LabVIEW® for Electric Circuits, Machines, Drives, and Laboratories

ISBN 0-13-061886-1

NATIONAL INSTRUMENTS | VIRTUAL INSTRUMENTATION SERIES

Jeffrey Y. Beyon
- Hands-On Exercise Manual for LabVIEW Programming, Data Acquisition, and Analysis

Jeffrey Y. Beyon
- LabVIEW Programming, Data Acquisition, and Analysis

Mahesh L. Chugani, Abhay R. Samant, Michael Cerra
- LabVIEW Signal Processing

Nesimi Ertugrul
- LabVIEW for Electric Circuits, Machines, Drives, and Laboratories

Rahman Jamal • Herbert Pichlik
- LabVIEW Applications and Solutions

Shahid F. Khalid
- Advanced Topics in LabWindows/CVI

Shahid F. Khalid
- LabWindows/CVI Programming for Beginners

Hall T. Martin • Meg L. Martin
- LabVIEW for Automotive, Telecommunications, Semiconductor, Biomedical, and Other Applications

Bruce Mihura
- LabVIEW for Data Acquisition

Jon B. Olansen • Eric Rosow
- Virtual Bio-Instrumentation: Biomedical, Clinical, and Healthcare Applications in LabVIEW

Barry Paton
- Sensors, Transducers, and LabVIEW

Jeffrey Travis
- LabVIEW for Everyone, second edition

Jeffrey Travis
- Internet Applications in LabVIEW

LabVIEW for Electric Circuits, Machines, Drives, and Laboratories

▲ Nesimi Ertugrul

Prentice Hall PTR, Upper Saddle River, NJ 07458
www.phptr.com

Library of Congress Cataloging-in-Publication Data

Erturgrul, Nesimi.
  LabVIEW for electric circuits, machines, drives, and laboratories / Nesimi Etrugrul.
    p. cm. — (National Instruments virtual instrumentation series)
  Includes bibliographical references and index.
  ISBN 0-13-061886-1
    1. Electric circuits — Computer simulation. 2. Electric machinery — Computer simulation.
  3. LabView. I. Title. II. Series

TK454 .E59 2002
621.3 — dc21   2002025259

Editorial Production/Composition: *G & S Typesetters, Inc.*
Acquisitions Editor: *Bernard Goodwin*
Editorial Assistant: *Michelle Vincenti*
Cover Design: *Talar Agasyan-Boorujy*
Cover Director: *Jerry Votta*
Marketing Manager: *Dan DePasquale*
Marketing Manager: *Alexis R. Heydt-Long*
Project Coordinator: *Anne R. Garcia*

  © 2002 by Prentice Hall PTR
A Division of Pearson Education, Inc.
Upper Saddle River, New Jersey 07458

Prentice Hall books are widely used by corporations and government agencies for training, marketing, and resale.

For information regarding corporate and government bulk discounts please contact:
Corporate and Government Sales (800) 382-3419 or corpsales@pearsontechgroup.com

All rights reserved. No part of this book may be reproduced, in any form or by any means, without permission in writing from the publisher.

Printed in the United States of America

10 9 8 7 6 5 4 3 2 1

ISBN 0-13-061886-1

Pearson Education Ltd.
Pearson Education Australia PTY, Ltd.
Pearson Education Singapore, Pte. Ltd.
Pearson Education North Asia Ltd.
Pearson Education Canada, Ltd.
Pearson Educación de Mexico, S.A. de C.V.
Pearson Education — Japan
Pearson Education Malaysia, Pte. Ltd.

Dedicated with love to Asuman, Atacan, and my parents, who were constant sources of inspiration and support.

Contents

Preface xv

1 Introduction 1
1.1 Some Features of Virtual Instruments 3
1.2 Hardware Suggestions 5

2 Basic Definitions and DC Circuits 13
2.1 Periodic Waveforms, and Average and RMS Values 14
    2.1.1 Virtual Instrument Panel 15
    2.1.2 Self-Study Questions 18
2.2 Periodic Waveforms and Harmonics 19
    2.2.1 Virtual Instrument Panel 21
    2.2.2 Self-Study Questions 21

Contents

- 2.3 DC Circuits 23
  - 2.3.1 Equivalent Resistance and Series/Parallel Resistance Circuits 23
- 2.4 Thevenin's and Norton's Equivalent Circuits 30
  - 2.4.1 Virtual Instrument Panel 32
  - 2.4.2 Self-Study Questions 34
- 2.5 References 35

3 AC Circuits 37
- 3.1 Fundamental Definitions 38
  - Per-Unit Values 42
- 3.2 AC Circuit Analysis 42
  - 3.2.1 Equivalent Impedances and Circuits 43
  - 3.2.2 A Reverse Study 48
- 3.3 Power and Power Triangles in AC Circuits 50
  - 3.3.1 Virtual Instrument Panel 53
  - 3.3.2 Self-Study Questions 53
- 3.4 Power Factor Correction 55
  - 3.4.1 Virtual Instrument Panel 57
  - 3.4.2 Self-Study Questions 57
- 3.5 Star-Delta and Delta-Star Conversion in Three-Phase AC Circuits 59
  - 3.5.1 Virtual Instrument Panel 60
  - 3.5.2 Self-Study Questions 60
- 3.6 Voltage and Currents in Star- and Delta-Connected Loads ... 62
  - 3.6.1 Virtual Instrument Panel 64
  - 3.6.2 Self-Study Questions 64
- 3.7 Voltage and Current Phasors in Three-Phase Systems 66
  - 3.7.1 Virtual Instrument Panel 67
  - 3.7.2 Self-Study Questions 67

Contents

- 3.8 Power in Three-Phase AC Circuits 69
  - 3.8.1 Virtual Instrument Panel 73
  - 3.8.2 Self-Study Questions 73
- 3.9 Three-Phase Power Measurement and Data Logging 75
  - 3.9.1 Virtual Instrument Panel 77
  - 3.9.2 Some Features of the VI and Operating Scenarios 81
- 3.10 References 84

4 Magnetic Circuits and Measurements 85
- 4.1 Background Information 86
- 4.2 Analysis of Magnetic Circuits 89
  - 4.2.1 Virtual Instrument Panel 92
  - 4.2.2 Self-Study Questions 92
- 4.3 BH Characteristics and Losses 95
  - 4.3.1 Virtual Instrument Panel 98
  - 4.3.2 Self-Study Questions 98
- 4.4 Measuring Magnetization Characteristics 100
  - 4.4.1 Principles of the Method 101
  - 4.4.2 Experimental Setup 103
  - 4.4.3 Virtual Instrument Panel 104
- 4.5 References 108

5 Electric Machines Laboratory 111
- 5.1 Introduction 112
- 5.2 Determining Moment of Inertia 115
  - 5.2.1 Virtual Instrument Panel 117
  - 5.2.2 Recommended Laboratory Hardware 120
- 5.3 Losses in DC Motors 121
  - 5.3.1 Virtual Instrument Panel 122
  - 5.3.2 Laboratory Hardware 134

- 5.4 Electromechanical Device Experiment 134
  - 5.4.1 Laboratory Hardware 145
- 5.5 Tests for AC Circuits 147
  - 5.5.1 Single-Phase AC Circuit Test 149
  - 5.5.2 Three-Phase AC Circuit Test 151
  - 5.5.3 Power Factor Correction Test 153
- 5.6 Transformer Test 153
  - 5.6.1 Virtual Instrument Panel 160
  - 5.6.2 Self-Study Questions 167
  - 5.6.3 Sample Results 169
- 5.7 Asynchronous (Induction) Motor Test 171
  - 5.7.1 Theory 173
  - 5.7.2 Virtual Instrument Panel 180
  - 5.7.3 Self-Study Questions 186
  - 5.7.4 Laboratory Hardware 187
- 5.8 Synchronization Observer 188
  - 5.8.1 Virtual Instrument Panel and Laboratory Hardware 190
- 5.9 Synchronous Machine Test 194
  - 5.9.1 Virtual Instrument Panel 204
  - 5.9.2 Self-Study Questions 214
  - 5.9.3 Sample Results 215
- 5.10 References 221

6 Introduction to Power Electronics Circuits 223
- 6.1 Diode Conduction 224
  - 6.1.1 Virtual Instrument Panel 226
  - 6.1.2 Self-Study Questions 226
- 6.2 SCR Conduction 228
  - 6.2.1 Virtual Instrument Panel 231
  - 6.2.2 Study Guides 236

6.3 Three-Phase Half-Way Diode Rectifier ... 236
- 6.3.1 Fundamental Theory ... 238
- 6.3.2 Virtual Instrument Panel ... 240
- 6.3.3 Implementation Details ... 243

6.4 Single-Phase AC Chopper ... 245
- 6.4.1 Virtual Instrument Panel ... 248
- 6.4.2 Features of the VI ... 251
- 6.4.3 Self-Study Questions ... 253

6.5 Cycloconverters ... 254
- 6.5.1 Virtual Instrument Panel ... 255
- 6.5.2 Features of the VI ... 256

6.6 PWM and Single-Phase Inverter (H-Bridge) Control Methods ... 258
- 6.6.1 Virtual Instrument Panel ... 260
- 6.6.2 Some Implementation Details of the Sub-VIs ... 272

6.7 References ... 276

7 Simulation of Electrical Machines and Systems ... 277

7.1 Rotating Field Simulation in AC Machines ... 278
- 7.1.1 Virtual Instrument Panel ... 280

7.2 Dynamic Simulation of Three-Phase Induction (Asynchronous) Motor ... 282
- 7.2.1 Virtual Instrument Panel ... 285
- 7.2.2 Self-Study Questions ... 285

7.3 Dynamic Simulation of Brushless Permanent Magnet AC Motor Drives ... 287
- 7.3.1 Virtual Instrument Panel ... 292
- 7.3.2 Self-Study Questions ... 297

7.4 Dynamic Simulation of Direct Current Motors ... 297
- 7.4.1 Virtual Instrument Panel ... 299
- 7.4.2 Self-Study Questions ... 301

7.5 Simulation of Stepper Motors 301
   7.5.1 Mathematical Model 304
   7.5.2 Control of Stepper Motors 306
   7.5.3 Virtual Instrument Panel 309
7.6 Steering and Control of Four-Wheel Direct-Drive
   Electric Vehicles 312
   7.6.1 Criteria Used to Develop a LabVIEW Simulation .. 314
   7.6.2 Virtual Instrument Panel 321
   7.6.3 Self-Study Questions 323
7.7 Fault-Tolerant Motor Drive for Critical Applications 324
   7.7.1 Fault-Tolerant Motor Drive System 324
   7.7.2 Virtual Instrument Panel 330
7.8 References 332

8 Real-Time Control of Electrical Machines 335
8.1 DC Motor Control 336
   8.1.1 Control of PM Brush DC Motors 337
   8.1.2 Hardware Implementation Details 341
   8.1.3 Details of the Virtual Instruments and Front Panels 346
8.2 Stepper Motor Control 352
   8.2.1 Virtual Instrument Panel 356
8.3 Brushless Trapezoidal PM Motor Control 358
   8.3.1 Virtual Instrument Panel 362
8.4 Starting Wound-Rotor Asynchronous Motors 366
   8.4.1 Principles of the Starting 366
   8.4.2 Hardware Details 370
   8.4.3 Virtual Instrument Panel 372
8.5 Switched Reluctance Motor Control 373
   8.5.1 Principles of Motor Control 374
   8.5.2 Virtual Instrumentation 380
8.6 References 382

Appendix 385
  For Chapter 1 385
  For Chapter 5 400
  For Chapter 8 406

Index 427

About the Author 435

Preface

*I hear and I forget; I see and I remember;
I do and I understand.*
Confucius

My strong belief is that education will have maximum value if knowledge and experience advance simultaneously regardless of quantity and complexity. This is the aim of *LabVIEW for Electric Circuits, Machines, Drives, and Laboratories*.

The groundwork for this book began about seven years ago, after I purchased my first copy of LabVIEW (Version 3.0) and a data acquisition card. Since then I have obtained more software and hardware components and integrated my teaching and research activities into LabVIEW-based modules, which have diverse areas of application varying from simple simulation and data acquisition to complex Motion Control. This book is a result of these studies. I now see *LabVIEW* as a verb as well as a noun, and many students who have utilized the end products say that they have improved their understanding of electrical circuits, machines, and drives.

It is a widely accepted fact that ac circuits, electrical machines, and drives are the most difficult subjects to teach and learn without some visual aids. Moreover, because in many institutions the time available for courses on electrical circuits, machines, and laboratories is severely limited, using custom-written virtual instruments (VIs) may conserve time by providing self-study tools and adding an extra dimension and visualization.

A first in its field, this book provides complete solutions for laboratory implementations and exposes the reader to manifold phenomena taking place in various electrical systems. It is an introduction to computer simulation of dc and ac electric circuits, electromechanical devices, and electric motors and drive systems and laboratory practices, all of which utilize VIs that are provided on the accompanying CD-ROM. The CD-ROM also contains a copy of an evaluation version of LabVIEW 6.0.

LabVIEW for Electric Circuits reinforces theoretical concepts and gives practical programming advice through examples of computer simulation in action and complete codes, which can also be utilized by potential developers. Furthermore, the book aims to fulfill the essential need for LabVIEW programming in this long-ignored area of electrical engineering—with hardware information, wiring diagrams, practical circuits, and printed circuit layouts all provided in the Appendix.

It should be emphasized here that discussion of the details of LabVIEW programming is beyond the scope of this book. Instead, basic or more complex theory of selected subjects is transformed in ways that learners and educators perceive to be useful and effective, utilizing a number of custom-written VIs. The book effectively employs a theoretical, a real-time, and a multidisciplinary approach to give readers a broader understanding of each topic, and it provides open-ended and highly interactive VIs that can be studied using computers.

While not intended as a detailed manual, *LabVIEW for Electric Circuits* provides sufficient basic knowledge to study the VI modules with a minimum background in LabVIEW. I have endeavored to make the front panels of the virtual instruments as simple and as close as possible to real-world operation. Furthermore, a considerable amount of time and effort was spent in developing cost-effective, high-performance hardware, details of which are all provided in the Appendix.

Each subsection is dedicated to a particular concept in an electric circuit and is accompanied by a VI. At the end of most subsections, a set of self-study questions is structured to introduce study guidelines and trigger further reflection. I provide a balanced coverage of fundamental definitions, various dc and ac electrical circuits, and magnetic circuits with animated operating modes of the circuits.

A considerable part of the book is necessarily devoted to laboratory experiments utilizing custom-written VIs, as the experimental modules are developed and currently used by the students at Adelaide University. Special attention is devoted to power electronics circuits involving basic converter topologies and to system simulations, which are related to earlier and successive chapters.

A number of dynamic motor and drive simulations are also provided. The essential principles of such systems, which underlie the performance of electrical machines and their applications, are discussed.

Using computer simulations without encountering the real devices is an incomplete experience since it will be only a toy (data), not a tool (information). Therefore, the book concludes with a number of real-time experimental modules that are intended to help students gain real-life experience.

The chapter topics cover a wide spectrum of areas in electric circuits, machines, and laboratory work. The chapters are organized to briefly introduce the fundamental theory and to provide alternative self-study tools (VIs). Note that the complexity of the topics gradually increases in successive chapters.

The CD-ROM that accompanies the book provides a wealth of supplements. The VIs supplied with the book form a veritable Virtual Electrical Circuits, Machines, and Drives Laboratory. Use of the VIs is mandatory for understanding the concepts covered in each study module. *LabVIEW for Electric Circuits* may well become a formidable partner for self-study and an enduring teaching assistant.

The intention of this book is to provide correct information and error-free virtual instruments. Considerable effort has been spent on testing every VI, yet you may still find an error. Therefore, I will be pleased to receive feedback, including suggestions for improvements. For feedback, use the following addresses: Dr. Nesimi Ertugrul, Adelaide University, Department of Electrical and Electronic Engineering, 5005, Adelaide, Australia; nesimi.ertugrul@adelaide.edu.au or nesimi@eleceng.adelaide.edu.au.

Acknowledgments

I am grateful to many people whose efforts have gone into the making of this book. I take pleasure here in expressing my thanks to Ravi Marawar of National Instruments, Inc., for the initial suggestion of writing this book; to Bernard Goodwin of Prentice Hall, Inc., for his recommendations and patience; and to three people who put great effort into preparing this book for printing: Anne Garcia of Prentice Hall, for coordination of in-house production; Joshua Goodman of G&S Typesetters, for coordination of production services; and Kathy Deselle, for copyediting the text.

I would like to acknowledge the support of my students, who worked closely with me in the development stages of various VI modules. Moreover, I would like to thank the electronic and mechanical workshop staff (specifically Geoff Pook for making PCBs) in our department, who contributed to the development of the hardware sections.

Finally, I am greatly indebted to my perceptive, supportive reviewers, Tony Parker (specifically for his continuous encouragement) and Wen Soong at Adelaide University, and Preston Johnson of National Instruments, Inc., who all have directly enriched the book.

Introduction

1

Within the last decade, electrical, electronic, and computer engineering have become increasingly intertwined. Although it is difficult to define what separates these disciplines in terms of theory, the applications and experimental work related to each discipline have distinctive features and, therefore, require special attention.

Most conventional textbooks provide a limited number of worked examples, which, in general, are used merely to illustrate facts that have previously been discussed. The computer-based framework, however, can provide students with alternative study questions along with meaningful, contemporary, and relevant practical experience, while not being restricted by limitations on resources such as infrastructure and laboratory hardware.

This book successfully introduces interactive LabVIEW-based virtual instruments (VIs) into the curriculum of an Electrical Circuits, Machines, and Drives course. The target audience is first- or second-year electrical and electronic engineering and mechatronics engineering students. In addition, students may find many of the VIs useful later in the engineering curriculum, when they study various drive system simulations.

LabVIEW programming can be defined as layers of software and hardware, added to a personal computer in such a way that the computer acts as a custom-designed instrument. LabVIEW software enables simulation, data capture, data analysis, and circuit animation. A VI consists of two main elements: a front panel and a block diagram. The front panel is a graphical user interface used for data presentation and control inputs and can be highly customized by the user in a particular application.

The overall mission of this book is to engage in both dedicated and interdisciplinary self-learning and teaching facilities while providing a lifelong experimental practice for students. The prime benefits of the virtual instruments provided here are the deep understanding of electrical circuits, electrical machines, and electromechanical devices; experiencing real-time signals and controls; and observing the limitations of the theory. With the VIs provided on the accompanying CD-ROM, students can monitor hardware of selected physical systems and can integrate animated computer simulations into the theory via custom-written VIs.

However, it should be emphasized here that this mission cannot be achieved without a fundamental knowledge and active participation in the practical experiments. Therefore, preliminary preparation, including the study of the theory sections in this book, is necessary. You should also study the reference books listed at the end of each chapter.

The VIs usually have step-by-step instructions and are designed to assist students methodically. Each section begins with a brief theory, which introduces concise background knowledge. The front panels of the VIs are designed as simple and as close as possible to the real world operation, which includes waveforms, indicators, controls, and animations. In the book, brief user guides for the main front panels are provided, which may be sufficient for a quick start. In addition, self-study questions at the end of most subsections guide the reader toward the operation of the custom-written VI. Self-study questions allow students to explore the concepts in depth and are organized in increasing difficulty.

The fundamental approach used in this book is to keep the VIs as flexible as possible, enabling future and alternative developments using the principles presented here. Note that the pure computer simulations and analysis sections of the experiments can run directly without any data acquisition board and experimental hardware. The remaining VIs, however, require application-specific hardware and signal conditioning devices. In the following subsections, the specifications of the hardware utilized during the development stage of the VIs are given for your reference. Furthermore, the experiments presented here can be made available to remote area users via an Internet link using LabVIEW's Internet Developers' Toolkit.

An additional VI is provided in the Appendix, which is used to implement multiple-choice quiz sessions that can run prior to each practical experiment.

These can be a part of the assessment process. Note that the questions that appear in the sample quiz are based on the electrical machines set available in our laboratory. Alternative questions may be easily customized.

Note that SI units are used throughout this book in equations and in the LabVIEW VIs.

1.1 Some Features of Virtual Instruments

The VIs provided in this book have a varying degree of complexity and special requirements. Therefore, it is beyond our coverage here to include all the implementation details. Instead, some unique approaches used in the VIs will be highlighted in the following paragraphs.

The discussions will be further extended in the relevant chapters to explain the structural features of some selected VIs. Our principal aim is that the approach and the methodology used in the VIs may provide insights into the area of study and may inspire users for further improvements and alternative study modules. The primary approach used to design virtual instrumentation is that "even complex problems are easy if they are divided into small sub-VIs."

As a common practice in LabVIEW programming, the VIs in this book are built by designing a front panel layout that shows the inputs and outputs to the function, with a diagram layer showing the interconnections and functions implemented in the VI. An icon equipped with a set of relevant input and output terminals is then used to provide an interface, allowing this VI to be called by other VI(s).

The data structures and control algorithms are developed in accordance with the theory. The formulas' nodes are often used to reduce the complexity of a VI. As complexity increases, such as in the discrete-time integration, multiple sub-VIs are incorporated to form a sub-VI.

Three integration methods are implemented in the VIs, which numerically solve the differential equations of a dynamic system:

- Runge-Kutta method available in LabVIEW's G-Math Toolkit
- Runge-Kutta method implemented separately using the weighted average values and the corresponding coefficients (which utilize the shift registers in LabVIEW)
- Trapezoidal integration method implemented using the shift registers

The implementation of each simulation in LabVIEW requires several embedded loops. A large portion of these loops enable the program to be restarted, paused, quit, and started from the front panel. This is necessary for when the program is repackaged, and the controls used during programming are not available to the user.

As a fundamental example, let's consider the simulations given in Chapter 6 and summarize the important implementation issues. To support the following explanations, refer to the VIs provided on the accompanying CD-ROM.

First, the programs ensure that the chart is cleared. Then a sub-VI is used to normalize the parameters to be viewed on the same chart. The first loop of the main VI is a case loop that allows the simulation to continue until the program is stopped. This case has been set to true. So the program is stopped by other means. The main purpose of this loop is to provide the required per-unit (normalized) value array in a shift register so that when the program is reset it does not go through the whole program initialization sequence again.

The second loop that the program enters contains the sequence frames. This sequence consists of initializing the graph and the running of the program. Initializing the graph frame itself contains a second set of sequence frames to initialize the simulation. The simulation initialization sequence first waits for the start button to be pressed, after which it initializes the graph values and implements the per-unit calculation for the plot of the graph. After the initialization frame has completed the program carries out the simulation.

The simulation sequence frame contains a while loop that runs until the reset button is pressed. Inside the while loop is a case loop, which is dependent on the reset button. When the reset button has been pressed, the case is true, and the program is allowed to reset and restart. However, while it has not been pressed, the case is false. In this case the program calculates the values to simulate the circuit and the determinants of the circuit are set. These are the input values that each simulation allows the user to set. These values are then entered into a for loop. As these values are read outside the for loop, they remain the values for the entire loop. It is only when the for loop has completed that these values are re-read from the front panel. This for loop calculates one entire period. Hence the circuit specifications are used for an entire period.

The for loop that calculates an entire period contains the controls for the speed of the simulation and for the animation of the circuit and provides the ability to change the per-unit values. However, the main part of this loop is the sequence frames. The first sequence frame carries out the calculations for the simulation including the supply voltages, the number of states, what state the simulation is in, and all the measured values that are plotted on the chart. All the values that are plotted are calculated in this sequence. The second sequence frame implements the pause function, the displaying of the per-unit values, and the graph selection.

The practical experiments provided in Chapter 5, for example, have a programming structure that has two main control buttons. These buttons can select a subsection or return to the previous level. In addition, the main VI includes menu rings, which are linked to the subsections of the experimental modules. The principal aim of the programming structure is easy navigation between the subsections of each experiment.

Another key feature of the VIs available in this book is the animation of an electric circuit and an electromechanical system, which may be linked to the operating states of the electrical circuit. The animations are achieved by using a picture ring. The visible item of this object is a single picture at any one time, out of many pictures, which are drawn, imported to the relevant front panels, and linked to the programming diagrams. Each item of the picture ring is associated with an (unsigned) integer value. By making the ring as an indicator, wiring it to a different integer value displays a different item that corresponds to a state of the electric or mechanical system.

1.2 Hardware Suggestions

Experience has shown that changes in laboratory practices are necessary because many of the things we do can be done better with the help of technology. Although the initial cost may be high, if the right technology for a job is selected, continuous improvement can be achieved with minimal cost.

The cost of such development may be reduced further if a portion of the existing hardware is utilized and integrated with an updated system. Furthermore, the selection criteria for suitable software and hardware are the major issue. Essential features are a long life cycle, easy interface with the hardware products, and compatibility with existing development tools. Because LabVIEW has these qualities, it was chosen in this book as an enabling technology.

From a technical point of view, all the engineering problems deal with some physical quantities such as temperature, speed, position, current, voltage, pressure, force, torque, and so on. The principal approach that is also used in the development stage of the laboratory tests is that a computer equipped with the suitable interface circuits, data acquisition systems, and software can give a visual look to these quantities, and then the acquired data can be analyzed in detail.

Depending on the target aim, the experimental system may contain various interface circuits and analog/digital I/O ports. Therefore, the final aim of the

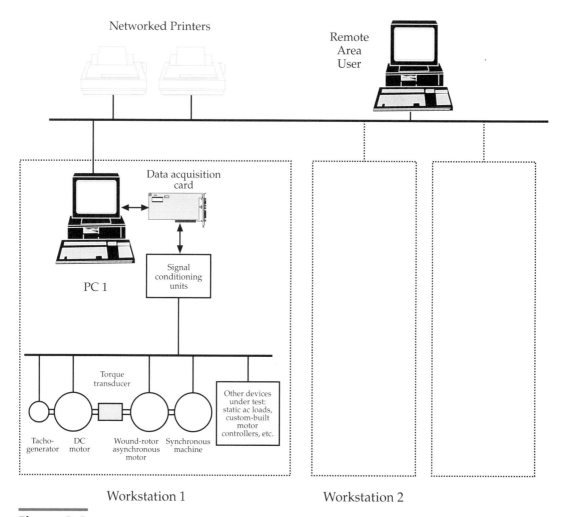

Figure 1-1
The system diagram of the computer-assisted laboratory.

teaching/learning technology should be identified to determine the subunits required.

Although the details of each piece of hardware used during the development stage are given later in the relevant sections and in the Appendix, let's provide an overview of the system used. The system diagram of the computer-assisted real-time experimental modules used in the laboratory component of this book is shown in Fig. 1-1.

Figure 1-2
A view of the Electrical Machines and Drives Teaching Laboratory, showing four workstations.

A view of our laboratory in Fig. 1-2 illustrates its layout and shows only four of the ten workstations. We took an existing laboratory with the traditional switchboards, analog measuring devices, and electrical machines and equipped it with computers, interface units, and custom-written VIs.

The essential parts of a real-time experimental system consist of five principal units: a device under test, signal conditioning devices, a PC, a data acquisition card, and a custom-written VI.

First, since the voltage and current levels present in ac circuits and machines are high, the primary requirement of a hardware component is to provide electrical isolation. Hence, measured analog signals have to be attenuated and isolated to a level that is safe for the data acquisition (DAQ) systems (which may accommodate customized signal conditioning devices) and for the operators (students). In addition, the frequency bandwidth of the signal conditioning devices has to be considered for accurate measurement.

Second, a suitable data acquisition system has to be selected. The main criteria are the maximum number of analog inputs and digital I/O ports, sampling rate, and resolution. Minimum specifications of the data acquisition system used in this book are at least eight differential A/D channels, 12-bit resolution, 100 kHz sampling frequency, and at least nine digital I/O ports. The minimum number of A/D channels is determined by the maximum number of parameters to be measured to perform the most complex experiment in the laboratory, such as three-phase voltages, three-line currents, speed, and torque. National Instruments' AT-MIO-16E-10 data acquisition card is used in the development system.

Note that the ratings of the real laboratory machines are not critical in the virtual instruments presented in this book. However, the details of the laboratory equipment may provide some insight about the development and experimentation environment and may help potential developers.

In the laboratory, there are 10 Pentium-based PCs running Windows NT and LabVIEW 5.0 full development system, 2 laser printers networked to the computers, 10 custom-built torque transducers to measure the instantaneous shaft torque in the machines (up to 50 Nm), 120 custom-built current transducers (50 A, and 100 A, DC to 100 kHz), 120 custom-built voltage isolation amplifiers (1000 V rms, 50 kHz), 10 benches and switchboards, 10 static starting circuits and interfaces for the slip-ring induction motors, and wiring.

Each workstation contains identical, medium-power, mechanically coupled rotating electrical machines: DC machine, slip-ring induction motor, and synchronous machine. Table 1-1 indicates the ratings of these machines. The measurement of shaft speed is achieved by a DC tacho-generator that is also attached to the common shaft.

The static devices under test and the custom-built motor control circuits are supplied on the bench where external interface terminals are available via custom-built voltage and current transducers.

Table 1-1 *The ratings of the electrical machines used in the laboratory.*

DC Machine	Wound-Rotor Induction Machine	Synchronous Machine
5.5 kW	415 V	3-phase
1250/1500 rpm	3-phase, Y-connected	8 kW
220 V	11 A	415 V
27.6 A	1410 rpm	10.5 A
Shunt field:	5.5 kW	1500 rpm
210 V	cos ϕ: 0.85	50 Hz
0.647 A	50 Hz	
	Rotor:	
	Y-connected	
	170 V, 22 A	

As previously mentioned, each workstation accommodates a number of transducers that are used to measure high voltages and currents even at high switching frequencies. Noise immunity and personal safety are always an issue in such systems. Therefore, the workstations are designed to achieve total isolation.

First, voltage dividers are used to attenuate the high voltages. Then the attenuated voltages are isolated using isolation amplifiers. Each isolation amplifier is also powered from a separate DC/DC converter. Each group of three transducers is equipped with separate floating power supplies for additional safety. The voltage transducers' boards are also physically guarded against potential danger, which may occur due to an arc.

Fig. 1-3 shows voltage and current transducers built in-house. Twenty-four of these transducers (12 current and 12 voltage transducers) are housed inside a box located behind the main switchboard of the workstations. The transducers are permanently wired to the switchboard. The circuit diagrams and the printed circuit board (PCB) layouts are provided in the Appendix.

In order to create a buffering circuit and to take full advantage of the resolution of the A/D conversion, additional amplifiers are used to amplify the

Figure 1-3
Custom-built current and voltage transducers: There are three transducers on each card powered by an onboard floating power supply.

(a)

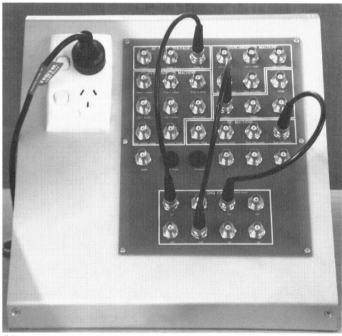

(b)

Figure 1-4
(a) A PC and (b) a BNC connector panel available in each workstation.

signals obtained via the Hall-effect current transducers and the isolation amplifiers. The amplified signals are transferred to BNC terminal panels via coaxial cables to eliminate the noise.

As stated earlier, the transducers are permanently wired to the main switchboard, and their outputs are made available on the BNC connector board shown in Fig. 1-4a. The BNC connector board (Fig. 1-4b) is used to select and assign any voltage and current signals to the analog inputs of the data acquisition card.

Basic Definitions and DC Circuits

2

This chapter's main objective is to highlight some of the commonly used definitions and fundamental concepts in electric circuits, which are supported by a set of custom-written VIs. These VIs enable students to examine various scenarios in circuits or control panels and, hence, provide an excellent tool for interactive studying. For example, a circuit can be modified easily by varying its controls on the front panel—a series resistance can be zeroed and a parallel resistance can be set to a very high value to introduce a short circuit and an open circuit, respectively. Or a dc offset can be introduced to a programmed waveform to obtain a desired average or root mean square (rms) value, which is supported by the visual display of the waveform.

This chapter is divided into five sections with accompanying custom-written VIs. The first two sections offer some basic explanations about common electrical waveforms and their distinguishing features. We then develop the concept further by studying harmonics in nonsinusoidal waveforms.

Section 2.3, DC Circuits, covers basic circuit topologies and mesh analysis. Section 2.4 presents Thevenin's and Norton's equivalent circuits. In addition, each subsection includes a set of self-study questions that are structured to assist learning and to encourage students to investigate alternative settings on the VIs.

Educational Objectives After completing this chapter, students should be able to

- understand the basic concepts in dc circuits, periodic waveforms, and harmonics.
- state the meaning of the terms *periodic* and *rectified waveforms*; *average*, *rms*, and *maximum values*; and *equivalent resistance*.
- solve for unknown quantities of resistance, current, voltage, and power in series, parallel, and combination circuits.
- examine the concepts of open and short circuits and describe their effects on dc circuits.
- understand Thevenin's and Norton's equivalent circuits.
- create various scenarios with the provided circuits, and verify the results analytically.
- gain skills in virtual instrumentation to create more complex and alternative systems by analyzing the programming block diagrams.

2.1 Periodic Waveforms, and Average and RMS Values

The electric power used for most industrial and household applications is generated and transmitted in the form of a fixed frequency (either 50 Hz or 60 Hz) sinusoidal voltage or current. These signals generated by alternators are time-dependent periodic signals that satisfy the equation

$$x(t) = x(t + nT) \tag{2.1}$$

where t is the time, T is the period of $x(t)$, and n is an integer.

Such signals are usually expressed as a perfect sine wave and known as ac quantities. Hence, a representation of an arbitrary ac sinusoidal voltage signal is given as

$$v(t) = V_m \sin \omega t \tag{2.2}$$

where V_m is the amplitude, and ω is the angular frequency in radians/s, which is equal to $2\pi f$. The frequency of a periodic signal f refers to the number of times the signal is repeated in a given time. The period is the time it takes for one cycle to be repeated. The frequency f and the period T are reciprocals of each other. (Note that one can represent a sine wave in terms of a cosine wave simply by introducing a phase shift of $\pi/2$ radians.)

Furthermore, there are other periodic waveforms observed in electrical circuits that can be approximated by time-varying ideal signals. Such approximation is usually done using exponential, linear, or logarithmic functions.

In the real world, however, two definitions for voltage and current waveforms are used to quantify the strength of a time-dependent electrical signal: the average (mean or dc) value and the root mean square (rms or effective) value.

The average value of a voltage signal corresponds to integrating the signal waveform over a period of time, which is given for the voltage signal by

$$V_{ave} = \frac{1}{T}\int_0^T v(t)\,dt \qquad (2.3)$$

The average value of a time-varying waveform may be considered as the dc voltage equivalent of a battery, which does not vary with time and will be used as a voltage source in dc circuits.

The root mean square value of a signal takes into account the fluctuations of the signal about its average value and is defined for the voltage signal as

$$V_{rms} = \sqrt{\frac{1}{T}\int_0^T v^2(t)\,dt} \qquad (2.4)$$

For an ideal sinusoidal voltage waveform $V_{rms} = V_m/\sqrt{2}$.

Note: True rms meters should be utilized to measure the rms value of any nonsinusoidal waveform. A custom-written LabVIEW VI equipped with a DAQ system can also provide a true rms measurement.

2.1.1 Virtual Instrument Panel

The custom-written VI for this section is named `Waveform Generator.vi` and is located on the accompanying CD-ROM. The objective of the VI is to study the concepts and definitions just introduced using a comprehensive waveform generator that can generate twelve different periodic waveforms (Fig. 2-1): sine wave, programmed harmonics, clipped sine wave, chopped sine wave, triangular wave, trapezoidal wave, rectangular wave, square wave, two different ramp waves, logarithmic wave, and exponential wave. These cover the majority of the practical waveforms featured in electrical and electronic engineering courses.

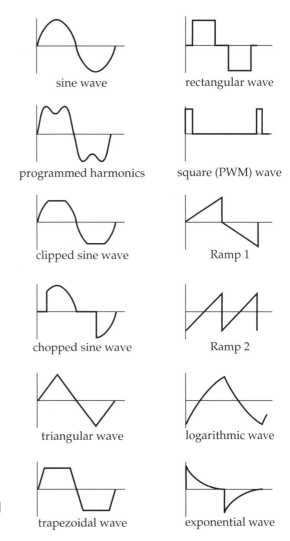

Figure 2-1
The waveforms that can be generated by the Waveform Generator.vi.

It should be noted here that at least three cycles of a waveform must be contained in the time-domain record for a valid estimate of rms and average values provided in the VI.

As seen on the front panel of the VI (Fig. 2-2a), various parameters of the periodic waveforms (such as frequency, phase, amplitude, dc offset, noise, etc.) can be controlled by the user. The VI can generate various outputs such as waveform graphs and average and rms values of the waveforms that are located next to the associated graph area. Furthermore, some subcontrols are

Chapter 2 • Basic Definitions and DC Circuits

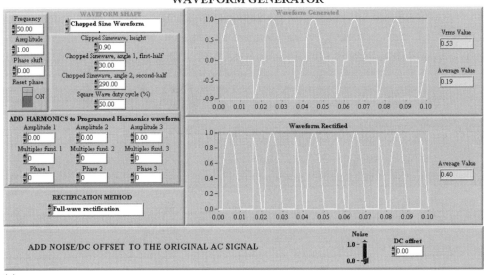

(a)

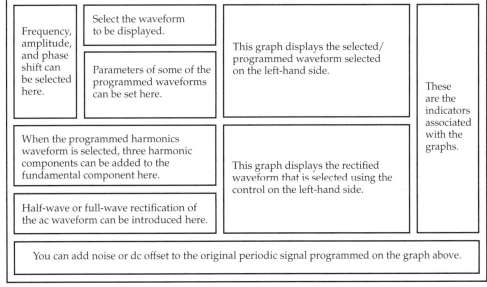

(b)

Figure 2-2
(a) Front panel and (b) brief user guide of `Waveform Generator.vi`.

provided to set additional parameters for the specific waveforms, such as duty cycle, chopping angle, clipped angle, and so on. The brief user guide in Fig. 2-2b explains the various features of the front panel.

In the subpanel named "Add Harmonics to Programmed Harmonic Waveform," the user can add three harmonic components onto a fundamental sine wave whose frequencies can be entered as multiples of the fundamental frequency. In addition, phase shifts can be introduced to each harmonic component if desired.

As it is implemented in the VI, practically any waveform can be generated using Formula Node in LabVIEW. I suggest you refer to the block diagram of the VI for the implementation details of the waveforms. After obtaining a basic understanding of LabVIEW programming, I encourage you to use this VI as a starting tool to develop more complex waveforms. In addition, remember that a complex-looking waveform may easily be obtained using a combination of two or three of the waveforms provided.

2.1.2 Self-Study Questions

Open and run the VI named `Waveform Generator.vi` in the `Chapter 2 VIs` folder on the accompanying CD-ROM, and investigate the following questions. Remember that the degree of difficulty varies in the following questions. You should verify your findings analytically.

1. Study the rms and average values (both for half-wave and full-wave rectifications) of the waveforms provided in the VI using the default values.
2. Investigate the effect of varying the amplitude, frequency, and phase shift of the waveforms on the calculated rms and average values. Verify that the periods of two waveforms of your choice displayed on the graph are correct. Remember that the period is $T = 1/f$.
3. Introduce some dc offset on a periodic waveform of your choice, and observe the change of the full-wave rectified waveform when the dc offset is introduced.
4. Verify that the rms value of a function $v = 50 + 30 \sin \omega t$ is 54.3 V.
5. Determine a chopped sine wave angle α that satisfies $V_{ave} = 0.5\, V_m$ for the waveform given in Fig. 2-3.
6. For the waveforms given in Table 2-1, verify that the values are correct.
7. Select the waveform option Programmed Harmonics, and introduce third, fourth, and fifth harmonics (one at a time). Vary their amplitudes

Chapter 2 • Basic Definitions and DC Circuits

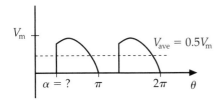

Figure 2-3
Example of a chopped sine wave for question 5.

Table 2-1 *Some selected waveforms and their average values.*

Waveform	Average
V_m half-wave, π, 2π	$0.32V_m$
V_m full-wave, π, 2π	$0.64V_m$
50 sawtooth, 2, 4, $t(s)$	25
10 decaying exp, .05, .10, $t(s)$	1
10 pulse, .01, .02, $t(s)$	3.33
$.707V_m$ clipped sine, π, 2π	$0.54V_m$
100 / −20 square, $\pi/4$, 2π	10
100 pulse sine, $\pi/4$, π, 2π	27.2

gradually and observe the effect of the harmonics on the original pure sine wave.

2.2 Periodic Waveforms and Harmonics

In practical electric circuits, voltage and current signals are not pure sine waves. Due to the nonideal behavior of electrical circuits, these signals are usually distorted.

The distortion of the signals in ac circuits can be due to various reasons, such as nonlinear loads (electric arc furnaces, etc.), magnetic saturation in the cores of transformers, or equipment containing switching devices or power supplies. Specifically, due to the switching action in adjustable speed motor drive systems, both the voltage and the current waveforms are highly distorted.

Table 2-2 *Classification of 50 Hz supply harmonics.*

Name	Funda-mental	2nd	3rd	4th	5th	6th	7th	Etc.
Frequency (Hz)	50	100	150	200	250	300	350	...

The distortion of dc signals, however, is mainly due to the rectification process. Rectification from an ac source involves various electronic converter circuits and supply transformers that generate ripples.

Nonsinusoidal, distorted waveforms (as illustrated in Table 2-1) can be represented by a series of harmonic components. Each harmonic has a name and frequency (see Table 2-2).

A special case in ac systems occurs when the positive and negative parts of the waveform have negative symmetry, that is, $f(t) = -f(t + T/2)$, where T is the period of the waveform. Hence, there is no dc component, and even harmonics (2nd, 4th, 6th, etc.) will not be generated.

It should be noted that in three-phase ac systems the harmonics are also defined with reference to their sequence, which refers to the direction of rotation with respect to the fundamental. For example, in an induction motor, a positive sequence harmonic generates a magnetic field that rotates in the same direction as the fundamental, while a negative sequence harmonic rotates in the reverse direction. Negative sequence voltages can produce large rotor currents, which may cause the motor to overheat. Zero sequence harmonics are known as Triplens (3rd, 9th, etc.), and they do not rotate but add in the neutral line of the three-phase four-wire system.

In ac circuits, fundamental power (which is produced by fundamental voltage and fundamental current) produces the useful power. The product of a harmonic voltage and the corresponding harmonic current produces a harmonic power. This is usually dissipated as heat in the ac circuits, and, consequently, no useful work is done.

Furthermore, harmonics can cause many other undesirable effects in electric motors, such as torque ripple, noise, vibration, reduction of insulation life, presence of bearing currents, and so on.

Waveforms with discontinuities, such as the ramp and square wave, often have high harmonics, which have amplitudes of significant value compared with the fundamental component. This can be visualized in the VI provided in this section.

The principal solution to reduce or eliminate the harmonics is to add harmonic filters at the source of the harmonics or to use various other techniques, such as programmed switching in motor control applications.

Although the level of distortion in a waveform can be seen by observing the real waveform, the distortion of the signals can be traced to the harmonics it contains using harmonic analysis techniques, one of which will be covered in this section. The tool presented here should provide an insight into the harmonics and enable you to take preventative measures to avoid distortion.

2.2.1 Virtual Instrument Panel

A number of periodic waveforms typically encountered in the study of electrical circuits are simulated in the virtual instrument provided in Section 2.1.1. This section develops the concept further and integrates the `Waveform Generator.vi` and the harmonic analysis module, providing a flexible user interface. In this section, we can decompose a given periodic wave into its fundamental and harmonic components.

The output of the `Waveform Generator.vi` is applied to the `Waveform and Harmonic Analyser.vi` (Fig. 2-4a) either as an ac signal or as a dc signal after rectification (half-wave or full-wave). The switch named AC Input or DC Input can be used to achieve the selection.

The well-known Fourier series expresses the periodic wave that is analyzed. Hence the original signal can be reconstructed using a number of terms of the trigonometric series, including the fundamental component of the signal. With more terms included in this reconstruction, the result more nearly resembles the original signal.

2.2.2 Self-Study Questions

Open and run the custom-written VI named `Harmonics.vi` in the `Chapter 2 VIs` folder, and investigate the following questions.

Note: When studying a specific case, unless otherwise stated, leave all the control values on the harmonic spectrum analysis panel in their default settings.

1. Certain functions contain a constant term, a fundamental, and a third harmonic. From the given signals available in the `Waveform Generator.vi`, list the signals, which have these features in their harmonic spectrum.
2. Demonstrate that an ac square wave with an amplitude of 100 V and a frequency of 50 Hz has the following harmonic contents:

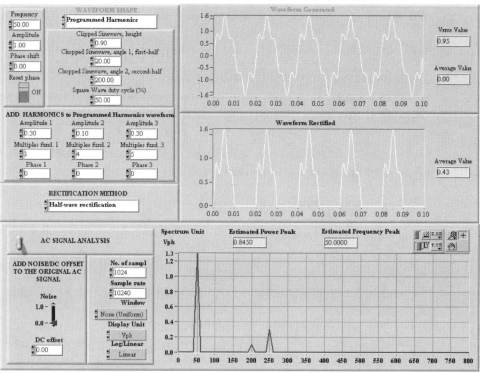

(a)

(b)

Figure 2-4
(a) Front panel of the complete VI, `Waveform and Harmonic Analyser.vi` and (b) brief user guide for the additional harmonics front panel.

Harmonic:	Fund.	3rd	5th	7th	9th
Frequency (Hz):	50	150	250	350	450
Amplitude (V):	127.3	42.4	25.5	18.5	14.1

Hint: An ac square wave can be obtained by introducing a dc offset to the square waveform with 50% duty cycle.

3. Select a ramp waveform and find the trigonometric Fourier series using the first three harmonic components displayed on the harmonic spectrum graph.
4. Waveform synthesis is a combination of the harmonics so as to form the actual waveform. Demonstrate that the ramp generated in question 3, which utilized three harmonic components, is not sufficient to form the actual waveform. Propose a solution.
5. The output of a full-wave rectified sine wave consists of a series of harmonics. Demonstrate that the Fourier representation of such a periodic wave is

$$f(t) = \frac{2V_m}{\pi}\left(1 - \tfrac{2}{3}\cos 2\omega t - \tfrac{2}{15}\cos 4\omega t - \tfrac{2}{35}\cos 6\omega t - \cdots\right)$$

6. Select a clipped sine wave with a clipped height of $0.2V_m$, and compare its harmonic contents to a trapezoidal waveform with an amplitude of 0.2.
7. Demonstrate that the average value of a waveform displayed in the corresponding indicator is equal to the magnitude of the dc component observed on the harmonic analysis graph.

2.3 DC Circuits

2.3.1 Equivalent Resistance and Series/Parallel Resistance Circuits

The basic circuit element we will use in this section is an ideal resistor, R. The current in an ideal resistor is linearly related to the voltage across it, and it has a value, which is time-invariant.

$$V = iR \tag{2.5}$$

Resistors can be connected in series or in parallel in electric circuits. When resistors are connected in series, they share the same current, and the voltages

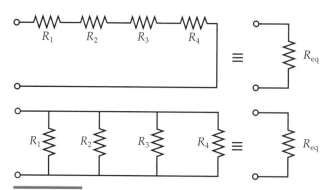

Figure 2-5
Series and parallel resistance circuit.

across them add to give the total voltage. The opposite is true in parallel resistance circuits; that is, parallel components share the same voltage, and their currents add to give the total current.

The equivalent resistances of a series and a parallel circuit (Fig. 2-5) can be calculated using the following formulas. These illustrate the case involving four elements.

$$R_{eq(series)} = R_1 + R_2 + R_3 + R_4 \tag{2.6}$$

$$R_{eq(parallel)} = \frac{1}{(1/R_1 + 1/R_2 + 1/R_3 + 1/R_4)} \tag{2.7}$$

Furthermore, the three basic circuits given in Fig. 2-6 will be used to study voltage and current division in the resistance circuits.

In Fig. 2-6a (voltage divider circuit), two resistors are connected in series across a voltage source. As seen in the figure, the resistors share the same current, and the voltages across them are proportional to their resistances. In addition, the power dissipated in each resistor can be calculated as follows.

$$i = \frac{V_{dc}}{(R_1 + R_2)} \tag{2.8}$$

$$V_2 = R_2 i = V_{dc} \frac{R_2}{(R_1 + R_2)} \tag{2.9}$$

$$V_1 = R_1 i = V_{dc} \frac{R_1}{(R_1 + R_2)} \tag{2.10}$$

$$P_{R1} = V_1 i = i^2 R_1 \tag{2.11}$$

$$P_{R2} = V_2 i = i^2 R_2 \tag{2.12}$$

Chapter 2 • Basic Definitions and DC Circuits

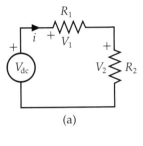

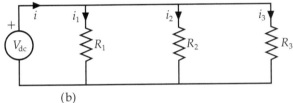

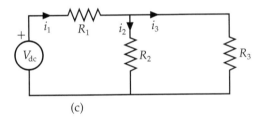

Figure 2-6
Voltage and current division circuits: (a) voltage divider circuit (series resistance circuit), (b) current divider circuit (parallel resistance circuit), and (c) series/parallel circuit.

The current divider circuit is studied using the circuit given in Fig. 2-6b. Since the resistors are in parallel, they share the same voltage. The current division between the resistors is inversely proportional to their resistances or directly proportional to their conductance, G. Note that the conductance is the reciprocal of resistance ($G = 1/R$).

$$i = i_1 + i_2 + i_3 = V_{dc}\left(\frac{1}{R_1} + \frac{1}{R_2} + \frac{1}{R_3}\right) = V_{dc}(G_1 + G_2 + G_3) \quad (2.13)$$

$$i_1 = V_{dc}\frac{1}{R_1} = i\frac{(1/R_1)}{(1/R_1 + 1/R_2 + 1/R_3)} = i\frac{G_1}{(G_1 + G_2 + G_3)} \quad (2.14)$$

$$i_2 = V_{dc}\frac{1}{R_2} = i\frac{(1/R_2)}{(1/R_1 + 1/R_2 + 1/R_3)} = i\frac{G_2}{(G_1 + G_2 + G_3)} \quad (2.15)$$

$$i_3 = V_{dc}\frac{1}{R_3} = i\frac{(1/R_3)}{(1/R_1 + 1/R_2 + 1/R_3)} = i\frac{G_3}{(G_1 + G_2 + G_3)} \quad (2.16)$$

A combination circuit is given in Fig. 2-6c. As seen in the figure, two parallel resistors are connected in series with a single resistor. Therefore, an equivalent circuit can be obtained that is similar to the circuit studied in Fig. 2-6a. Hence, both voltage and current divider rules can be applied to the original circuit, where

$$i_1 = i_2 + i_3 \tag{2.17}$$

$$i_1 = \frac{V_{dc}}{(R_1 + R_{eq})}, \quad R_{eq} = \frac{1}{(1/R_1 + 1/R_2)} \tag{2.18}$$

$$V_{eq} = R_{eq} i_1 = V_{dc} \frac{R_{eq}}{(R_1 + R_{eq})} \tag{2.19}$$

$$i_3 = i_1 \frac{(1/R_3)}{(1/R_2 + 1/R_3)} = i_1 \frac{G_3}{(G_2 + G_3)} = \frac{V_{eq}}{R_3} \tag{2.20}$$

2.3.1.1 Virtual Instrument Panel

The LabVIEW VI implemented in this section (Fig. 2-7) contains six different circuit options covering the previously discussed circuit topologies, including a circuit study of the concept of mesh analysis, which is discussed in the following section. The desired circuit can be selected from the library file, which contains all the circuits discussed.

The VIs of the circuits here calculate the voltage, current, and power of each circuit element and present them in the same format as given in conventional textbooks. My intention is to emphasize the effect of changing certain circuit parameters on the current and the power of the other circuit elements.

Moreover, you can experiment with open-circuit and short-circuit concepts in any branch of the circuit by varying the circuit parameters depending on their connection.

2.3.1.2 Self-Study Questions

Open and run the custom-written VIs located in Resistance Circuits .11b, in the Chapter 2 VIs folder. Remember that if a circuit contains more components than the circuits presented here, you can subdivide your circuit into small subsections that are similar to the other circuits analyzed. Furthermore, if the value of a resistance has to be set to 0 Ω, it means a short circuit of that branch. However, if the value of a resistance is very large (compared with the resistances of the other components), the branch can practically be assumed open circuit.

Chapter 2 • Basic Definitions and DC Circuits

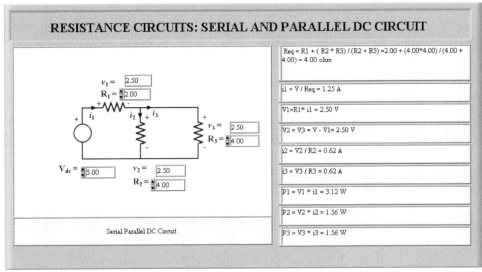

(a)

(b)

Figure 2-7
(a) A sample front panel and (b) brief user guide for the VIs in the `Resistance Circuits.llb`.

1. Select the options Series Resistance Circuit and Parallel Resistance Circuit, respectively, and vary the values of the resistances in the circuit to estimate the equivalent resistance. First, start with one resistance only, then gradually add more resistances, and verify your results analytically.
2. One common application of the voltage divider circuit is to reduce a high voltage to the low levels that are used in signal conditioning circuits. First, select the Voltage Divider Circuit option from the Menu Ring of the VI. We would like to measure a 200 V voltage that should be scaled down to 5 V to be linked to a computer. Find the values of the resistances to make sure that their powers do not exceed 1 W.
3. Select the Current Divider Circuit, and set the parameters as $V_{dc} = 12$ V, $R_1 = 2\ \Omega$, $R_2 = 2\ \Omega$, $R_3 = 10{,}000\ \Omega$ (to introduce an open circuit), and estimate the currents in each branch. First double and then halve the values of the resistances and compare your results.
4. Select the Series/Parallel Circuit and set the parameters as $V_{dc} = 12$ V, $R_1 = 2\ \Omega$, $R_2 = 2\ \Omega$, $R_3 = 2\ \Omega$. Comment on the powers of each component. What is the power taken from the supply V_{dc}?
5. Select the Series/Parallel Circuit and set the parameters as $V_{dc} = 12$ V, $R_1 = 2\ \Omega$, $R_2 = 2\ \Omega$, $R_3 = 10{,}000\ \Omega$, and record the current values I_1, I_2, and I_3. Change R_3 to 20,000 Ω, and find the new values of I_1, I_2, and I_3. Comment on your findings.

2.3.2 Mesh Analysis

In solving electric circuits, Kirchhoff's laws, mesh analysis (unknowns are currents), and nodal analysis (unknowns are voltages) can be utilized. These can provide all the independent current and voltage equations.

Kirchhoff's Current Law (KCL) states that the algebraic sum of currents entering a node (where two or more elements have a common terminal) is equal to zero. In standard notation, the ingoing currents are considered negative, and the outgoing currents are considered positive.

Kirchhoff's Voltage Law (KVL), on the contrary, states that the algebraic sum of voltages around a loop (consisting of nodes and branches, which form simple closed paths) is equal to zero. In standard notation, an arrow or $+/-$ signs are used to indicate the sign of the voltage potential. The + sign is equivalent to the head of the arrow, which is an arbitrary choice.

Mesh analysis starts by defining a current circulating around each mesh (the loops corresponding to the open areas in the circuit without any crossovers). The element currents are then the algebraic sums of the mesh currents

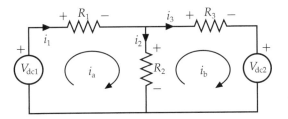

Figure 2-8
A sample electric circuit used in mesh analysis.

that pass through them. Since each mesh current enters and leaves a node, KCL is automatically satisfied. The resulting equations are the result of KVL applied to each mesh, hence the unknowns are the mesh currents. The remaining unknowns in the circuit (elements' currents and voltages) can be calculated using the mesh currents.

For the simple circuit given in Fig. 2-8, if KVL is written for each mesh using the standard notation in relation to the mesh currents i_a and i_b,

$$-v_{R2} - v_{R1} + v_{dc1} = -R_2(i_a - i_b) - R_1 i_a + V_{dc1} = 0 \qquad (2.21)$$

$$-v_{dc2} - v_{R3} + v_{R2} = -V_{dc1} - R_3 i_b + R_2(i_b - i_a) = 0 \qquad (2.22)$$

If equations 2.21 and 2.22 are rearranged, the unknown currents, i_a and i_b, can be calculated.

$$V_{dc1} = i_a(R_1 + R_2) - i_b R_2 \qquad (2.23)$$

$$V_{dc2} = i_a R_2 - i_b(R_2 + R_3) \qquad (2.24)$$

Hence, the currents and voltages of the resistance elements are

$$i_1 = i_a, \qquad v_{R1} = R_1 i_a \qquad (2.25)$$

$$i_2 = i_a - i_b, \qquad v_{R2} = R_2(i_a - i_b) \qquad (2.26)$$

$$i_3 = i_b, \qquad v_{R3} = R_3 i_b \qquad (2.27)$$

2.3.2.1 Virtual Instrument Panel

The front panel of the `Mesh Analysis.vi` illustrated in Fig. 2-9 can be accessed via `Resistance Circuits.llb`.

2.3.2.2 Self-Study Questions

Open and run the VI named `Resistance Circuits.vi`, in the `Chapter 2 VIs` folder, and select the option Mesh Analysis.

1. Consider the circuit parameters as $V_{dc1} = 42$ V, $V_{dc2} = 10$ V, $R_1 = 6\,\Omega$, $R_2 = 3\,\Omega$, $R_3 = 4\,\Omega$, and confirm that the mesh currents I_a and I_b are 4.889 A and 0.667 A, respectively. Verify your findings by manual calculations.

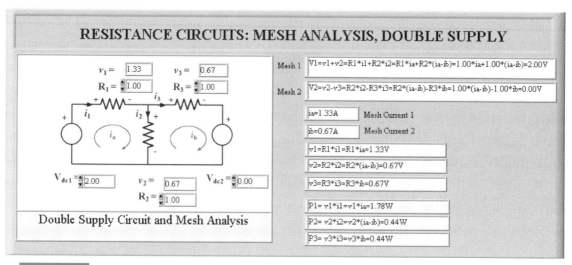

Figure 2-9
Front panel of the `Mesh Analysis Circuit.vi`.

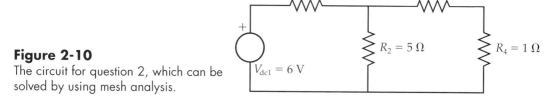

Figure 2-10
The circuit for question 2, which can be solved by using mesh analysis.

2. Consider the circuit given in Fig. 2-10, and confirm that the voltage across the resistances R_2 and R_4 is 3.333 V and 0.667 V, respectively.

Hint: Use an equivalent resistance for R_3 and R_4 in Fig. 2-10, and set $V_{dc2} = 0$ V in the front panel circuit.

2.4 Thevenin's and Norton's Equivalent Circuits

Thevenin's and Norton's equivalent circuits are used to transform complex circuits to simple circuits, voltage sources into current sources, or current sources to voltage sources (which are also known as source transformations), provided that there is an appropriate resistance in series with the voltage source or in parallel with the current source.

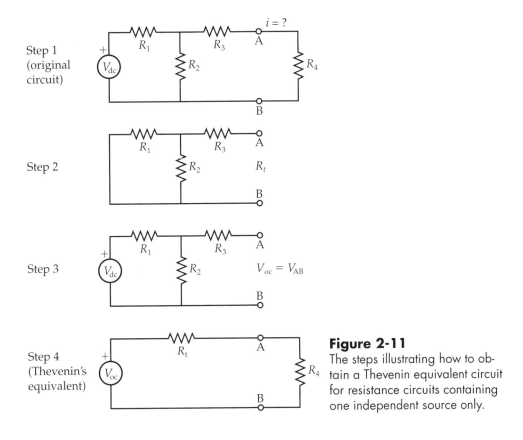

Figure 2-11
The steps illustrating how to obtain a Thevenin equivalent circuit for resistance circuits containing one independent source only.

Consider a circuit with resistance elements and a voltage source with identified output terminals A and B. A Thevenin's equivalent circuit can be constructed by a series combination of an ideal voltage source V_{oc} and a resistance R_t, where V_{oc} is the open-circuit voltage at the identified terminals and R_t is the Thevenin's equivalent of the resistor.

The resistor R_t is the ratio of the open-circuit voltage to the short-circuit current at the terminals A–B. The steps followed to obtain this transformation are visually illustrated in Fig. 2-11, which is also used in the LabVIEW simulation.

As shown in Fig. 2-11, the principal aim is to find the current that flows through the resistor R_4. This can easily be estimated if the circuit on the left-hand side of the terminals A–B is transformed to a simple circuit given in Step 4 that contains the Thevenin equivalent circuit.

In Step 2, short-circuiting the source terminals deactivates the voltage source and allows the equivalent Thevenin resistance R_t to be calculated. Note that if a current source is present in a circuit, it should be open-circuited in Step 2.

Table 2-3 *Three methods of finding Thevenin and Norton equivalent circuits.*

Features of the Resistance Circuits	Steps to Obtain **Thevenin** Equivalent Circuit	Steps to Obtain **Norton** Equivalent Circuit
With independent sources	• Deactivate the sources and find R_t • Find open-circuit voltage v_{oc} with the sources included *Sample:* Fig. 2-12a	• Deactivate the sources and find R_t • Find short-circuit current i_{sc} with the sources included *Sample:* Fig. 2-12d
With independent and dependent sources or With independent sources	• Find open-circuit voltage v_{oc} with the sources included • Find short-circuit current i_{sc} by short circuiting the terminals A and B. • Calculate $R_t = v_{oc}/i_{sc}$ *Sample:* Fig. 2-12b	• Find short-circuit current i_{sc} with the sources included • Find open-circuit voltage v_{oc} at the terminals A and B. • Calculate $R_t = v_{oc}/i_{sc}$ *Samples:* Fig. 2-12e and 2-12f
With dependent sources where $v_{oc} = 0$ or $i_{oc} = 0$	• Where $v_{oc} = 0$ • Connect a 1 A current source to the terminals A and B, and calculate v_{AB}. • Estimate $R_t = v_{AB}/1$ A *Sample:* Fig. 2-12c	• Where $i_{sc} = 0$ • Connect a 1 A current source to the terminals A and B, and calculate v_{AB}. • Estimate $R_t = v_{AB}/1$ A

In Step 3, the open-circuit voltage v_{oc} across the terminals A–B is calculated, and the Thevenin equivalent circuit is replaced with the original circuit in the final step, Step 4.

The Norton's equivalent circuit can also be constructed with a single current source equal to the short-circuit current at terminals A–B, in parallel with a single resistance. The resistance in the Norton equivalent is the same as the Thevenin resistance.

As summarized in Table 2-3, three methods can be identified for the distinct electrical circuits, which can be used to determine Thevenin and Norton equivalent circuits. Six distinct electric circuits illustrated in Fig. 2-12 are used to study Thevenin and Norton equivalent circuits.

2.4.1 Virtual Instrument Panel

The objective of this section is to study Thevenin's and Norton's equivalent circuits in sufficient detail, which requires the custom-written `Thevenin Norton.vi`. Two front panels given in Fig. 2-13 illustrate the layout of the Thevenin and Norton equivalent circuits with the associated steps.

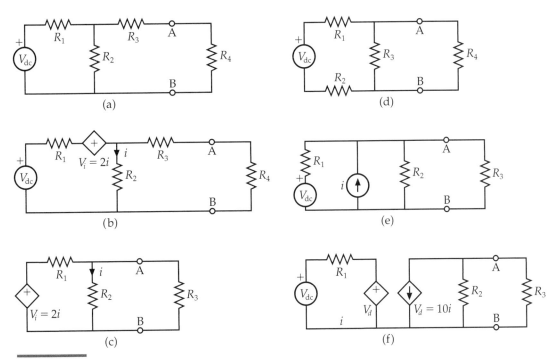

Figure 2-12
(a), (b), and (c) Typical electric circuits used to obtain Thevenin equivalent circuits. (d), (e), and (f) Typical electric circuits used to obtain Norton equivalent circuits.

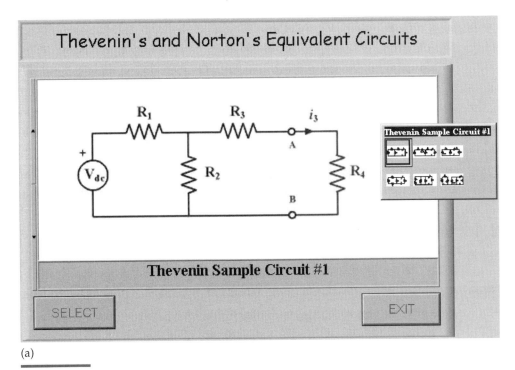

(a)

Figure 2-13
(a) The main front panel of `Thevenin Norton.vi` and (b) a sample front panel. *(cont.)*

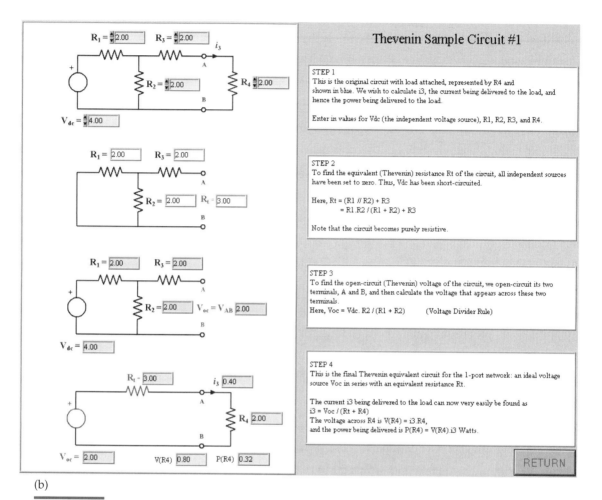

(b)

Figure 2-13
Continued

2.4.2 Self-Study Questions

Open and run the custom-written VI named `Thevenin Norton.vi` in the `Chapter 2 VIs` folder, and study the following questions.

1. Consider the circuit parameters for the circuit given in Part Thevenin (1) as $V_{dc} = 12$ V, $R_1 = 3\,\Omega$, $R_2 = 6\,\Omega$, $R_3 = 7\,\Omega$, $R_4 = 3\,\Omega$. Using the Thevenin's equivalent of the circuit, find the voltage across the output resistor R_4.
   Answer: $V_{AB} = 2$ V (for $V_{oc} = 8$ V, $R_t = 9\,\Omega$)

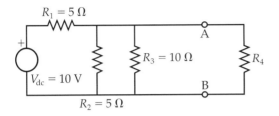

Figure 2-14
The circuit diagram for question 5.

2. Consider the circuit parameters for the circuit given in Part Thevenin (2) as $V_{dc} = 20$ V, $R_1 = 6\ \Omega$, $R_2 = 6\ \Omega$, and $R_3 = 10\ \Omega$. Determine the Thevenin's equivalent circuit.
   Answer: $V_{oc} = 12$ V, $R_t = 13.6\ \Omega$

3. Consider the circuit given in Part Thevenin (3), where $R_1 = 3\ \Omega$ and $R_2 = 6\ \Omega$. Determine the Thevenin's equivalent circuit.
   Answer: $V_{oc} = 0$ V, $R_t = 3.85\ \Omega$

4. Consider the circuit parameters for the circuit given in Part Norton (1) as $V_{dc} = 15$ V, $R_1 = 8$ kΩ, $R_2 = 4$ kΩ, and $R_3 = 6$ kΩ. Determine the Norton's equivalent circuit.
   Answer: $i_{sc} = 1.25$ mA, $R_n = 4$ kΩ

5. Consider the circuit given in Fig. 2-14 and determine the Thevenin's equivalent circuit. *Hint:* You can use the circuit given in Part Thevenin (1) of the custom-written VI. However, you have to include an equivalent resistance into the control box of R_2 and input $R_3 = 0$.
   Answer: $V_{oc} = 4$ V, $R_t = 2\ \Omega$

2.5 References

Davis, B. R., and B. E. Bogner. "Electrical Systems A, Electric Circuits Lecture Notes." Department of Electrical and Electronic Engineering. University of Adelaide, 1998.

Dorf, R. C., and J. A. Svoboda. *Introduction to Electric Circuits*. New York: Wiley, 1996.

Edminister, J. A. *Theory and Problems of Electric Circuits*. Schaum's Outline Series. New York: McGraw-Hill, 1972.

Ertugrul, N. "Electric Power Applications Lecture Notes." Department of Electrical and Electronic Engineering. University of Adelaide, 1997.

Wildi, T. *Electrical Machines, Drives, and Power Systems*. Englewood Cliffs, NJ: Prentice Hall, 1991.

AC Circuits

3

Electric power is generated, distributed, and used as sinusoidal voltages and currents in a great variety of commercial and domestic applications. Furthermore, in the industrial world a vast number of small-, medium-, or large-size ac power generators and loads are interlinked. Therefore, the design, operation, maintenance, and management of such systems very much depend on a good understanding of ac circuit theory.

The analysis of ac circuits involves the study of the behavior of the system under both normal and abnormal operating conditions. However, this book is not intended to include abnormal conditions. Instead, it focuses on the foremost fundamental issues and normal conditions and presents visual aids and interactive tools. Additionally, it is assumed that a steady-state sinusoidal condition is reached, which means that all transient effects in ac circuits have disappeared.

This chapter's topics are of practical importance in ac systems. The fundamental terminologies used in ac circuits are introduced, and a number of custom-written VIs, considering both single-phase and three-phase ac circuits, are provided.

The chapter begins with essential definitions of such terms as *power factor, phasor, impedance,* and *per-unit value.*

Section 3.2 describes the topological analysis of five basic ac electric circuits containing impedances and ac supplies. Then a reverse study is presented where an unknown impedance is determined by user-defined ac voltage and currents.

Section 3.3 is concerned with a description and visual demonstration of the powers in ac circuits and is followed by a discussion of power factor correction in Section 3.4.

The remaining sections describe various other aspects of three-phase circuits, accompanied by interactive VI modules. The chapter ends with a comprehensive study of real-time three-phase data logging.

Educational Objectives The chapter develops appropriate relationships and visual aids for describing ac systems using ac voltage, ac current, impedance, ac power, and phasors. After completing this chapter, students should be able to

- plot and interpret the characteristics of ac voltage, current, and power waveforms.
- understand the definitions of peak to peak; peak and rms values; and phase/line voltage and current, phase angle, power factor, complex impedance, phasor and base (per-unit) values in ac circuits.
- state the equations for series, parallel, and combination ac circuits that contain impedances and ac supplies, understand the effects on current caused by changes in impedances, and state the meaning of the term *equivalent impedance.*
- analyze the sinusoidal steady-state behavior of single- and three-phase ac circuits using phasors, and study the effect of resistive, inductive, and capacitive loads in single-phase ac circuits.
- understand the concept of complex power, power measurement methods, and power factor correction.
- recognize the data logging techniques associated with ac circuits.

3.1 Fundamental Definitions

Steady-state sinusoidal time-varying voltage and current waveforms can be given by

$$v(t) = V_m \sin(\omega t) \quad (3.1)$$
$$i(t) = I_m \sin(\omega t + \theta) \quad (3.2)$$

where v and i are the time-varying voltage and current, and V_m and I_m are the peak values (magnitudes or amplitudes) of the voltage and current waveforms. In equation 3.2, θ is known as the phase angle, which is normally defined with reference to the voltage waveform.

The term $\cos \theta$ is called a power factor. Remember that we assumed a voltage having a zero phase. In general, the phase of the voltage may have a value other than zero. Then θ should be taken as the phase of the voltage minus the phase of the current.

In a linear circuit excited by sinusoidal sources, in the steady-state, all voltages and currents are sinusoidal and have the same frequency. However, there may be a phase difference between the voltage and current depending on the type of load used.

The three basic passive circuit elements, the resistor (R), the inductor (L), and the capacitor (C), are considered in this chapter. An ac load may be a combination of these passive elements, such as R + L or R + C.

Note that the current and voltage waveforms in the resistor are in phase, while inductances and capacitors both have a 90° phase shift between voltage and current. The inductor current waveform lags the inductor voltage waveform by 90°, while in the capacitor, the current leads the voltage by 90°.

The peak-to-peak value is also used in the analysis of ac circuits; it is the difference between the highest and lowest values of the waveform over one cycle. This can easily be visualized in the ac waveforms generated.

It might seem difficult to describe an ac signal in terms of a specific value, since an ac signal is not constant. However, as shown in Chapter 2, these sinusoidal signals are periodic, repeating the same pattern of values in each period. Therefore, when a voltage or a current is described simply as ac, we will refer to its rms or effective value, not its maximum value, which simplifies the description of ac signal.

Power Factor

In general, for nonsinusoidal systems (distorted waveforms), the power factor PF is equal to

$$\text{PF} = \frac{P_{\text{average}}}{S_{\text{total}}} \quad (3.3)$$

where S_{total} is the total apparent (or complex) power in VA, which is equal to $V_{rms} \cdot I_{rms}$.

If the voltage and current waveforms of an ac system are measured in real time, it is much easier to calculate the power factor simply by using the definitions of average and rms values (given in equations 2.3 and 2.4).

However, in many utility applications, the distortion in ac voltage is usually small, hence the voltage waveform can be assumed as an ideal sine wave at fundamental frequency. This assumption simplifies the analysis, which results in an analytical solution of power factor for the nonsinusoidal systems as

$$\text{PF} = \frac{I_{s1}}{I_s} \cos \theta \tag{3.4}$$

where I_s is the rms value of the current (as in equation 2.4 for sinusoidal quantities), I_{s1} is the rms Fourier fundamental component of the current, and $\cos \theta$ is the power factor in linear circuits with sinusoidal voltages and currents.

From equation 3.4, we can conclude that a large distortion in the current waveform will result in a small value of the current ratio, I_{s1}/I_s, which means a small value of power factor, even if $\theta = 0$ (a unity power factor, $\cos \theta = 1$).

As seen, since the Fourier analysis is involved for the estimation of I_{s1}, a quantity called total harmonic distortion (THD) can be defined in the current, which simplifies the estimation of PF.

$$\text{THD} = \frac{\sqrt{I_s^2 - I_{s1}^2}}{I_{s1}} \tag{3.5}$$

$$\text{PF} = \frac{1}{\sqrt{1 + \text{THD}^2}} \cos \theta \tag{3.6}$$

Note that the definitions given for the PF will be utilized in the transformer experiment later since the excitation current of transformers is highly distorted.

Phasors

In most ac circuit studies, the frequency is fixed, so this feature is used to simplify the analysis. Sinusoidal steady-state analysis is greatly facilitated if the currents and voltages are represented as vectors in the complex number plane known as phasors. The basic purpose of phasors is to show the rms value (or the magnitude in some cases) and phase angle between two or multiple quantities, such as voltage and current.

The phasors can be defined in many forms, such as rectangular, polar, exponential, or trigonometric:

$$Z = R + jX, \quad Z = Z\angle\theta, \quad Z = |Z|e^{j\theta}, \quad Z = |Z|(\cos\theta + j\sin\theta) \quad (3.7)$$

where R is the real part and X is the imaginary part of the complex number, and $|Z|$ is the absolute value of Z.

To understand the phasors theoretically, let us consider a sinusoidal voltage function. If the rotating term at angular frequency ω is ignored, the phasor function can be given by using the real part of a complex function in polar form.

$$v(t) = V_m \cos(\omega t + \theta) = \text{Re}\{V_m e^{j(\omega t + \theta)}\} = \text{Re}\{(V_{rms}e^{j\theta})(\sqrt{2}e^{j(\omega t)})\} \quad (3.8)$$

and the voltage phasor function is

$$V = V_{rms}e^{j\theta} = V_{rms}\angle\theta \quad (3.9)$$

This phasor is visualized as a vector of length V_{rms} that rotates counterclockwise in the complex plane with an angular velocity of ω. As the vector rotates, its projection on the real axis traces out the voltage as a function of time. In fact, the "phasor" is simply a snapshot of this rotating vector at $t = 0$ and will be shown in the phasor graphs of the VIs implemented.

In addition, when the periodic voltage, current, and power waveforms are considered, each data point in these waveforms can be represented in the complex plane. This feature is also demonstrated in the chapter's phasor VIs.

Impedance

The impedance Z in ac circuits is defined as the ratio of voltage function to current function. Hence, the impedance is a complex number and can be expressed in the rectangular form as

$$Z = \frac{V_{rms}\angle 0°}{I_{rms}\angle\theta°} = R \pm jX \quad (3.10)$$

The real component of the impedance is called the resistance R and the imaginary component is called the reactance X, both of which are in ohms. The reactance is a function of ω in L and C loads, and for an inductive load, X is positive, whereas for a capacitive load, X is negative (Fig. 3-1).

The impedance can also be displayed in the complex plane as phasors, the voltage, and the current waveforms. However, since the resistance is never negative, only the first and fourth quadrants are required. This restriction is implied in the associated VIs by limiting the upper and lower values of the controls.

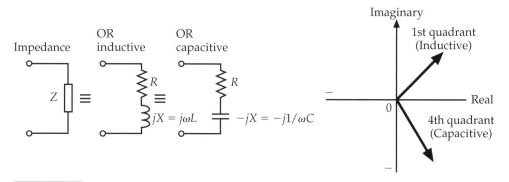

Figure 3-1
Equivalent circuits of an impedance and impedance phasor in the complex plane.

Per-Unit Values

The per-unit system of measurement and computation is used in electrical engineering for two principal reasons:

- To display multiple quantities (such as voltages and currents) on the same scale for comparison purposes
- To eliminate the need for conversion of the voltages, currents, and impedances in the circuits and to avoid using transformation from three-phase to single-phase, and vice versa

The quantity subject to conversion is normalized in terms of a particularly convenient unit, called the per-unit base of the system. Note that whenever per-unit values are given, they are always dimensionless. To calculate the actual values of the quantities, the magnitude of the base of the per-unit system must be known. In electrical circuits, voltage, current, impedance, and power can be selected as base quantities.

3.2 AC Circuit Analysis

The topological analysis of basic ac electric circuits containing impedances and ideal ac supplies are presented in the following subsections. As will be demonstrated, using phasors greatly simplifies the analysis, and the VIs provide a flexible self-learning tool allowing users to create different circuit scenarios. Although Thevenin and Norton equivalent circuits as discussed for dc circuits in Chapter 2 can also be applied to ac circuits, they are not covered here.

3.2.1 Equivalent Impedances and Circuits

An ac circuit may contain a number of series and/or parallel branches. As will be studied in the following paragraphs, however, it is possible to divide any complex ac circuit into subcircuits that include simple circuit combinations. Therefore, five ac circuit combinations are identified and studied here: equivalent impedance circuit, voltage divider circuit, current divider circuit, series/parallel (combination) circuit, and circuit with dual ac supply.

The equivalent impedance of any number of impedances in series or in parallel (Fig. 3-2) is the sum of the individual impedances or the sum of the admittances that is equal to $1/Z$, respectively.

$$Z_{eq(series)} = Z_1 + Z_2 + Z_3 + Z_4 \tag{3.11}$$

$$Z_{eq(parallel)} = \frac{1}{(1/Z_1 + 1/Z_2 + 1/Z_3 + 1/Z_4)} \tag{3.12}$$

Since the current phasors and voltage phasors are related by complex impedances, Kirchhoff's Voltage and Current Laws can be applied to ac circuits containing sinusoidal sources operating in steady-state.

The voltage divider circuit in Fig. 3-3a indicates that if two impedances are connected in series and share the same current (meaning no other element is connected to the node where Z_1 and Z_2 join), the voltages across each of the

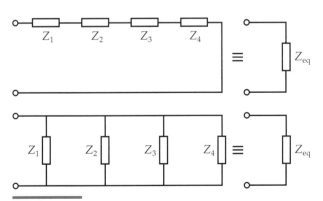

Figure 3-2
Series and parallel impedance circuits.

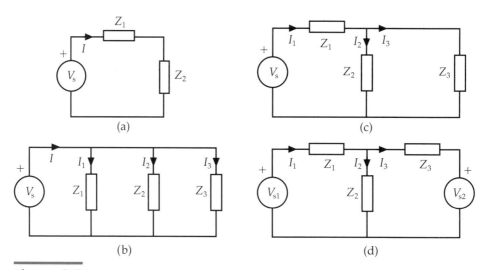

Figure 3-3
Circuits with impedances: (a) voltage divider, (b) current divider, (c) series/parallel circuit, and (d) circuit with dual supplies.

elements are proportional to their impedances. The current and the voltages across each impedance can be given as

$$I = \frac{V_s}{(Z_1 + Z_2)} \tag{3.13}$$

$$V_1 = Z_1 I \tag{3.14}$$

$$V_2 = Z_2 I \tag{3.15}$$

Let's consider the current divider circuit in Fig. 3-3b now. Since the voltages across the impedances are equal to the supply voltage,

$$V_s = V_1 = V_2 = V_3 \tag{3.16}$$

then the currents in each impedance branch can be calculated easily.

$$I_1 = \frac{V_s}{Z_1}, \quad I_2 = \frac{V_s}{Z_2}, \quad I_3 = \frac{V_s}{Z_3} \tag{3.17}$$

Hence the supply current I is simply the sum of all three currents.

$$I = I_1 + I_2 + I_3 \quad \text{or} \quad I = \frac{V_s}{Z_{eq}} \tag{3.18}$$

The series/parallel circuit of Fig. 3-3c can be analyzed easily if the equivalent impedance is calculated first.

$$Z_{eq} = Z_1 + \frac{Z_2 \cdot Z_3}{Z_2 + Z_3} \tag{3.19}$$

All remaining unknown parameters in the circuit can be derived as

$$I_1 = \frac{V_s}{Z_{eq}}, \quad I_2 = \frac{V_2}{Z_2}, \quad I_3 = \frac{V_3}{Z_3} \quad (3.20)$$

$$V_1 = I_1 Z_1, \quad V_2 = V_3 = V_s - V_1, \quad V_3 = I_3 Z_3 \quad (3.21)$$

The circuit given in Fig. 3-3d is selected since it represents a commonly used equivalent circuit in electrical engineering: single-phase transformer, and asynchronous motor. Note that the circuit given in Fig. 3-3c can be obtained easily by eliminating the supply voltage V_{s2} in Fig. 3-3d.

In the analysis of this dual supply circuit, the mesh analysis can be used as described in Chapter 2. Hence the supply voltages are given as

$$V_{s1} = I_1(Z_1 + Z_2) - I_3 Z_2 \quad (3.22)$$

$$V_{s2} = I_1 Z_2 - I_3(Z_2 + Z_3) \quad (3.23)$$

The currents and voltages of the impedance elements can be determined after solving equations 3.22 and 3.23, in which the known parameters are V_{s1}, V_{s2}, Z_1, Z_2, and Z_3, and the unknowns are I_1 and I_3. To solve for I_1 and I_3, the back substitution method is used in the VI provided in the following section. Then the remaining unknowns, I_2, V_1, V_2, and V_3, in the circuit are calculated as

$$I_2 = I_1 - I_3 \quad (3.24)$$

$$V_1 = Z_1 I_1, \quad V_2 = Z_2 I_2, \quad V_3 = Z_3 I_3 \quad (3.25)$$

3.2.1.1 Virtual Instrument Panel

The virtual instruments provided in this section contain six circuit options. The desired circuits can be selected from the available folders. When a VI is selected, the corresponding circuit and its associated controls are displayed with their default values. Then students can run the VI and vary the complex quantities of the circuit and observe the calculated values all in the complex forms.

As demonstrated, the parameters in the circuits are given in rectangular form, which can easily be related to the practical values. In addition, the rectangular form allows the user to omit either a real or an imaginary component of the complex numbers. See Fig. 3-4 for a sample front panel.

3.2.1.2 Self-Study Questions

Open and run the custom-written VI named `Impedance Circuits.vi` in the `Chapter 3` folder, and investigate the following questions.

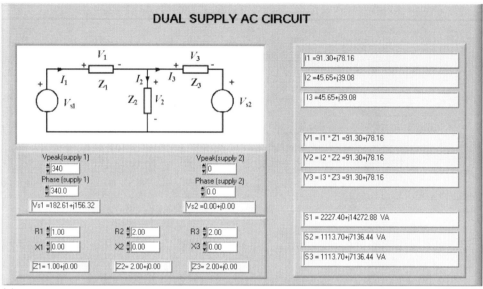

(a)

(b)

Figure 3-4
A sample front panel and brief user guide.

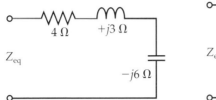
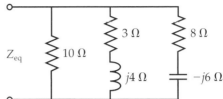

Figure 3-5
Example circuits for question 1.

1. In the impedance circuits shown in Fig. 3-5, calculate the equivalent impedances.
   Answers: $Z_{eq(series)} = 5\angle-36.9°\ \Omega$, $Z_{eq(parallel)} = 3 + j1\ \Omega$

2. Consider $V_s = 10\angle 0°$, $Z_1 = +j100\ \Omega$, $Z_2 = 100\ \Omega$, and $Z_3 = -j100\ \Omega$ for the circuit given in Figure 3-3b. Determine the currents in each impedance branch, and verify the results.

3. For the identical settings in question 2, vary the frequency of the voltage and observe its effect on the complex values of the calculated values.

4. Consider the voltage divider circuit, and set the supply voltage to $V_s = 10\angle 90°$. Set the impedances to $Z_1 = 6$ and $Z_2 = j8$, and determine the voltage across Z_2.
   Note: Remember to convert the complex voltage to the format that is acceptable by the custom-written VI, `Impedance Circuits.vi`.
   Answer: $8.00\angle 126.87°$

5. Compute the currents in each element for the circuit given in Fig. 3-6. Assume that the supply voltage is equal to $10\angle -90°$. Then verify the results analytically.
   Answers: $I_L = 0.114\angle -135°$, $I_R = 0.1\angle -180°$, $I_C = 0.1\angle -90°$

6. Consider the ac circuit in Fig. 3-7, and calculate the value of the capacitor if $\omega = 2$ rad/s and $v_s(t) = 90 \sin 2t$.

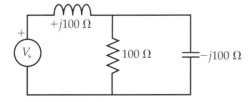
Figure 3-6
Example circuit for question 5.

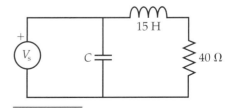
Figure 3-7
Example circuit for question 6.

Hint: You can ignore one of the impedance branches shown in Fig. 3-3b or 3-3c. Remember that capacitive reactance is equal to $1/\omega C$.
Answer: 6 mF

3.2.2 A Reverse Study

A reverse study is presented here, where an unknown impedance is determined utilizing ac voltage and current waveforms that are set by the user.

Fig. 3-8 illustrates the block diagram of the VI model, where the ideal sine wave current i_s and voltage v_s waveforms are defined by the user, and the corresponding equivalent impedance seen from the two terminals A–B is determined.

The complex value of the equivalent impedance is displayed on the front panel, which is used to interpret the nature of the load (R, L, C, R + L, or R + C). The phasor diagram is also provided in the VI to illustrate the voltage and impedance phasors.

3.2.2.1 Virtual Instrument Panel

Fig. 3-9 illustrates the front panel and the explanations diagram of the VI under investigation. The VI provided here (`Single Phase AC Definitions .vi`) uses the voltage and current as the base of the per-unit system, mainly for display purposes. The equivalent impedance is calculated and displayed on the front panel (as a per-unit value and a real value).

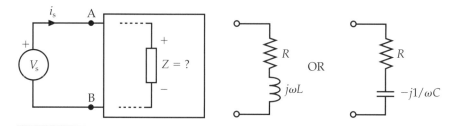

Figure 3-8
The circuit simulated to identify the equivalent impedance of an unknown load Z.

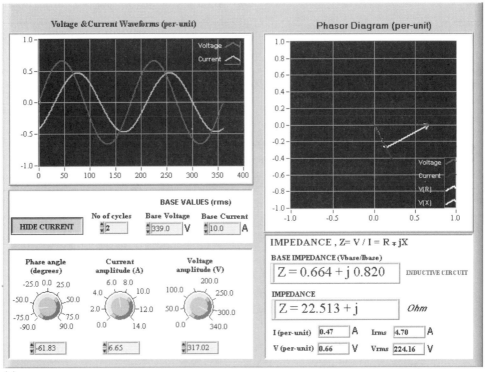

(a)

(b)

Figure 3-9
The front panel and the brief user guide of `Single Phase AC Definitions.vi`.

3.2.2.2 Self-Study Questions

Open and run the custom-written VI named `Single Phase AC Definitions.vi` in the `Chapter 3` folder, and investigate the following questions.

1. Vary the values of the phase angle, the current amplitude, and the voltage amplitude by using the knobs provided. Observe the relative positions of the voltage and current waveforms on the graph (in terms of phase angle).
2. Set the amplitudes of the voltage and the current equal to the base values, and analyze the estimated per-unit values of the voltage and current and the rms values.
3. Set the voltage and current waveforms to $V_S = 50\angle 0°$ and $I_S = 15.8\angle -18.45°$, and verify that the equivalent impedance estimated in the front panel ($Z_{eq} = 3 + j1\ \Omega$) is correct.
4. Without referring to the phasor diagram on the graph, plot the phasor diagram for the voltage and the current waveform that are shown on the waveform graph, and compare them with the displayed phasors. Is the load resistive, capacitive, or inductive, and why?
5. After making graphical observations of the voltage and current waveforms under three typical conditions (in phase, with lagging angle, and leading phase angle), observe and record the impedances, and verify the results analytically. In addition, observe that the impedances of both the pure inductor and the pure capacitor are pure imaginary numbers.
6. In the case of R + L or R + C load, calculate the value of the inductor L or the capacitor C. Assume that the supply frequency is 60 Hz.

3.3 Power and Power Triangles in AC Circuits

This discussion will focus on single-phase ac circuits and develop the definitions for ac powers: complex power, active (real) power, and reactive power. In addition, the graphical method for ac powers, the power triangles, will be presented.

The instantaneous power delivered to a load can be expressed as

$$p(t) = v(t)i(t) \qquad (3.26)$$

The instantaneous power may be positive or negative depending on the sign of $v(t)$ and $i(t)$, which is related to the sign of the signal at a given time. A positive power means that power flows from the supply to the load, and a negative value indicates that power flows from the load to the supply.

In the case of sine wave voltage and current, using trigonometric identities, the instantaneous power may be expressed as the sum of two sinusoids of twice the frequency.

$$v(t) = V_m \cos(\omega t) \tag{3.27}$$

$$i(t) = I_m \cos(\omega t + \theta) \tag{3.28}$$

$$p(t) = V_m I_m \cos(\omega t)\cos(\omega t + \theta) \tag{3.29}$$

$$p(t) = V_{rms} I_{rms} \cos\theta \cdot (1 + \cos 2\omega t) + V_{rms} I_{rms} \sin\theta \cdot \cos(2\omega t + \pi/2) \tag{3.30}$$

In equations 3.27 through 3.30, the phase angle θ can be any value varying from $-90°$ to $+90°$ for general resistance, inductance, and capacitance loads (RLC).

The first term on the right-hand side of the power equation is known as instantaneous average power, real power, or active power, and is measured in watts (W), kW, or MW.

The second term on the right-hand side is called instantaneous reactive power, and its average value is zero. The maximum value of this term is known as the reactive power, and it is measured in volt-ampere reactive (VAR), kVAR, or MVAR.

The instantaneous powers corresponding to various loads are graphically illustrated in Fig. 3-10. Note that if a single phase of the system is concerned, our previous discussions can be applied to the three-phase balanced ac circuits as well.

Hence, the active power and the reactive power are given by

$$P = V_{rms} I_{rms} \cos(\theta) \tag{3.31}$$

$$Q = V_{rms} I_{rms} \sin(\theta) \tag{3.32}$$

As studied earlier, the cosine of the phase angle θ between the voltage and the current is called power factor.

The apparent power S can be calculated from P and Q as

$$S = V_{rms} I_{rms} = \sqrt{P^2 + Q^2} \tag{3.33}$$

The apparent power is measured in volt-amperes (VA), kVA, or MVA.

The complex power in ac circuits can be given as

$$S = P \pm jQ = V_{rms} I_{rms} \cos(\theta) + jV_{rms} I_{rms} \sin(\theta) \tag{3.34}$$

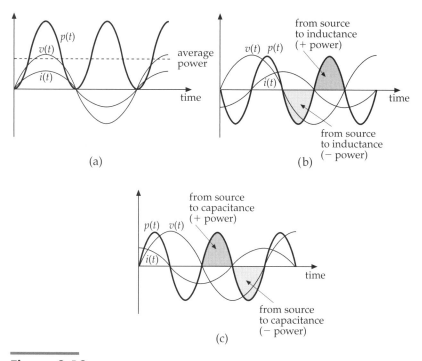

Figure 3-10
The waveforms showing the features of the instantaneous powers: (a) pure resistance load, (b) pure inductance load, and (c) pure capacitance load.

Here S indicates a complex number. As indicated, the real part of the complex power equals the active power P, and the imaginary part is the reactive power Q.

Hence, from these definitions, the equations associated with the active, reactive, and apparent power can be developed geometrically on a right triangle called a power triangle. The power triangle is shown in Fig. 3-11 for an inductive and a capacitive load.

The impedance angle θ is also called a power angle. If the current lags the voltage, the load is inductive, the angle θ is positive, and the case is said to have a lagging power factor. Conversely, if the current leads voltage, the load is capacitive, the angle θ is negative, and the case is said to have a leading power factor. Remember that the power factor $\cos\theta$ does not actually lag or lead; the current lags or leads the voltage.

Chapter 3 • AC Circuits

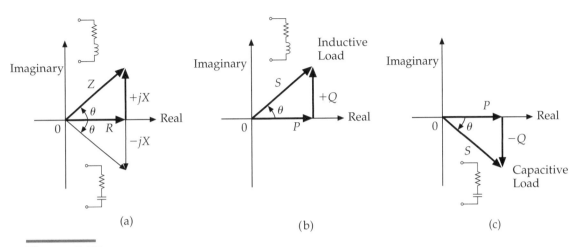

Figure 3-11
Derivation of power triangles in ac circuits: (a) impedance triangles, (b) power triangles in inductive loads, and (c) in capacitive loads.

3.3.1 Virtual Instrument Panel

The power triangle using the phasors is illustrated on the front panel of the VI in Fig. 3-12. In the phasor graph, the horizontal axis represents the active power and the vertical axis represents the reactive power.

3.3.2 Self-Study Questions

Open and run the custom-written VI named `Power Triangles.vi` in the `Chapter 3` folder, and investigate the following questions.

1. Set Voltage Amplitude = 339 V, Base Voltage = 339 V, Base Current = 10 A, R_{load} = 10 Ω, X_{load} = 10 Ω, f = 50 Hz, and observe the waveforms of the voltage, the current, and the power triangle graph. What are the values of the active, reactive, and apparent powers and the power factor of the load? Verify the displayed values analytically, and estimate the power values by using the power triangle graph.
2. Keep the identical settings as in question 1. Then vary the values of the impedance as

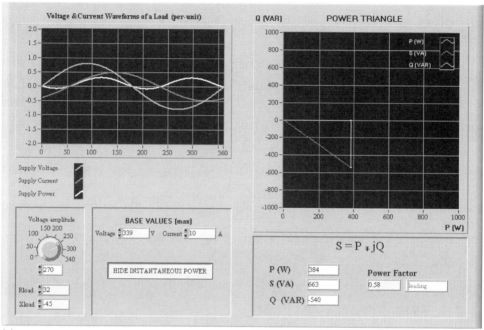

Figure 3-12
The front panel and brief user guide of `Power Triangles.vi`.

Chapter 3 • AC Circuits

$$R_{load} = 0\ \Omega,\ X_{load} = j10\ \Omega$$
$$R_{load} = 0\ \Omega,\ X_{load} = -j10\ \Omega$$
$$R_{load} = 10\ \Omega,\ X_{load} = 0\ \Omega$$

respectively, and observe the waveforms of the voltage, the current, the power, and the power phasors.

3. A single-phase motor winding has a resistance of 10 Ω and an inductive reactance of 25 Ω at 50 Hz. Calculate the current, the phase angle, the power factor, the apparent power, the active power, and the reactive power.
Answers: 9.28 A, 68.13°, 0.371, 2320 VA, 860 W, 2155 VAR

4. A load connected to an ac supply consists of a lamp load of 10 kW at a unity power factor, a motor load of 80 kVA at a power factor of 0.8 lagging, and another motor load of 40 kVA at a power factor of 0.7 leading. What are the values of the total active power supplied by the source and the power factor of the source?
Hints: To find the total active power you may consider each load separately and sum its active powers and reactive powers. The power factor of the source is equal to the combined power factors of the loads.
Answers: $P_{total} = 102$ kW, PF = 0.98

3.4 Power Factor Correction

When the complex power definition is analyzed, it will be seen that, if a pure inductive or a pure capacitive load is connected to an ac source, the source will be fully loaded while the active power delivered will be zero.

A practical load, however, absorbs both active power and reactive power. As illustrated in Fig. 3-10, the active power does the useful work. On the other hand, the reactive power only represents oscillating energy; however, it is required in many practical loads (for example, in inductive devices, an ac motor or a transformer absorb reactive power to produce the ac magnetic field).

Both the active and the reactive powers place a burden on the conductor (or on the transmission line). Nevertheless, the power company must provide the current to the load whether it is pure inductive or pure capacitive, and this current generates the power losses in the transmission lines. However, customers wish to pay only for the active power since it does the useful work, and the power company wishes to reduce the power losses in the transmission line.

Referring to the power triangle (Fig. 3-11), the hypotenuse S is a measure of the loading on the source, and the side P is a measure of the useful power delivered. Therefore, it is desirable to have the apparent power as close as possible to the active power, which makes the power factor approach unity (1.0).

The process of making the power factor approach 1.0 (or below 1.0 but above the existing power factor) is known as power factor correction or power factor compensation.

In practice, power factor correction is performed simply by placing a capacitor or an inductor across the existing load, which itself may be an inductive or a capacitive load, respectively. Although a more complex solution may be employed to shape the nonsinusoidal input current of a load (which mainly occur due to the power electronics converters and arc furnaces in the system), and hence to improve the power factor, this concept will not be covered here.

During the power factor correction process, the voltage across the load remains the same and the active power does not change. However, the current and the apparent power drawn from the supply decrease. This means that the amount of decrease in supply current/power can be utilized somewhere else (may be used by additional loads) without increasing the capacity of the supply.

As an example (Fig. 3-13), if the existing powers and the power factor of a single-phase ac circuit are $P = 1200$ W, $Q = 1600$ VAR, $S = 2000$ VA, and

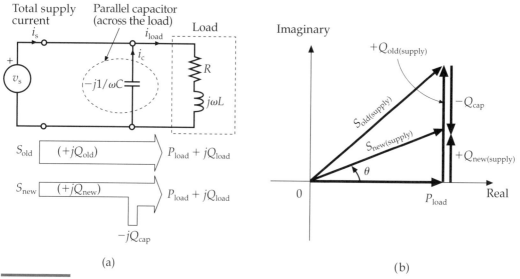

Figure 3-13
Representation of the power factor correction concept.

PF = cos θ = 0.6 lagging, and if we wish to correct the power factor to 0.9 lagging, a capacitor must be added across the load. After the correction is introduced, the active power remains unchanged but the apparent power delivered by the source is reduced to 1333 VA, and the reactive power of the capacitor equals 1015 VAR leading, which can be used to determine the ratings of the capacitor for a given voltage and supply frequency.

3.4.1 Virtual Instrument Panel

The front panel of Power Factor Correction.vi is given in Fig. 3-14. This VI provides a highly flexible virtual instrument to study the complex power definitions and the power factor correction concept in single-phase ac circuits.

3.4.2 Self-Study Questions

Although there can be many combinations of settings in the VI, the following studies are sufficient to understand the power factor correction in single-phase ac circuits. Open and run the custom-written VI named Single Phase Power and Power Factor Correction.vi in the Chapter 3 folder, and investigate the following questions.

1. Set Voltage Amplitude = 339 V, Base Voltage = 339 V, Base Current = 10 A, the value of the impedance to $R_{load} = 0\ \Omega$, $X_{load} = 10\ \Omega$, and observe the waveforms of the voltage, the current, the power, and the power triangle. Then gradually increase the Desired Power Factor to the unity power factor 1.0, and observe the Power Triangle graph. What difference(s) have you noticed?
2. Consider the operating conditions in question 1, and determine the values of the active, the reactive, the apparent power, and the power factor of the load before and after the power factor is corrected.
3. Repeat questions 1 and 2 for new values of the load impedance: $R_{load} = 10\ \Omega$, $X_{load} = -10\ \Omega$.
4. Verify the results of questions 1, 2, and 3 analytically.
5. A single-phase ac motor draws a current of 40 A at a power factor of 0.7 lagging from a 400 V, 50 Hz supply. What value must a parallel capacitor have to raise the power factor to 0.9 lagging? Note that the motor's active power remains unchanged.

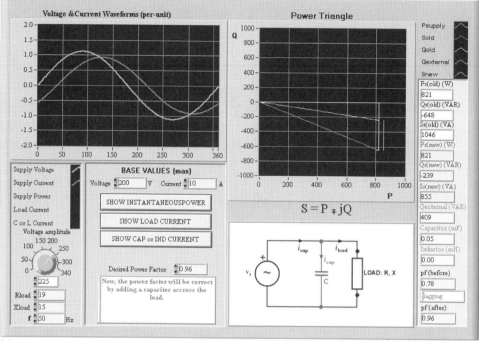

Figure 3-14
Front panel and brief user guide of `Single Phase Power and Power Factor Correction.vi`.

3.5 Star-Delta and Delta-Star Conversion in Three-Phase AC Circuits

In this book, the three-phase ac systems are considered as a balanced circuit, made up of a balanced three-phase source, a balanced line, and a balanced three-phase load. Therefore, a balanced system can be studied using only one-third of the system, which can be analyzed on a line to neutral basis.

The star-delta (Y-Δ) or delta-star (Δ-Y) conversion (Fig. 3-15) is required in three-phase ac systems to simplify the circuits and ease their analysis. If a three-phase supply or a three-phase load is connected in delta, it can be transformed into an equivalent star-connected supply or load. After the analysis, the results are converted back into their original delta equivalent.

The complex delta-star or star-delta conversion formulas are given next. These are based on the electric circuits shown in Fig. 3-15.

$$Z_A = \frac{Z_1 \cdot Z_2 + Z_2 \cdot Z_3 + Z_3 \cdot Z_1}{Z_3} \tag{3.35}$$

$$Z_B = \frac{Z_1 \cdot Z_2 + Z_2 \cdot Z_3 + Z_3 \cdot Z_1}{Z_2} \tag{3.36}$$

$$Z_C = \frac{Z_1 \cdot Z_2 + Z_2 \cdot Z_3 + Z_3 \cdot Z_1}{Z_1} \tag{3.37}$$

$$Z_1 = \frac{Z_A \cdot Z_B}{Z_A + Z_B + Z_C} \tag{3.38}$$

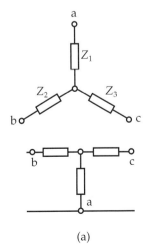
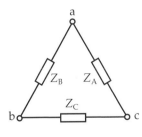
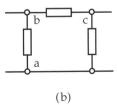

Figure 3-15
Impedance circuits that are equivalent in relationship to terminals a, b, and c: (a) star-connected and T-connected impedances, and (b) delta-connected and π-connected impedances.

$$Z_2 = \frac{Z_A \cdot Z_C}{Z_A + Z_B + Z_C} \tag{3.39}$$

$$Z_3 = \frac{Z_B \cdot Z_C}{Z_A + Z_B + Z_C} \tag{3.40}$$

where Z is the complex impedance, $Z = R \pm jX$.

Since the load is balanced, the impedance per phase of the star-connected load will be one-third of the impedance per phase of the delta-connected load. Hence the equivalent impedances can be given by

$$Z_A = Z_B = Z_C = Z \qquad Z_1 = Z_2 = Z_3 = \frac{Z}{3} \tag{3.41}$$

One of the common uses of these transformations is in power system transmission line modeling and in three-phase transformer analysis. Circuit analysis involving three-phase transformers under balanced conditions can be performed on a per-phase basis. When Δ-Y or Y-Δ connections are present, the parameters refer to the Y side. In Δ-Δ connections, the Δ-connected impedances are converted to equivalent Y-connected impedances.

3.5.1 Virtual Instrument Panel

The objective of the following VI is to study these transformation concepts and provide an easy calculation tool using the complex impedances. The front panel of `Star Delta Transformations.vi` is given in Fig. 3-16 and is capable of transforming balanced or unbalanced three-phase impedance loads.

3.5.2 Self-Study Questions

Open and run the custom-written VI named `Star Delta Transformations.vi` in the `Chapter 3` folder, and investigate the following questions.

1. Set all impedances equal and perform Δ-Y and Y-Δ transformations, then repeat the transformations for unequal impedances, and verify the results analytically.

2. The circuit shown in Fig. 3-17 is called an unbalanced Wheatstone Bridge. Find the equivalent resistance between terminals A and D, which then can be used to calculate the source current for a given supply voltage.
   Answer: 20.94 Ω
   Hint: Use Δ-Y transformation to simplify the circuit.

Chapter 3 • AC Circuits

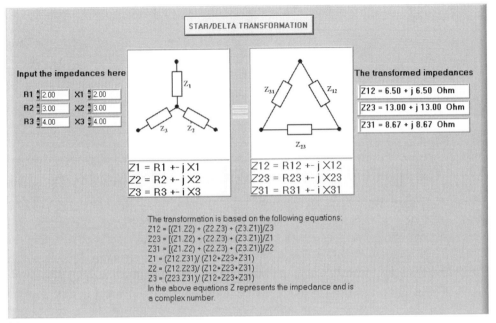

(a)

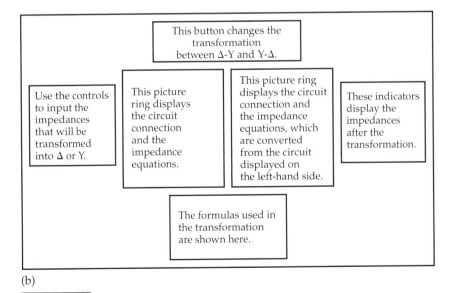

(b)

Figure 3-16
Front panel and brief user guide of `Star Delta Transformations.vi`.

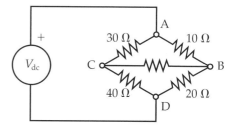

Figure 3-17
Sample circuit for question 2.

3.6 Voltage and Currents in Star- and Delta-Connected Loads

A three-phase ac system consists of three voltage sources that supply power to loads connected to the supply lines, which can be connected to either delta (Δ) or star (Y) configurations as stated previously.

In three-phase systems, the voltages differ in phase 120°, and their frequency and amplitudes are equal. If the three-phase loads are balanced (each having equal impedances), the analysis of such a circuit can be simplified on a per-phase basis. This follows from the relationship that the per-phase real power and reactive power are one-third of the total real power and reactive power, respectively.

It is very convenient to carry out the calculations in a per-phase star-connected line to neutral basis. If Δ-Y, Y-Δ, or Δ-Δ connections are present, the parameters on Δ side(s) are transformed to Y-connection, and computations are carried out.

Two three-phase load connections that are commonly used in the ac circuits were given in Fig. 3-15. In this section, the voltage and the current functions are examined while the three-phase loads are connected to the star-connected three-phase supplies, shown in Fig. 3-18.

In Fig. 3-18, v_{1s}, v_{2s}, v_{3s}, v_{1p}, v_{2p}, v_{3p}, v_{12p}, v_{23p}, and v_{31p} are the phase voltage functions, and v_{12}, v_{23}, and v_{31} are the line-to-line voltages (or simply line voltages). Similarly, i_{1p}, i_{2p}, i_{3p}, i_{12p}, i_{23p}, and i_{31p} are the phase currents, and i_{1L}, i_{2L}, and i_{3L} are the line currents.

The phase voltages of a three-phase supply can be given as

$$v_{1s}(t) = V_m \sin(\omega t) \qquad (3.42)$$
$$v_{2s}(t) = V_m \sin(\omega t - 2\pi/3) \qquad (3.43)$$
$$v_{3s}(t) = V_m \sin(\omega t - 4\pi/3) \qquad (3.44)$$

In the case of sinusoidal steady-state operation, similar expressions can be written for the current waveforms with identical phase difference θ, which depend on the phase angle of the balanced load inductances.

Chapter 3 • AC Circuits

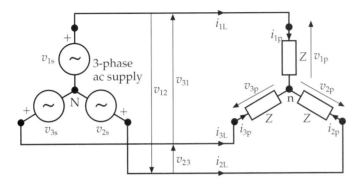

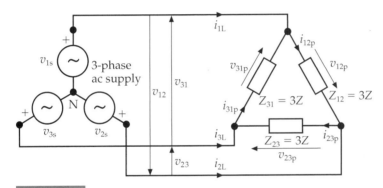

Figure 3-18
Two common balanced-load connections in three-phase ac circuits.

$$i_{1s}(t) = I_m \sin(\omega t - \theta) \quad (3.45)$$
$$i_{2s}(t) = I_m \sin(\omega t - 2\pi/3 - \theta) \quad (3.46)$$
$$i_{3s}(t) = I_m \sin(\omega t - 4\pi/3 - \theta) \quad (3.47)$$

A three-phase load is balanced when the line voltages are equal in magnitude and mutually displaced in phase by $2\pi/3$ in radians and the line currents are equal. In a balanced three-phase system, there is a very simple relationship between the line and phase quantities, which can be obtained from the phasor quantities or the time-varying expressions of the voltages and the currents.

The voltage and current relationships in three-phase ac circuits can be simplified by using the rms values (I and V) of the quantities. Refer to Fig. 3-18, and study Table 3-1.

Table 3-1 *Voltage and current relationships in three-phase circuits.*

Star-Connected Balanced Load	Delta-Connected Balanced Load
Phase current: $I_{1p} = I_{1L}$, $I_{2p} = I_{2L}$, $I_{3p} = I_{3L}$ Line current: $I_L = I_{1L} = I_{2L} = I_{3L}$	Phase current: $I_p = I_L/\sqrt{3}$ Line current: $I_L = I_{1L} = I_{2L} = I_{3L}$ and $I_p = I_{12p} = I_{23p} = I_{31p}$
Phase voltage: $V_p = V_L/\sqrt{3}$ Line voltage: $V_L = V_{12} = V_{23} = V_{31}$	Phase voltage: $V_{12} = V_{12p}$, $V_{23} = V_{23p}$, $V_{31} = V_{31p}$ Line voltage: $V_L = V_{12} = V_{23} = V_{31}$ and $V_p = V_{1p} = V_{2p} = V_{3p}$

The voltages across the impedances and the currents in the impedances are 120° out of phase.

3.6.1 Virtual Instrument Panel

Fig. 3-19 shows the front panel of the VI named `Voltage and currents in delta/star loads.vi`. The VI provides a visual aid to understanding the definitions of phase and line voltages and phase and line currents in the delta- and the star-connected ac systems that contain the loads as well as the ac supplies. In addition, the instantaneous voltage and currents are displayed in the front panel of the VI.

3.6.2 Self-Study Questions

Open and run the custom-written VI named `Voltage and currents in delta/star loads.vi` in the `Chapter 3` folder, and investigate the following questions.

1. Show that the line voltage V_{line} in the three-phase system is $\sqrt{3}$ times the phase voltage V_{phase}, and verify the result by using the VI for a given phase voltage.
2. Study the concept in question 1 this time for the line currents and the phase currents in the case of a delta-connected three-phase load.
3. In question 2, find out the angles in degrees between the phase and the line quantities on the supply side and the load side.

Figure 3-19
The front panel and brief user guide of `Voltage and currents in delta/star loads.vi`.

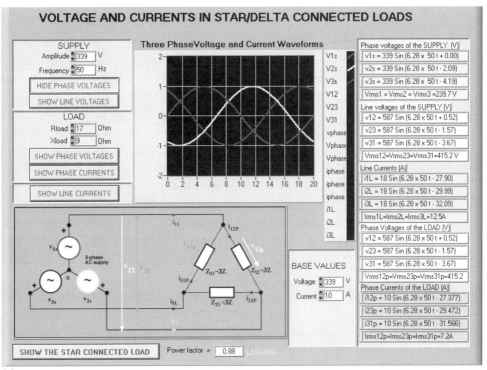

4. Use the single-phase equivalent circuit in each load configuration and calculate the phase currents for given values of the voltage and the load impedance.
5. Three incandescent lamps rated 60 W, 120 V (rms) are connected in the delta form. What line voltage is needed so that the lamps burn normally (at rated conditions)? What are the line and phase currents in the circuit? *Hint:* First calculate and set the resistance of the lamps using the controls provided.
6. Three load resistors are connected in the delta form. If the line voltage is 415 V (rms) and the line current is 100 A (rms), calculate the current in each resistor, the voltage across the resistors, and the resistance of each resistor. Verify the results analytically.
7. Each phase of a delta-connected load comprises a resistor of 50 Ω and a capacitor of 50 μF in series. The three-phase load is connected to a 440 V (rms, line voltage) and 50 Hz three-phase star-connected supply. Calculate the phase and line currents.
Answer: 5.46 A (rms), 9.46 A (rms)

3.7 Voltage and Current Phasors in Three-Phase Systems

As can be seen in equations 3.42, 3.43, and 3.44, the voltage in Phase 1 reaches a maximum first, followed by Phase 2 and then Phase 3 for sequence 1, 2, 3. This sequence should be evident from the phasor diagram of the three-phase source where the phasors should pass a fixed point in the order 1-2-3, 1-2-3, ...

In this section, the variation of the phasors in a three-phase ac circuit will be examined. The phasors are obtained by selecting one phase voltage as a reference with a phase angle of zero and determining the phase angles of the other two phases. Since the amplitudes and the frequencies of the voltage sources are equal, the phasors have equal lengths.

The phase and the line voltage phasors in the three-phase system can also be represented in polar form. Note that similar phasor representation can be given for the current waveforms if the phase angle between the voltage and the current is known.

$$V_1 = V_{Ph}\angle 0°, \quad V_2 = V_{Ph}\angle -120°, \quad V_3 = V_{Ph}\angle -240° \quad (3.48)$$

$$V_{12} = \sqrt{3}V_{Ph}\angle 30°, \quad V_{23} = \sqrt{3}V_{Ph}\angle -90°, \quad V_{31} = \sqrt{3}V_{Ph}\angle -210° \quad (3.49)$$

where V_{Ph} is the phase voltage.

Chapter 3 • AC Circuits

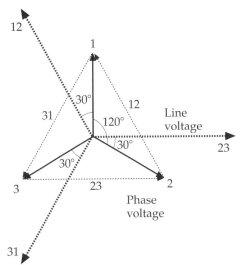

Figure 3-20
The three-phase phasor representation.

Fig. 3-20 shows the sequence and position of the voltage phasors in a three-phase star-connected ac circuit. As shown in the figure, there are two different ways to illustrate the phase and line voltages in the phasor form. The dotted lines in the figure demonstrate the line voltage phasors starting from origin, which is also used in this study.

3.7.1 Virtual Instrument Panel

The objective of this section is to understand the phasors and the phase sequences in the three-phase balanced ac circuits.

In the VI named `3phase phasors.vi`, two types of phasors are studied: for phase voltages and for line voltages.

As demonstrated in the front panel in Fig. 3-21, since one phase is always the reference, changing the phase angle affects all phasors equally and they rotate in the same direction as expected. On the front panel of the VI, the phase and the line voltages are also displayed in the time domain.

3.7.2 Self-Study Questions

Open and run the custom-written VI named `3phase phasors.vi` in the `Chapter 3` folder, and investigate the following questions.

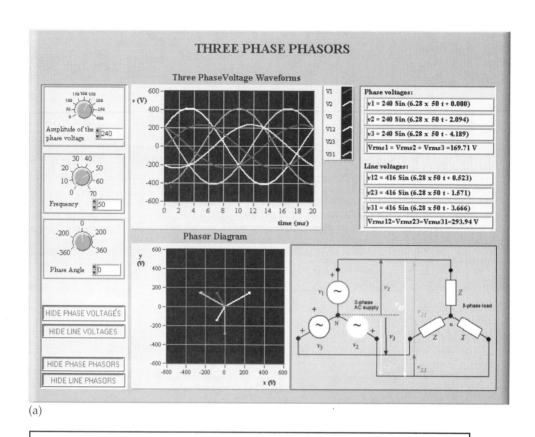

(a)

(b)

Figure 3-21
The front panel and brief user guide of `3phase phasors.vi`.

1. Use the knobs provided on the front panel to vary the voltage amplitude, frequency, and phase angle. Observe the changes in the voltage and current phasors and the waveforms.
2. Vary the phase angle clockwise and counterclockwise and observe the direction of the rotation of the phasors. Did the phase angles change, and if so, why?
3. Read the magnitude of each phasor from the phasor graph and compare with the values displayed in the time domain.
   Note: To make the comparison easy, you may display either phase or line quantities simultaneously, but not both.

3.8 Power in Three-Phase AC Circuits

Since the phase impedances of a balanced star- or delta-connected load contain equal currents, the phase power is one-third of the total power. As a definition, the voltage across the load impedance and the current in the impedance can be used to compute the power per phase.

Let's assume that the angle between the phase voltage and the phase current is θ, which is equal to the angle of the impedance. Considering the load configurations given in Fig. 3-22, the phase power and the total power can be estimated easily.

In the case of Fig. 3-22a, the total active power is equal to three times the power of one phase.

$$P_1 = P_2 = P_3 = P = V_{line} I_{phase} \cos \theta \qquad (3.50)$$
$$P_{total} = 3 \cdot P = 3 V_{line} I_{phase} \cos \theta \qquad (3.51)$$

Since the line current $I_{line} = \sqrt{3} I_{phase}$ in the balanced delta-connected loads, if this equation is substituted into equation 3.51, the total active load becomes

$$P_{total} = \sqrt{3} V_{line} I_{line} \cos \theta \qquad (3.52)$$

In Fig. 3-22b, however, the impedances contain the line currents I_{line} (= phase current, I_{phase}) and the phase voltages V_{phase} (= $V_{line}/\sqrt{3}$). Therefore, the phase active power and the total active power are

$$P_1 = P_2 = P_3 = P = V_{phase} I_{line} \cos \theta \qquad (3.53)$$
$$P_{total} = 3 \cdot P = 3 V_{phase} I_{line} \cos \theta \qquad (3.54)$$

If the relationship between the phase voltage and the line voltage ($V_{phase} = V_{line}/\sqrt{3}$) is used, the total active power becomes identical to the equation

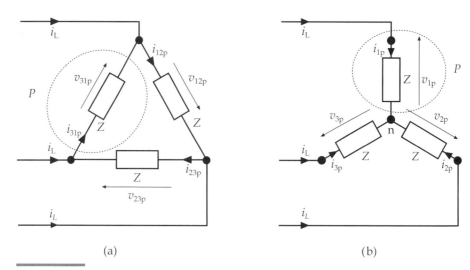

Figure 3-22
Per-phase powers in (a) a delta-connected load and (b) a star-connected load.

developed in equation 3.52. This means that the total power in any balanced three-phase load (Δ- or Y-connected) is given by equation 3.52.

Similarly, the total reactive and the total apparent power in the three-phase balanced ac circuits can be given by

$$Q_{total} = \sqrt{3} V_{line} I_{line} \sin \theta \tag{3.55}$$

$$S_{total} = \sqrt{3} V_{line} I_{line} \tag{3.56}$$

Power Measurement Techniques

In the three-phase power systems, one, two, or three wattmeters can be used to measure the total power. A wattmeter may be considered to be a voltmeter and an ammeter combined in the same box, which has a deflection proportional to $V_{rms} I_{rms} \cos \theta$, where θ is the angle between the voltage and current. Hence, a wattmeter has two voltage and two current terminals, which have + or − polarity signs. Three power measurement methods utilizing the wattmeters are described next, and are applied to the balanced three-phase ac load.

1. Two-Wattmeter Method

This method can be used in a three-phase three-wire balanced or unbalanced load system that may be connected Δ or Y. To perform the measurement, two wattmeters are connected as shown in Fig. 3-23.

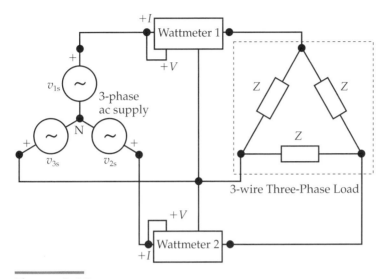

Figure 3-23
Two-wattmeter method in star- or delta-connected load.

In the balanced loads, the sum of the two wattmeter readings gives the total power. This can be proven in a star-connected load mathematically using the power reading of each meter as

$$P_1 = V_{12}I_1 \cos(30° + \theta) = V_{\text{line}}I_{\text{line}} \cos(30° + \theta)$$
$$P_2 = V_{32}I_3 \cos(30° - \theta) = V_{\text{line}}I_{\text{line}} \cos(30° - \theta)$$
$$P_{\text{total}} = P_1 + P_2 = \sqrt{3} V_{\text{line}} I_{\text{line}} \cos \theta \tag{3.57}$$

If the difference of the readings is computed,

$$P_2 - P_1 = V_{\text{line}}I_{\text{line}} \cos(30° - \theta) - V_{\text{line}}I_{\text{line}} \cos(30° + \theta)$$
$$= V_{\text{line}}I_{\text{line}} \sin \theta \tag{3.58}$$

which is $1/\sqrt{3}$ times the total three-phase reactive power. This means that the two-wattmeter method can also indicate the total reactive power in the three-phase loads and also the power factor (see Fig. 3-24).

2. Three-Wattmeter Method

This method is used in a three-phase four-wire balanced or unbalanced load. The connections are made with one meter in each line as shown in Fig. 3-25. In this configuration, the total active power supplied to the load is equal to the sum of the three wattmeter readings.

$$P_{\text{total}} = P_1 + P_2 + P_3 \tag{3.59}$$

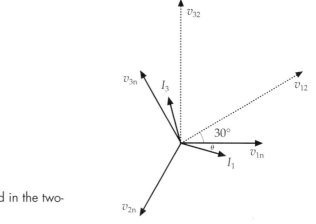

Figure 3-24
Three-phase voltage phasors used in the two-wattmeter method.

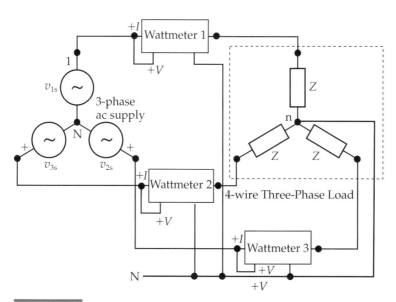

Figure 3-25
The wattmeter connections in the three-phase four-wire loads.

3. One-Wattmeter Method

This method is suitable only in three-phase four-wire balanced loads. The connection of the wattmeter is similar to the drawing given in Fig. 3-25. The total power is equal to three times the reading of only one wattmeter that is connected between one phase and the neutral terminal.

3.8.1 Virtual Instrument Panel

The objective of this section is to understand the powers and the power measurement techniques associated with the three-phase ac circuits. Fig. 3-26 illustrates the front panel of the VI named `Three phase power measurements.vi`.

3.8.2 Self-Study Questions

Before studying this VI, make sure that you study and understand the power concept presented in Section 3.2.2. Open and run the custom-written VI named `Three phase power measurements.vi` in the `Chapter 3` folder, and investigate the following questions.

1. A balanced three-phase, three-wire star-connected load is connected to a three-phase supply. The line voltage is 400 V. The load comprises an impedance of $100 + j100$ Ω per phase. Set these parameters and select the suitable circuit to determine the total active, reactive, and apparent power by using the VI provided.
2. Assume that the load in question 1 is a four-wire circuit. Use three power measurement methods and confirm your findings manually.
3. Three 10 μF capacitors are connected in star (Y) across a 2300 V (rms, line voltage), 60 Hz line. Calculate the line current, the active power, the reactive power, and the apparent power by using the VI.
4. A three-phase heater dissipates 15 kW when connected to a 208 V, three-phase line. Determine the value of each resistor if they are connected as star.
5. An industrial plant draws 600 kVA from a 2.4 kV line at a power factor of 0.8 lagging. What is the equivalent line-to-neutral impedance of the plant?
6. An electric motor having a power factor 0.82 draws a current of 25 A from a 600 V three-phase ac supply. Find the active power supplied to the motor.
7. Each phase of a delta-connected load comprises a resistor of 50 Ω and a capacitor of 50 μF in series. The three-phase load is connected to a 440 V (rms, line voltage) and 50 Hz three-phase star (Y)-connected supply. Calculate the phase and line currents, the power factor, the total active power, and total apparent power. Observe the phasor diagrams using the VI given in the previous section, `3phase phasors.vi`.

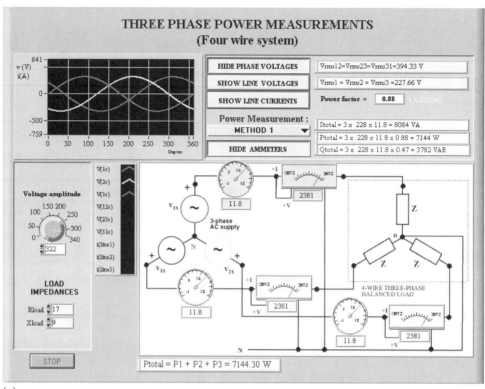

(a)

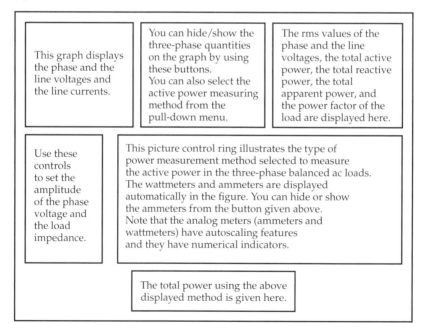

(b)

Figure 3-26
The front panel and brief user guide of `Three phase power measurements.vi`.

Hint: First, transform the delta-connected load to a star-connected equivalent (see Section 2.5). Then perform the power calculations.
Answers: 5.46 A (rms), 9.46 A (rms), 0.62, 4480 W, 7240 VA

3.9 Three-Phase Power Measurement and Data Logging

Three-phase power calculation techniques are well established. However, a number of factors have precipitated the need for flexible and reliable measurement and data logging systems for use in power system applications.

Power monitoring may be needed to analyze the power demand. Furthermore, observing the quality of the power helps us to analyze the source of disturbance that may cause important stability problems in power systems, or may seriously affect the operation of a device, or may cause catastrophic failure of a device.

In this section, a simple and cost-effective measurement tool that is based on LabVIEW is presented. This tool measures three-phase voltages and currents in real time using voltage and current transducers, and calculates all aspects of power (such as active, reactive, apparent power and power factor, and harmonic distortion). In addition, it records various parameters to view the trend over a period of time (from minutes to months).

Measurement Equations

In this study, three-phase measurement is done assuming that the load is star-connected and the star point is accessible. The three-phase system setup is illustrated in Fig. 3-27, which also illustrates the correct channel assignment for the VI.

It should be emphasized here that the voltage and current interfaces should be selected (or designed) to operate the system safely and accurately. In addition, note that in high voltage measurements, insulating via differential amplifiers or via differential input channels of the data acquisition card should not be considered since such methods do not provide real isolation, and thus are not safe.

To achieve complete isolation, I suggest that Hall-effect or clamp-based current transducers for current measurement and high-performance isolation amplifiers (or voltage transformers at low frequency) for voltage measurement should be employed (as provided in the Appendix).

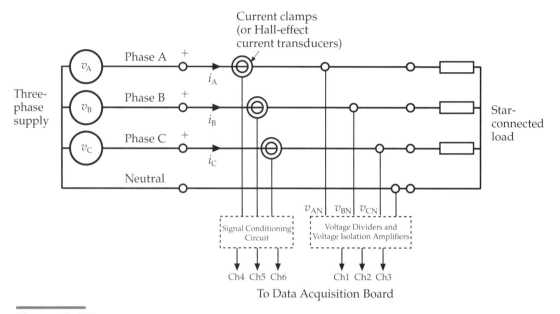

Figure 3-27
Three-phase measurement block diagram and channel assignments for four-wire system (with neutral).

The phase angle (or power factor) calculation is achieved using a single line-to-neutral voltage and a line current of the same phase (Ch1 and Ch4 in Fig. 3-27).

The following power calculations per phase were employed in the VI. Instantaneous, average, and rms powers per phase are estimated by

$$p(t) = v(t)i(t) \tag{3.60}$$

$$P_{ave} = \frac{1}{T}\int_0^T p(t)\,dt \tag{3.61}$$

$$P_{rms} = \sqrt{\frac{1}{T}\int_0^T [p(t)]^2\,dt} \tag{3.62}$$

Hence, the total apparent, active, and reactive powers in this four-wire three-phase system are calculated as

$$S_{total} = V_{rms(1)}I_{rms(1)} + V_{rms(2)}I_{rms(2)} + V_{rms(3)}I_{rms(3)} \tag{3.63}$$

$$P = S_{total}\cos\theta \tag{3.64}$$

$$Q = S_{total}\sin\theta \tag{3.65}$$

In the VI, the phase angle θ is calculated measuring the time instances of the first peaks of the phase voltage and the phase current. Then the time difference is calculated as a reference to the period of the waveform and converted to degrees.

The interval between the two consecutive peaks of the same waveform is used to calculate the period T of a selected waveform, and then the frequency of the waveform is calculated, $f = 1/T$.

If the voltage and current waveforms of an ac system are measured in real time, it is much easier to calculate the power factor simply by using the definitions of average and rms values, which were given earlier in equation 3.3 for nonsinusoidal systems (distorted waveforms).

Note that since the majority of three-phase power systems are not balanced, the VI developed here does not consider a balanced system. Hence, all of the three-phase line-to-neutral voltages and three line currents are measured for correct estimation of the power.

3.9.1 Virtual Instrument Panel

Open the VI named `Data Logging.vi`. Fig. 3-28 and Fig. 3-29 illustrate the front panel and the brief user guide for this VI. The descriptions of the front panel items are all available in the VI's help menu. The features of the VI are summarized below.

The following information can be displayed on the graphs:

- Voltages (rms and instantaneous) and their harmonics
- Line currents (rms and instantaneous) and their harmonics
- Period, frequency, phase angle, and power factor (leading/lagging)
- Instantaneous power and calculated values of active (W), reactive (VAR), and apparent (VA) powers
- Harmonics, total harmonic distortion (THD), THD + noise, peak power, peak frequency, and the amplitude and frequency of the selected number of harmonics

The following settings can be made:

- Per-unit values and dc off-set for voltages and currents
- Gains of the measured signals
- Triggering based on the reference parameters (rms or instantaneous), triggering channel selection, and triggering bandwidth selection for data logging

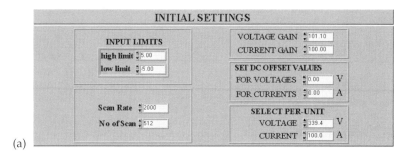

(a)

(b)

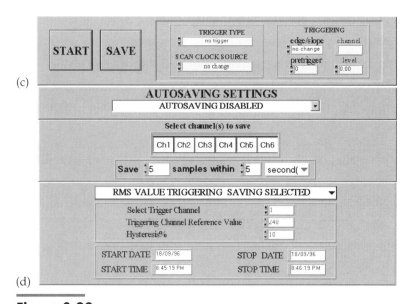

(c)

(d)

Figure 3-28
The front panel of Data Logging.vi (LabVIEW 4.0). *(continues)*

(e)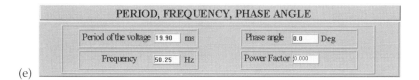

(f)

(g)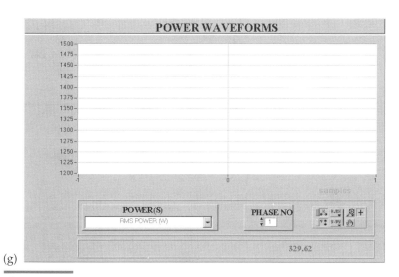

Figure 3-28
Continued

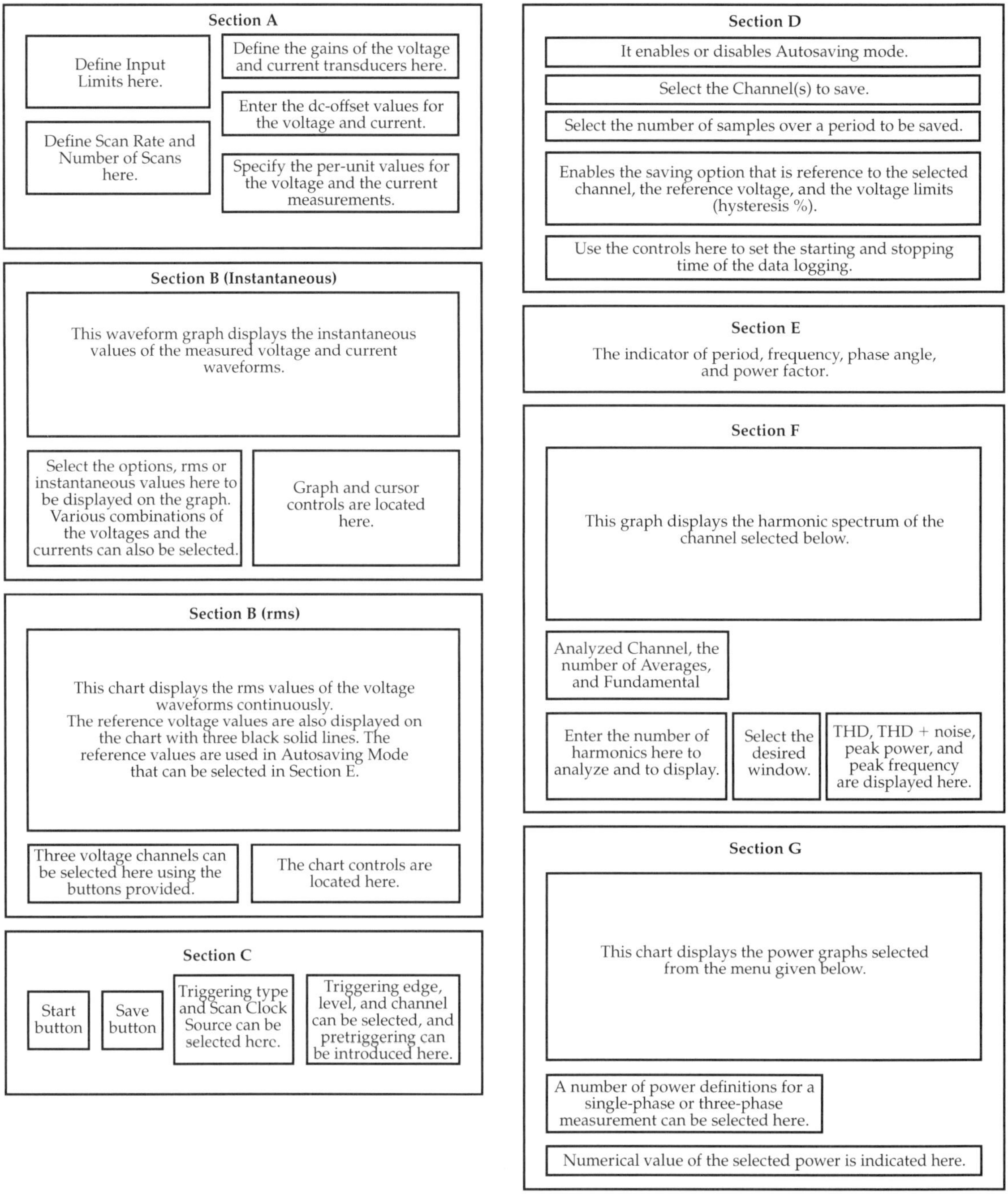

80 LabVIEW for Electric Circuits, Machines, Drives, and Laboratories

Section A

- Define Input Limits here.
- Define the gains of the voltage and current transducers here.
- Enter the dc-offset values for the voltage and current.
- Define Scan Rate and Number of Scans here.
- Specify the per-unit values for the voltage and the current measurements.

Section B (Instantaneous)

This waveform graph displays the instantaneous values of the measured voltage and current waveforms.

- Select the options, rms or instantaneous values here to be displayed on the graph. Various combinations of the voltages and the currents can also be selected.
- Graph and cursor controls are located here.

Section B (rms)

This chart displays the rms values of the voltage waveforms continuously.
The reference voltage values are also displayed on the chart with three black solid lines. The reference values are used in Autosaving Mode that can be selected in Section E.

- Three voltage channels can be selected here using the buttons provided.
- The chart controls are located here.

Section C

- Start button
- Save button
- Triggering type and Scan Clock Source can be selected here.
- Triggering edge, level, and channel can be selected, and pretriggering can be introduced here.

Section D

- It enables or disables Autosaving mode.
- Select the Channel(s) to save.
- Select the number of samples over a period to be saved.
- Enables the saving option that is reference to the selected channel, the reference voltage, and the voltage limits (hysteresis %).
- Use the controls here to set the starting and stopping time of the data logging.

Section E

The indicator of period, frequency, phase angle, and power factor.

Section F

This graph displays the harmonic spectrum of the channel selected below.

- Analyzed Channel, the number of Averages, and Fundamental
- Enter the number of harmonics here to analyze and to display.
- Select the desired window.
- THD, THD + noise, peak power, and peak frequency are displayed here.

Section G

This chart displays the power graphs selected from the menu given below.

- A number of power definitions for a single-phase or three-phase measurement can be selected here.
- Numerical value of the selected power is indicated here.

Figure 3-29
The brief user guide for Fig. 3-28, `Power and Data Logging.vi`.

- Logging period (from one minute to a day/a month/a year)
- Number of samples to be saved (from seconds to hours)

3.9.2 Some Features of the VI and Operating Scenarios

The details of the measurement program and some of the principal calculations are explained in the following paragraphs.

Channel number in the VI starts from zero, "0". Hence, for six analogue inputs use the numbers 0, 1, 2, 3, 4, and 5 to set the channels to Ch1, Ch2, Ch3, Ch4, Ch5, or Ch6, respectively.

The power factor calculation is based on the phase specified on the front panel, Phase No.

The data is saved into a spreadsheet file with an extension of `.dat` and can be read by Microsoft Excel. Only the rms values of the voltages, currents, power factor, and frequency are saved in a column format. However, a path named `c:\labview\datafile\` has to be created first.

To distinguish the data files in future use, a unique name is given to each data file that is created automatically by the VI. All data files created the same day are saved into a single folder.

For example, the name of a file on 17 December 1997 will be located in a folder named `97D17M12` (year+D+day+M+month). If this file is created at 12:33:12 AM, then the data file will have the following name:

`12AM3312.dat`   (hour+AM(or PM)+minute+second)

The data format used to set DATE/TIME TRIGGERING SAVING mode is shown below.

DATE FORMAT: 17/09/96   or   7/09/96
TIME FORMAT: 3:56:12 AM   or   11:5:12 AM

It should be noted here that sampling should be synchronous with the system frequency. Therefore, the settings for Scan Rate and No. of Scan are important in the harmonic analysis.

When Pretrigger scans are not zero, data is acquired continuously while waiting for the start trigger. When the trigger occurs, the specified pretrigger scans are kept, and the rest of the scans are acquired after the trigger.

The LED of Trigger Timeout, which is in the harmonic analysis section, turns yellow if a trigger timeout condition occurs, with an error message of "Trigger did not occur." Therefore acquisition timed out. Under this condition, if the Save button is pressed, the software keeps trying.

Choose either a Software Analog Trigger or a hardware Analog Trigger for the acquisition. With a Software Analog Trigger, for example, the acquisition starts immediately but the data is not retrieved by the program until the level and slope conditions are met. With a hardware Analog Trigger, the acquisition does not start until the level and slope conditions are met. Furthermore, remember that the hardware Analog Trigger is not available on all data acquisition devices.

Two autosaving modes are possible in the VI:

- by using a pre-set rms value
- by using a specified Date and Time

To start autosaving, enable the autosaving function by selecting the mode AUTOSAVING ENABLED. Remember that all the settings in this part of the front panel have to be done before running the software.

Select either RMS VOLTAGE TRIGGERING SAVING SELECTED or DATE/TIME TRIGGERING SAVING SELECTED mode. In both modes of operation, the number of samples can be selected using the settings that are located just below the triggering selection.

In the RMS mode of operation, Reference Voltage and Hysteresis % should also be specified.

Let us assume that Reference Voltage is 240 V and Hysteresis % is 10. The operation of the RMS triggering is illustrated in Fig. 3-30. As illustrated, the saving starts if the measured rms value of any one of the phase voltages is outside the specified hysteresis bandwidth.

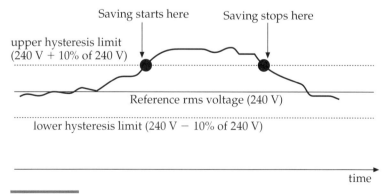

Figure 3-30
An illustration of RMS VOLTAGE TRIGGERING.

The selection of DATE/TIME TRIGGERING SAVING is similar to the one just described for RMS VALUE TRIGGERING.

When DATE/TIME TRIGGERING SAVING is selected, Start Time, Start Date, Stop Time, and Stop Date should also be specified. In this mode of operation, autosaving starts when the specified date and time match with the actual date and time of the computer, and saving stops when Stop Date and Stop Time are reached.

Example Setting 1

To save the rms values into a spreadsheet file, make the following settings on the front panel and run the VI:

- Autosaving Enabled
- RMS Voltage Triggering Saving Selected
- Trigger Channel, Triggering Channel Reference Value, and Hysteresis %

Saving will start when the measured rms voltage is outside the hysteresis bandwidth that was determined earlier. Then the data will be saved into a spreadsheet file that is named and placed in a folder created automatically.

A new directory is created if the data logging continues next day. This automatic file generation eases the retrieving of data for further processing.

It should be noted here that percentage of the reference value is used to determine the hysteresis bandwidth, which also determines the triggering condition in the auto-saving mode.

Example Setting 2

Alternatively, if DATE/TIME TRIGGERING SAVING is selected, triggering will start when the current date and the current time match the starting date and start time. The saving will stop when the current time and the date match the settings STOP DATE and STOP TIME.

The settings for this example are

- Autosaving Enabled
- Date/Time Triggering Saving Selected
- Start Date, Start Time, Stop Date, Stop Time

Example Setting 3

If Analog Trigger is selected, the triggering channel and the triggering signal level have to be specified. The level in the Triggering Level box indicates the input value of the actual analog signal before it is multiplied by the gain. Therefore, the level setting should be scaled accordingly. In addition, the other triggering features (Rising edge, positive or Falling edge, negative) have to be specified; these can be accessed from the Edge/Slope menu.

Let's assume that Rising slope is selected, and the triggering level is set to 0.4. If the gain is 100, the triggering will occur at a voltage level of +40 V.

Example Setting 4

Let's assume that the option Save 5 samples within 5 second(s) is selected. Under this condition, a maximum of one sample per second will be saved into a spreadsheet file after the rms triggering condition is met. In this mode of operation, the program waits 1 second between the two consecutive time intervals, regardless of the speed of the execution. However, remember that the minimum sampling interval is limited by the specifications of the DAQ card and the software.

3.10 References

Edminister, J. A. *Theory and Problems of Electric Circuits.* Schaum's Outline Series. New York: McGraw-Hill, 1972.

Ertugrul, N. "Electric Power Applications Lecture Notes." Department of Electrical and Electronic Engineering. University of Adelaide, 1997.

Ertugrul, N. "Electrical Circuits and Machines Laboratory with LabVIEW." June 2000 ed. National Instruments. Part no. 322765A-01.

Shotton, A. C. *Worked Examples in Electrotechnology.* London: George G. Harrap and Co. Ltd., 1957.

Wildi, T. *Electrical Machines, Drives, and Power Systems.* Englewood Cliffs, NJ: Prentice Hall, 1991.

Yamaee, Zia A., L. Juan, and J. R. Bala. *Electromechanical Energy Devices and Power Systems.* New York: Wiley, 1994.

Magnetic Circuits and Measurements

4

In practice, many practical devices use some form of magnetic circuit that usually contains coils wound on magnetic materials (such as iron cores). The coil currents flow in associated electrical circuits.

The magnetic effect of a current in a magnetic circuit is utilized in numerous applications, such as in electromagnets for lifting loads, in electric motors to produce torques, and in generators to generate electromotive force (emf).

Therefore, it has become essential for engineers to understand the basic theoretical standpoints and practical applications of magnetic circuits, which are covered in the first three sections of this chapter.

Furthermore, many electromechanical devices used in modern technology present highly nonlinear magnetization characteristics. To design, verify, and predict the performance of electromechanical devices under various operating conditions, an accurate knowledge of the magnetization characteristics is required. We will study these characteristics in Section 4.4.

Educational Objectives As in previous chapters, I recommend that students study each section in the order given, and answer the associated questions with the help of the custom-written VIs. After completing this chapter, students should be able to

- understand the basic concepts in magnetic circuits.
- analyze some typical magnetic circuit topologies using the simulation tools provided.
- study the effects of changing the physical dimensions, the number of turns, and the fringing effect in magnetic circuits.
- observe and visualize the influence of current waveforms in magnetic circuits, and study the BH (hysteresis) characteristics and associated losses.
- understand the real-time measurement method in magnetic circuits with graphical programming that allows a high degree of software modularity.

4.1 Background Information

Although there are some limitations in a magnetic circuit approach, such as saturation, nonlinearity, leakage, and fringing flux, in sizing the magnetic components during the design stage, the approach of simple dc circuit analysis can be utilized. For a more definite picture of magnetic field distribution in practical situations, which might have complicated geometries, it is necessary to use more advanced methods, such as Finite Element Analysis.

First, it's convenient to see the analogy between electrical and magnetic quantities. Hence the magnetic circuits can be simplified to take into account some of the difficulties in modeling some of the phenomena previously mentioned, such as fringing, which occur in the practical circuits.

To be able to apply electric circuits concepts to magnetic circuits, we need to make some assumptions.

- If the frequency is 60 Hz or less, you can apply Ampere's Law.

$$N \cdot I = \sum H_k l_{mk} \tag{4.1}$$

Here N is the number of turns, I is the current, H_k is the field intensity, l_{mk} is the mean length of the medium, and k indicates the number of the subsection in a closed path of the magnetic circuit, where the cross section of the magnetic circuit is uniform.

- Magnetic flux density B is uniform across the plane, therefore magnetic flux ϕ and magnetic field intensity H are constant, and H and ϕ have the same direction.

$$\phi = B \cdot A \tag{4.2}$$

Here A is the cross section of the uniform part of the magnetic section.

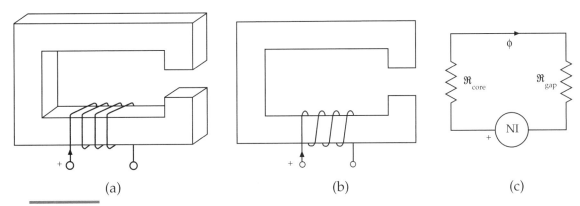

Figure 4-1
Representation of magnetic circuits: (a) 3-D representation; (b) schematic diagram, 2-D representation; and (c) magnetic equivalent circuit.

- Total flux entering a node is zero.
- A one-dimensional circuit is used instead of a three-dimensional magnetic field.
- Each section of the path has a constant cross-sectional area.
- The mean path length l_m is used in all calculations.

As shown in Fig. 4-1, the magnitude of the flux produced in a coil of the magnetic circuit depends on a quantity named *the reluctance,* which is a function of the mean length of the path of the magnetic flux, the cross-sectional area, and the magnetic properties of the medium in which the magnetic field is generated.

The reluctance, or magnetic resistance, of a section of the magnetic circuit is given by

$$\mathcal{R} = \frac{l_m}{\mu A} \tag{4.3}$$

where l_m is the mean length of the uniform path of the magnetic circuit, A is the cross-sectional area of the uniform magnetic section, and μ is known as permeability, which will be explained later.

In addition to these assumptions, if the fringing (spreading of flux lines in an air-gap) exists, the flux density in the air-gap becomes less than the core, and it is impossible to calculate it analytically. However, the fringing can be taken into account by adding the length of the gap to each of the dimensions of the air-gap cross section (assuming that the cross-sectional areas of each side of the air-gap are identical).

Once the magnetic equivalent circuit is obtained, apply Ampere's Law to determine the parameters of the circuit.

In a magnetic core, the magnetic flux density B is related to the magnetic field H according to the BH curve, which is nonlinear. The slope of this curve μ is known as the permeability of the material in henries per meter. The relationship between B and H is given by

$$B = \mu H \qquad (4.4)$$

However, the BH relationship of the ferromagnetic materials is highly nonlinear. Hence, the permeability depends on the operating value of magnetic flux density.

In most cases, electromagnetic devices are designed to operate within the linear section of the BH characteristic of the magnetic material. In magnetic circuits, the permeability of free space μ_0 is chosen as the reference datum, and every other material is quoted as a multiple of μ_0.

$$\mu = \mu_r \mu_0 \qquad (4.5)$$

Here μ_r is the dimensionless relative permeability of the core, and the value of μ_0 is a fundamental constant equal to $4\pi 10^{-7}$ Wb/A · m or H/m.

For most materials the relative permeability μ_r is close to unity 1.0. However, the ferromagnetic materials that are used for magnetic circuits (e.g., in rotating electric machines and transformers) have much higher values than unity. Some of the typical ferromagnetic materials comprise

Cobalt	μ_r	up to 70
Nickel	μ_r	up to 200
Iron and alloys	μ_r	up to 100,000

In the application of Ampere's Law, depending on the unknown parameter, one of the following methods can be used.

- Determine magnetomotive force (mmf) NI for a given flux density B
- Determine magnetomotive force for a given flux ϕ
- Determine flux for a given mmf (either by a trial-and-error method or a graphical method)

In the simulation tools provided in this section, the first two analytical methods are demonstrated.

Definitions of EMF and MMF Electromotive force (emf) is used with reference to potential difference and a source of electricity. It should be noted that current-carrying coils produce a magnetic field in a magnetic circuit. For uniform magnetic flux density B, the rate of change of flux linked by the N number of turns of a coil is equal to the emf, in volts.

$$e = N\frac{d\phi}{dt} \tag{4.6}$$

Similarly, the induced emf in a coil with N number of turns when the magnetic flux linked with it is changed is proportional to the same value defined by equation 4.6. However, remember that the direction of the current due to the induced emf opposes the variation of the magnetic flux linked with the coil, hence the right-hand side of equation 4.6 has a negative sign in a generator operation.

The emf concept is widely used in the measurement of the magnetic flux ϕ. For example, if a search coil is situated within the established magnetic flux with its plane perpendicular to the field, and if the induced voltage, emf, is measured, the integral of this voltage over a period of time will give the magnetic flux in the circuit.

Furthermore, let's consider the above coil, which has N number of turns carrying a current of I. The strength of a magnetic field depends on this current flowing through the individual turns. As a definition: Magnetomotive force (mmf) tends to drive the flux through the magnetic circuit and is analogous to emf in the electric circuit. The mmf is directly proportional to the ampere-turns NI of the electric circuit.

4.2 Analysis of Magnetic Circuits

Now let's examine two typical magnetic circuits (Fig. 4-2) while considering the simplifying assumptions given in the previous section. In addition to these assumptions, let's assume that the permeability of the magnetic material used in the core is constant at the operating point.

Magnetic Circuit 1

In the magnetic circuit shown in Fig. 4-2b, mmf (NI) produces magnetic flux ϕ. If Ampere's Law is applied to this circuit,

$$H_c l_c + H_g l_g = NI = \phi(\mathcal{R}_c + \mathcal{R}_g) \tag{4.7}$$

where l_c and l_g are the mean lengths of the core and the air-gap, respectively.

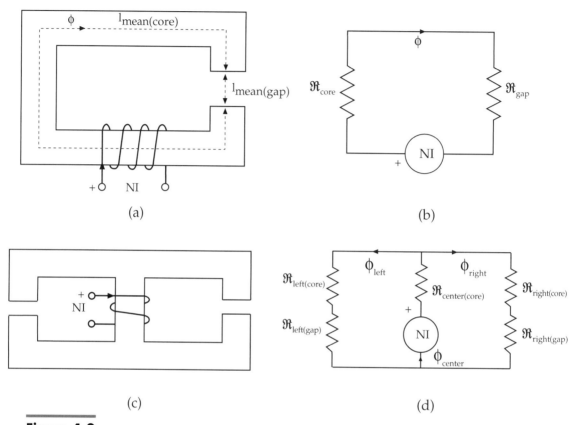

Figure 4-2
Sample magnetic circuits: (a, c) Schematic diagrams, and (b, d) magnetic equivalent circuits illustrating the reluctance $\mathcal{R}$ and the mmf source NI.

As seen in the equivalent circuit, the magnetic circuit contains two different magnetic reluctances, $\mathcal{R}_c$ and $\mathcal{R}_g$, which are

$$\mathcal{R}_c = \frac{l_c}{\mu_r \mu_0 A_c}, \quad \mathcal{R}_g = \frac{l_g}{\mu_0 A_g} \quad (4.8)$$

and the magnetic field intensities H_c and H_g are

$$H_g = \frac{B_g}{\mu_0}, \quad H_c = \frac{B_c}{\mu_r \mu_0} \quad (4.9)$$

In practical magnetic circuits, flux lines in an air-gap spread, making the flux density in the air-gap less than in the core. In other words, the effective cross section of the air-gap is larger than the cross section of the core. This is called fringing, and it is almost impossible to calculate analytically.

However, for simple magnetic circuits empirical formulas can be used to take the fringing into account. For example, for the circuit in Fig. 4-2a,

If there is no fringing $A_c = A_g = W \cdot T$

With fringing $A_c = W \cdot T$ and $A_g = (W + l_g) \cdot (T + l_g)$

where W and T are the width and the depth (thickness) of the magnetic core, respectively. Therefore, substituting all the unknowns and solving equation 4.7, we can obtain the required current for a given number of turns in the magnetic circuit to achieve a given flux.

Magnetic Circuit 2

In the magnetic circuit of Fig. 4-2d, however, the flux generated by NI has divided into two branches and provides flux in the two air-gaps, where

$$\phi_{center} = \phi_{right} + \phi_{left}$$

To determine the fluxes in the circuit, we can use the reluctance method. The reluctances of each flux path can be given as

$$\mathcal{R}_{center(core)} = \frac{l_{center(core)}}{\mu_r \mu_0 A_{center(core)}} \tag{4.10}$$

$$\mathcal{R}_{left} = \frac{l_{left(core)}}{\mu_r \mu_0 A_{left(core)}} + \frac{l_{left(gap)}}{\mu_0 A_{left(gap)}} \tag{4.11}$$

$$\mathcal{R}_{right} = \frac{l_{right(core)}}{\mu_r \mu_0 A_{right(core)}} + \frac{l_{right(gap)}}{\mu_0 A_{right(gap)}} \tag{4.12}$$

The flux in the center core can be estimated by using the total reluctance of the circuit, which is

$$\phi_{center} = \frac{NI}{\mathcal{R}_{total}} \tag{4.13}$$

where

$$\mathcal{R}_{total} = \mathcal{R}_{center(core)} + \frac{\mathcal{R}_{left(core)} \mathcal{R}_{right(core)}}{\mathcal{R}_{left(core)} + \mathcal{R}_{right(core)}} \tag{4.14}$$

Then the fluxes in other branches can be determined as

$$\phi_{left} = \phi_{center} \frac{\mathcal{R}_{right(core)}}{\mathcal{R}_{left(core)} + \mathcal{R}_{right(core)}} \tag{4.15}$$

$$\phi_{right} = \phi_{center} \frac{\mathcal{R}_{left(core)}}{\mathcal{R}_{left(core)} + \mathcal{R}_{right(core)}} \tag{4.16}$$

Furthermore, the flux densities in each section of the magnetic circuit are calculated using the associated cross-sectional areas as

$$B_{\text{left}} = \frac{\phi_{\text{left}}}{A_{\text{left}}}, \qquad B_{\text{right}} = \frac{\phi_{\text{right}}}{A_{\text{right}}}, \qquad B_{\text{center}} = \frac{\phi_{\text{center}}}{A_{\text{center}}} \qquad (4.17)$$

Finally, the fringing effect can be included in the estimation of the flux densities in the air-gaps by changing the effective area of the air-gaps as explained in the previous example.

4.2.1 Virtual Instrument Panel

Two VIs named `Magnetics 1.vi` and `Magnetics 2.vi` are provided in this section (Fig. 4-3 and Fig. 4-4). The sample magnetic circuits explained earlier are solved and their states are animated in the associated VIs. After running the VIs, the user can modify the input data (controlled by the control boxes provided) and view the corresponding calculated values on the indicators.

The brief problem definitions are given on the front panels of the VIs. In both VIs, the front panels have four distinct subsections.

The top-left section displays the 3-D view of the magnetic circuit together with the control boxes, where the physical dimensions can be set. A 2-D magnetic equivalent circuit is given on the top-right corner of the front panel. This section animates the direction of the flux in the circuit and shows the polarity of the mmf source. It displays the values of the reluctances and the mmf. The other two sections on the front panels provide all the controls and the indicators that display the calculated values in the VIs.

4.2.2 Self-Study Questions

Open and run the custom-written VIs named `Magnetics 1.vi` and `Magnetics 2.vi` (which are found in Section 4.2 of the Chapter 4 folder on the accompanying CD-ROM) in the continuous run mode, and investigate the following questions.

1. Use the default values on each VI and verify the values shown on the indicators analytically.
2. Vary the relative permeability of the core in question 1, observe the values of the reluctance of the core and the mmf, and comment.
3. Repeat similar observations as in question 2 for varying number of turns and the current level in the coils.

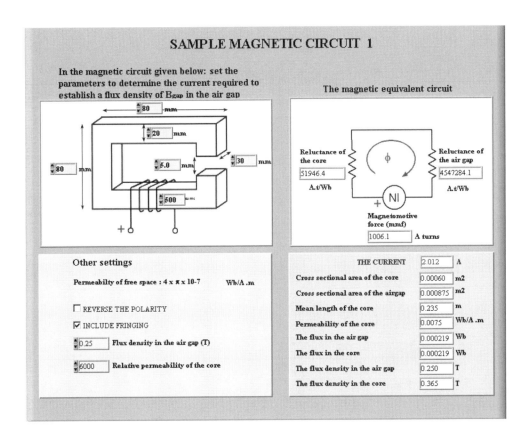

Figure 4-3
The front panel and the brief user guide of Magnetics 1.vi.

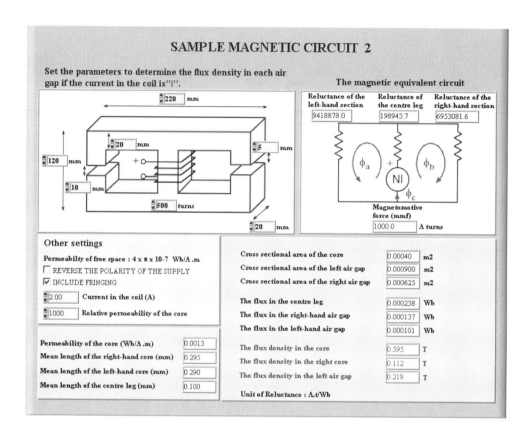

Figure 4-4
The front panel and brief user guide of `Magnetics 2.vi`.

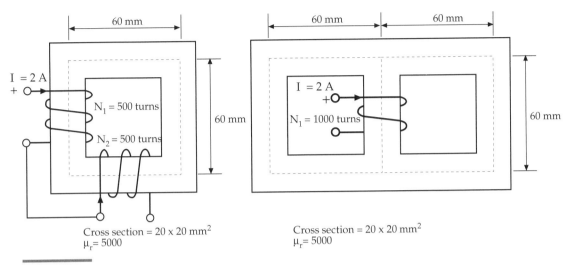

Figure 4-5
Magnetic circuits of question 7.

4. Alter the cross-sectional area(s) and the dimensions of the magnetic cores and observe the variation of the reluctances.

5. As stated earlier, fringing describes the spreading of flux lines in an air-gap of a magnetic circuit. While the simulation is running, use the control Include Fringing and observe the changes in the values of Cross-sectional area of the air-gap.

6. Observe the impact of the polarity change of the supply on the direction of the flux and the polarity of mmf source.

7. Compute the flux in each leg of the magnetic circuits shown in Fig. 4-5.

4.3 BH Characteristics and Losses

Magnetic hysteresis, caused by the cyclic reversals of magnetic flux in the alternating magnetic field, occurs in an ac current circuit surrounded by iron or other magnetic material. The alternating magnetic flux that is produced by the ac current induces a counter emf in the electric circuit. Ideally, if the emf imposed by the ac current is a sine wave, the counter emf and resulting magnetic flux follow a sine wave.

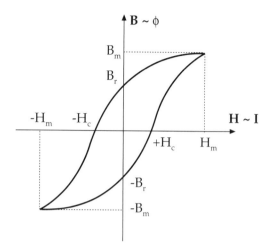

Figure 4-6
Typical BH curve of a magnetic circuit.

However, the ac current waveform is greatly distorted in the magnetic circuits due to a phenomenon called magnetic saturation in the cores of the magnetic circuit, and therefore it is not a sine wave.

As mentioned in the previous section, there is a definite relationship between the flux density B and the magnetic field intensity H of a magnetic material: $B = \mu H$. In addition, the relationship between H and B is usually nonlinear and is expressed graphically by the BH curve of the material.

Fig. 4-6 illustrates a typical BH curve of a magnetic material (also known as hysteresis loop) for a cyclic input current waveform.

In the hysteresis loop, when $H = 0$ the flux density is not zero but has a value $\pm B_r$, called the remnant magnetization or residual flux density. Similarly, when $B = 0$, the magnetic field is not zero, but is equal to $\pm H_c$, a parameter called the coercive force of the material. As can be seen in the hysteresis loop curve, the slope of the curve is the incremental permeability (also known as the dynamic permeability). It decreases with increasing B, which ultimately reaches the permeability of the free space μ_0. The state of low incremental permeability is known as saturation, and at this point, $B = B_m$ and $H = H_m$, to their maximum values.

Furthermore, a magnetic material absorbs energy during each cycle of the current, and this energy is dissipated as heat. It is proven that the energy lost as heat for each ac cycle in the magnetic circuit is equal to the area of the hysteresis loop. The hysteresis losses can be reduced by selecting a magnetic material that has a narrow hysteresis loop.

If the magnetic circuit is operated on ac current and if the flux is estimated, the BH curve shown in Fig. 4-6 can easily be obtained. Note that in the circuit,

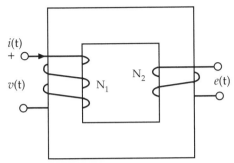

Figure 4-7
The sample magnetic circuit used in this study.

the magnetic field intensity H is proportional to the ac current i flowing in the winding, and B is proportional to the flux ϕ.

Two methods for determining BH characteristics are given next, based on the magnetic circuit provided in Fig. 4-7.

In Fig. 4-7, $v(t)$ and $i(t)$ are the ac voltage and currents applied to winding 1, which creates the magnetic flux in the circuit, and $e(t)$ is the induced emf on winding 2, which is known as a search coil.

Assuming that the leakage flux is negligible, the flux in the magnetic circuit links both windings. Hence the integral of the emf, $(v(t) - Ri(t))$ across winding 1 or the induced voltage $e(t)$ across winding 2 can be used to calculate the total flux linkage.

$$\lambda(t) = \int_0^t (v(t) - Ri(t))\, dt \quad \text{or} \quad \lambda(t) = \int_0^t e(t)\, dt \qquad (4.18)$$

where R is the resistance of winding 1, and λ is the total flux linkage, which is given by

$$\lambda = \phi N = NBA \qquad (4.19)$$

As stated earlier, due to the magnetic saturation, the ac current waveform that produces the magnetic field is greatly distorted. Any distorted waveform can be defined by the harmonics it contains. Since the harmonic is any voltage and current waveform whose frequency is an integral multiple of the line frequency, a distorted waveform can be represented by a set of sine waves whose frequency, magnitude, and phase angles are different from the fundamental waveform.

To study the effect of distorted current in the magnetic circuit, as it is provided in the VI, the distorted wave of current can be resolved into multiple components: a true sine wave of effective intensity and power equal to the distorted wave and higher harmonic components that primarily consist of a triple harmonic.

Note that a return to the initial point in the BH curve may not be obtained unless the integration is carried out over a number of cycles of the voltages and currents.

4.3.1 Virtual Instrument Panel

The principal aim of the VI `BH Hysteresis.vi` provided in Section 4.3 of the Chapter 4 folder on the accompanying CD-ROM is to explain the hysteresis phenomena using the ac voltage and the current waveform that exist in the magnetic circuits. Furthermore, the VI provides a flexible simulation tool for studying the impact of the higher harmonics on the voltage and current waveforms and for examining the hysteresis characteristics and estimating the hysteresis losses in the magnetic circuits.

The controls of the front panel of the VI (Fig. 4-8) can be utilized to construct a periodic voltage or current of any conceivable shape. To achieve this, use the fundamental component and add an arbitrary set of harmonic components (3rd and 5th harmonic components).

Furthermore, two different types of analysis states are provided: reading from a data file (`RealTimeVoltageCurrent.txt`, which contains a set of real measured data of the time, voltage, and current waveforms over multiple periods) and reading the simulated waveforms directly from the front panel settings. However, the correct path name should be defined to read the text file automatically via the VI. The current path name for the real data file is `c:\Magnetic Circuit\RealTimeVoltageCurrent.txt`.

4.3.2 Self-Study Questions

Open the VI named `BH Hysteresis.vi` (in Section 4.3 of the Chapter 4 folder on the accompanying CD-ROM) and assume that the cross-sectional area of the magnetic core illustrated on the front panel is A, the mean length of its magnetic circuit is l_{mean}, and the core is wound with two coils of N_1 and N_2 turns. The winding N_1 is excited by a distorted current varying cyclically from $-I_m$ to $+I_m$. In this circuit, as the current varies, the flux will vary from $-\phi_m$ to $+\phi_m$. The second winding N_2 (search coil) is used to measure the induced voltage e, which is equal to $-N_2(d\phi/dt)$. Then run the VI and investigate the following questions.

1. Use the default values for the settings and select the button Reading Real Data, which will read the associated data file that contains a set of real

Chapter 4 • Magnetic Circuits and Measurements

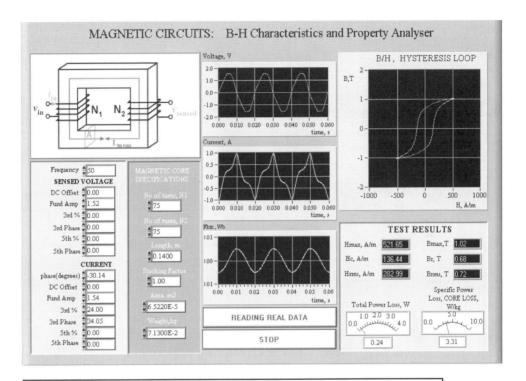

Figure 4-8
The front panel and brief user guide of `BH Hysteresis.vi`.

data in column format: sampling instants in seconds, the search coil voltage in volts, and the winding current in amperes. Then observe the shape of the current, the voltage, the flux linkage, and the *BH* characteristic, and identify which waveforms are close to an ideal sine wave.

2. In fact, the *BH* curve or hysteresis loop exhibits a lag between the magnetic induction (magnetic flux density) and the magnetizing field. This is somewhat analogous to the phase difference between voltage and current in the electrical circuits. Observe this analogy varying the phase angle of the current waveforms.
3. Determine the values of B_r, H_c, B_m, and H_m on the *BH* graph.
4. Suggest analytical solutions for the rms values of the voltage, the current, and the flux linkage waveforms.
5. While the VI is running, select a new mode of operation, Using Simulated Waveforms. Use the associated controls to set a current waveform that has a significant number of triple harmonics, and observe the corresponding hysteresis loop until you obtain a *BH* characteristic that is similar to the real *BH* graph studied earlier.
6. Under the operating condition stated in question 5, introduce some dc offset to the voltage and the current waveforms, and report the changes to the graph of the flux linkage and the *BH* characteristic. Explain why there is a change in the flux linkage graph.
7. Vary the stacking factor (which is equal to the net volume of the magnetic material/total volume) and the cross-sectional area of the core and observe the changes in the graphs and in the indicators. What conclusions can you draw from these changes?

4.4 Measuring Magnetization Characteristics

In this section, we describe a fully automated method of measuring the magnetization characteristics (flux versus current) of electromechanical devices. The measuring scheme was developed using a custom-written VI, which has a high degree of software modularity and provides the features needed for sensor zero adjustment, data acquisition, analysis, and automated presentation of results.

Even though the operation of many electromechanical devices (such as relays, actuators, Switched Reluctance motors, stepper motors) is easily understood qualitatively, because of the magnetic nonlinearity, it is very difficult to model and simulate such devices by analytical methods. This is primarily due to the saliency (nonuniform air-gap) of the devices and difficulty in modeling the magnetic materials. Moreover, the method of current excitation (such as using switching devices) combined with the nonlinearity of the circuit produces flux linkage and electromagnetic torque waveforms, which are highly nonsinusoidal in both space and time.

The performance analysis of electromechanical devices is usually evaluated through modeling the magnetic nonlinearity by either the geometrically based flux-current curves using finite elements or magnetic circuits or the experimentally determined flux-current curves as demonstrated in this section.

The determination of magnetization characteristics is necessary for design verification, performance evaluation, and even control of various electromechanical devices, such as Switched Reluctance (SR) motors.

The necessity of having accurate knowledge of the flux linkage characteristics in nonlinear devices requires an accurate and automated measuring method due to the laborious nature of obtaining the characteristics. In practice, a number of methods are available for determining B versus H or ϕ versus i characteristics:

- Using a ballistic galvanometer with search coil
- Using an application of the Hall-effect
- Using a flux meter

Most of these measurement methods, however, require modification of the specimen or expensive signal conditioning and measurement devices.

In addition, a number of computer-aided measuring techniques have been proposed in the literature. However, these methods usually ignore the secondary effects, such as the range of flux variation in practical sensors and the harmonic contents of the currents. In this section, various improvements on the previous methods are made, and an automated PC-based measurement system equipped with a custom-written VI is explained.

4.4.1 Principles of the Method

The method described and implemented here is based on indirectly measuring the flux linkage at standstill from the instantaneous voltage and the current measurements on the winding of a device that is connected to a supply. The voltage across the excited winding (or the source) is given by

$$v(t) = Ri(t) + \frac{d\lambda(t)}{dt} \tag{4.20}$$

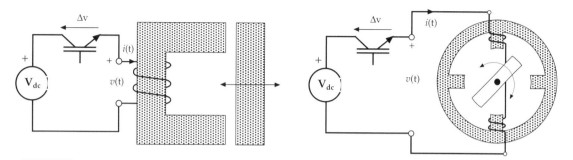

Figure 4-9
Sample device arrangements for measuring magnetization characteristics.

where $v(t)$ is the instantaneous voltage, R is the winding resistance, $i(t)$ is the instantaneous winding current, and $\lambda(t)$ is the instantaneous flux linkage of the winding.

The flux linkage characteristics of a device can be obtained by rearranging and integrating equation 4.20 over a time period. In addition, in a practical system (Fig. 4-9), if the total voltage drop of the external components, Δv, and the value of the initial flux linkage, $\lambda(0)$, are included, a general flux linkage equation can be given by

$$\lambda(t) = \int_0^t [v(t) - \Delta v - Ri(t)]\, dt + \lambda(0) \qquad (4.21)$$

where dt is the sampling time interval.

In equation 4.21, the voltage drop across the switching device and the resistance of the winding are usually assumed to be constant. The initial value of the flux linkage $\lambda(0)$ is very significant in electromagnetic circuits containing remnant fluxes, such as permanent magnets. However, the initial flux value can be assumed to be zero in other devices when there is no current.

Note that this integration can be performed easily by using analog or digital methods. Nevertheless, the analog integration method is not preferred due to the drift in the analog integrators and inaccurate components of the practical integrator.

Evaluating the flux linkage by a digital technique eliminates the error involved with analog integrators and has the following principal benefits:

- No drift in the integrator
- Greatly reduced measurement time
- An almost continuous range of data points available for further use, such as for performance prediction

However, in the digital integration method, the winding resistance must be known to a good degree of accuracy, and if a step voltage is applied to the winding and the instantaneous current and voltage are measured, the flux linkage at any instant can be estimated.

4.4.2 Experimental Setup

The experimental setup used to measure the flux linkage versus current characteristics consisted of a 4 kW, four-phase SR motor; a rotor clamping and positioning device; a transistor switching device; custom-written VI; and a computer and data acquisition (DAQ) card. Fig. 4-10 shows a complete block diagram of the automated experimental setup and a cross section of the SR motor under test.

The dc voltage supply that is switched onto the motor winding consists of a three-phase rectifier supplied via a three-phase autotransformer. The applied dc voltage is adjusted so that the peak current reached during the test should not exceed a preset level of current, which can be slightly higher than the rated current (to include the saturated operation).

The hardware used in the test facility includes a switching device, an IGBT transistor, which is used to switch on the dc supply to the motor phase winding. The switch is controlled by the VI `FluxLinkageCurrentTest.vi`, which provides a control signal via the digital output port of the data acquisition board.

In the test setup, the phase current and phase voltage are measured by using a Hall-effect device and an isolation amplifier, respectively, which present small amplitude voltage signals to the DAQ board.

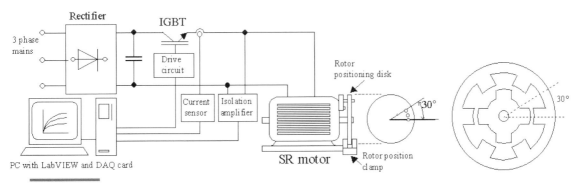

Figure 4-10
Hardware block diagram of the test system.

A PC bus based plug-in DAQ board fully compatible with the LabVIEW software is used to collect data, as well as to provide digital control signals. The board has a maximum sampling rate of 100,000 sample/s, and provides a 12-bit sampling A/D converter with up to eight differential inputs (AT-MIO-16E-10). The inputs of the board are software configured for either 0 to +10 V or −10 V to +10 V operation. In addition, each channel had software programmable gains of 1, 2, 5, 10, 20, 50, and 100. Also provided on the board are eight digital IO lines and two 16-bit D/A converters.

As shown in Fig. 4-10, the motor clamping and positioning device consists of a disc, attached to the motor's shaft, with small holes on the outer diameter spaced by one mechanical degree. This arrangement is used to fix the rotor position to a given angle by inserting a pin into the given angle pinhole, and to lock the disc with an attached clamp. Note that the clamping/disc device must be carefully designed and assembled so that errors in rotor position are minimized. In addition, the clamping action must be strong enough to hold the rotor through the impulse torque attempting to align the rotor with the rotor pole with the nearest stator pole, when current is applied.

Let me emphasize here that the hardware suggested in Section 5.4 can be utilized to perform this experiment.

4.4.3 Virtual Instrument Panel

The general steps carried out by the VI are shown in Fig. 4-11. The PC-based system initiates the test on the device and the following steps are performed:

- An offset error adjustment of the voltage and current sensors is performed.
- The winding voltage is switched on, and then switched off, and the instantaneous voltage and current are acquired.
- The flux linkage is calculated.
- The graphical output is produced for viewing and storage purposes.
- The rotor of the device under test is rotated by one mechanical degree (or more or less, depending on the accuracy of the setup), and the above process is repeated.

The front panel of the VI developed in this section (Fig. 4-12) allows the user to enter the resistance of the winding and displays the graphs of the current, voltage, and flux linkage versus time, and also displays flux linkage–versus-current graphs.

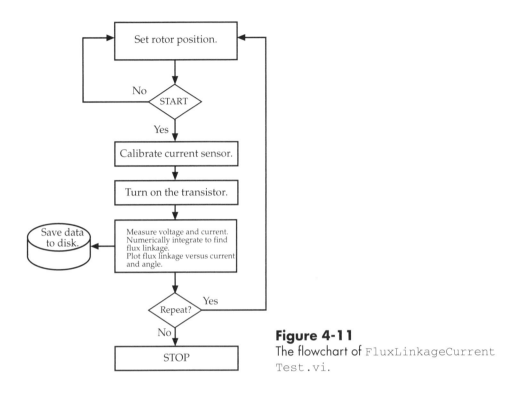

Figure 4-11
The flowchart of `FluxLinkageCurrent Test.vi`.

Four principal steps of the block diagram are followed in the designed algorithm.

In the first step, the user is asked to check that the rectified dc power supply is off, in order that the current sensor can be calibrated.

The second step of the block diagram performs the calibration (offset error zero adjustment) of the current sensor, and informs the user when this step is finished. This is required in practical systems since Hall-effect current sensors normally have some offset error, which should be removed for higher measurement accuracy. Furthermore, as the offset error can drift over time, the sensor must be calibrated periodically to obtain accurate results. The calibration sub-VI reads 50 random values from the current sensor when the applied current is zero and averages the values to determine the offset voltage of the Hall-effect current sensor. The front panel also displays a plot of the current values and their mean during the execution of this step. The voltage measurement circuit may also have some small offset error, due to offset in the ground plane of the voltage measurement circuit. Hence, a similar offset adjustment is also made with respect to the voltage measurement circuit.

In the third sequence of the block diagram, the user is then instructed to set a new rotor position (using the rotor clamping device), to turn on the main dc

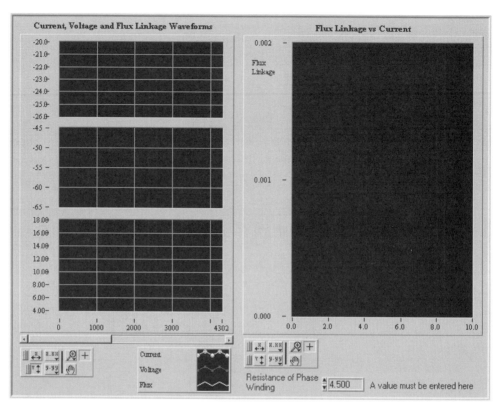
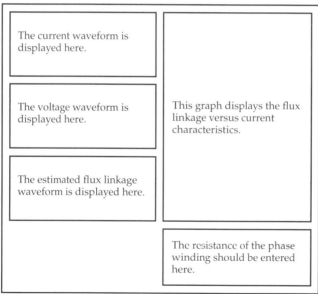

Figure 4-12
The front panel and brief user guide of FluxLinkageCurrentTest.vi (LabVIEW 5.0).

voltage supply to the motor, and to press the on-screen button to start the test. The user is requested to enter the angle that is being measured (for graphical plotting purposes) and the maximum current required.

In the final step, the transistor switch is turned on and off repeatedly for a short period. This is done so that the current will flow in the winding, and thus the winding will be at a warm operating temperature before the actual test is performed.

The transistor is turned on via a low-level sub-VI that controls the state of the digital output port. After this is performed, a sub-VI turns on the transistor again, the voltage and current are measured, and then the transistor is turned off. The measured voltage and current are integrated to calculate the flux linkage by a sub-VI that performs the integration. From this, the user is presented with on-screen plots of current, voltage, flux linkage, and flux linkage versus current. These values are also automatically stored numerically onto the computer disk drive. A separate file is created to store a set of data regarding the rotor position. The filename is also automatically created in such a way as to indicate the measured rotor position.

A safety feature has also been added to the software so that if any error occurs, the transistor is automatically turned off. This ensures against a sustained high current if the transistor is left on during some software or data overflow error.

Sample Test Results

As the test motor has eight stator poles and six rotor poles, the electrical angle is 30°, and therefore the complete characteristics of the motor are obtained by measuring up to 30°.

The flux linkage versus current curves for each test performed on a single rotor angle are automatically saved to a spreadsheet file at the end of each test. The current path names used in the VI are `c:\Magnetic Circuit\flux#` and `c:\Magnetic Circuit\volcur#`, where the # sign indicates the test (or position) number.

A set of experimental results is given in Fig. 4-13. The figure illustrates the measured instantaneous current, the estimated flux linkage waveforms (based on equation 4.21), and the flux linkage versus current curve for the rotor position of 30° mechanical. In this result, one can observe the distinct saturation seen on the magnetization characteristics, when the current is above approximately 6 A.

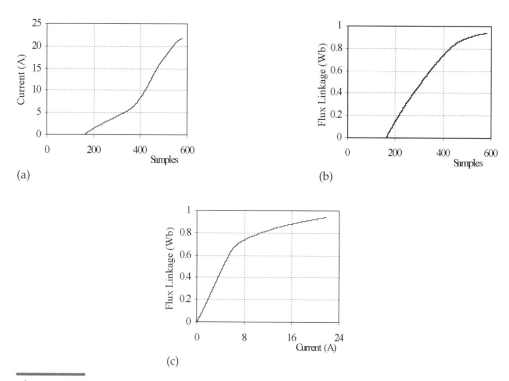

Figure 4-13
A set of experimental results for the rotor position of 30° mechanical obtained using the VI and the setup described in Fig. 4-10: (a) The waveforms of measured current, (b) estimated flux linkage, and (c) flux linkage versus current.

4.5 References

Cheok, A., and N. Ertugrul. "Computer-Based Automated Test Measurement System for Determining Magnetization Characteristics of Switched Reluctance Motors." *IEEE Transactions on Instrumentation and Measurements 50*, no. 3 (June 2001): 690–696.

Ertugrul, N., and A. Cheok. "An Automated Method of Determination of Magnetization Characteristics of Switched Reluctance Machines." Australasian Universities Power Engineering Conference (AUPEC '99), 17–22. Darwin, 26–29 September 1999.

Hambley, A. R. *Electrical Engineering: Principles and Applications.* Upper Saddle River, NJ: Prentice Hall, 1997.

Smith, R., and R. Dorf. *Circuits, Devices, and Systems.* New York: Wiley, 1992.

Wildi, T. *Electrical Machines, Drives, and Power Systems.* Englewood Cliffs, NJ: Prentice Hall, 1991.

Yamaee, Z. A., L. Juan, and J. R. Bala. *Electromechanical Energy Devices and Power Systems.* New York: Wiley, 1994.

Yarwood, J. *Electricity and Magnetism.* London: University Tutorial Press, 1973.

Electric Machines Laboratory

5

The main objectives of the laboratory experiments provided here are to help students in observing what happens to the principal electromechanical devices when they are linked to a power source by studying the real-time waveforms and how the devices can be modeled, facilities that are not fully present in conventional methods.

By conducting experiments and making observations in real time on the experimental systems, students can build a good model of the systems that can then be used to perform more detailed analyses.

Eight generic experiments are presented in this chapter, which utilize custom-written LabVIEW-based experiments. Table 5-1 lists these experiments. As a guide, the table also indicates the estimated time to complete each task.* The tasks are divided into two sections: measurement and analysis. The experiments described here are self-contained, with sufficient material included in the associated sections to describe the basic theory. However, outside references may be useful to expand the discussions further.

---

*Note that the estimated times in Table 5-1 do not take into account preparation time, quiz time, time to disseminate the information, or time to answer questions for report writing, which are based on the laboratory infrastructure where they are developed, as explained in the Appendix.

Table 5-1 *The list of experiments and estimated times for each test.*

Title	Estimated Time (min.)	
	Measurement*	Analysis**
Determination of Moment of Inertia	5	10
Losses in DC Motors	40	30
Electromechanical Device Experiment	30	30
Laboratory Tests for Basic AC Circuits	50	40
Transformer Test	30	30
Asynchronous Motor Test	30	40
Synchronization Observer	10	10
Synchronous Machine Test	60	30

*Includes the time to set up the experiment and collect the real-time data.
**This is the computer-aided analysis section and, if desired, can be done away from the laboratory.

5.1 Introduction

Fig. 5-1 illustrates the common hierarchical structure of the experiments. Each experiment consists of a main menu with a list of options indicating the parts of the experiment. After selecting from the list, a submenu displays two options: theory and experiment. The theory section provides the user an on-site reference containing some background information and wiring diagrams to guide the real-time implementations. This section should be followed by the real data capturing section.

Note that some of the experiments provided here contain a multiple-choice-type quiz section in our laboratory, which is delivered by PC and should be studied before initiating the practical sections. The quiz questions are mainly based on safety issues and the specific laboratory infrastructure. A sample quiz VI is given in the Appendix.

The real-time measurement section is followed by the analysis VIs where basic calculations are carried out and the results are recorded for further calculations. The analysis sections link the measured results to well-known textbook knowledge and verify the experimental results and provide insight into the operation of the electrical machines.

Real-Time Measurement Interfaces

It should be emphasized here that the specifications of the real-time interface circuits (such as for speed, torque, force, current, and voltage) vary consider-

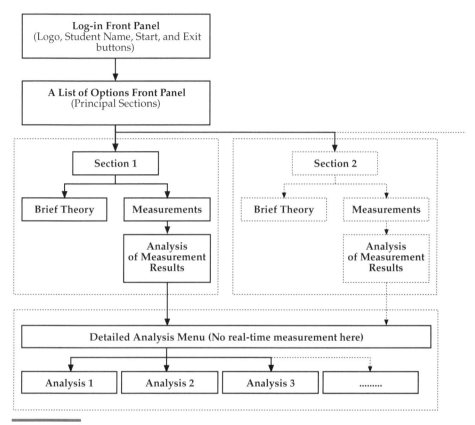

Figure 5-1
The common hierarchical structure of the VIs.

ably among educational institutions. Therefore, some minor adjustments have to be made in the block diagrams of the VIs for the correct real-time measurements. To achieve this, some initial tests have to be carried out on the site, and the correct measurement gains have to be entered into the relevant boxes in the block diagram as shown in Fig. 5-2.

The ratings of the real laboratory machines are not critical in the VIs, but the frequency bandwidths of the signal conditioning devices are. The measured signals have to be conditioned (attenuated or amplified, and isolated) before connecting to the DAQ card.

The essential parts of a real-time experimental system may consist of a number of units: a device under test, transducer(s), a PC and a DAQ card, and a custom-written VI, which are illustrated in Fig. 5-3. However, the details of the hardware for the individual experiments will be given later.

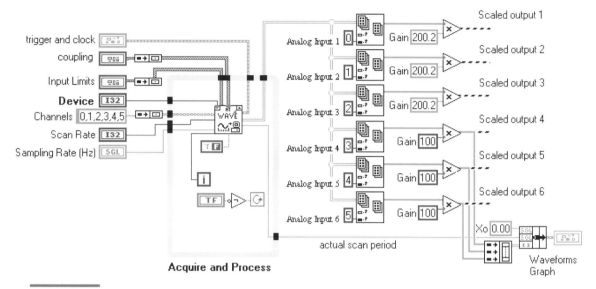

Figure 5-2
A sample block diagram used to acquire data from the multiple analog inputs.

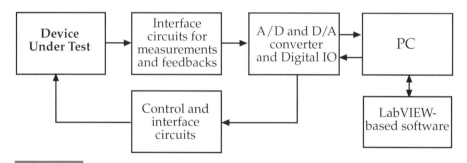

Figure 5-3
Common components of a real-time measurement system.

Precautions in Real-Time Measurement

Due to the nature of the experimental systems covered here, our view is that if the VIs are intended to be used in practical experiments, certain safety precautions have to be considered and implemented. Some of these precautions will be emphasized later in the Appendix. However, the following list re-emphasizes the safety issues:

- Make sure that you have sufficient background knowledge about the system to foresee potential dangers.

- Make sure that the correct wiring and isolation procedures are followed before starting the experiment.
- During the wiring, make sure that you use only one hand at a time to reduce the risk of electrocution.
- Do not use a faulty cable or equipment.
- Do not use a trial-and-error method to find the function of a certain unit.

5.2 Determining Moment of Inertia

The primary objective of this simulation is to demonstrate the transient speed profiles of rotating electrical machines, which occur during acceleration and deceleration. Furthermore, the experiment provides insight into how to measure the moment of inertia of rotating bodies. Moment of inertia may be necessary to determine, specifically, if the rotating body is inaccessible or if its dimensions are not known.

The moment of inertia is the name given to rotational inertia, the rotational analog of mass for linear motion. It appears in the relationships for the dynamics of rotational motion.

A rotating body possesses kinetic energy. This energy depends on the moment of inertia J, which can be calculated if the exact mass and dimensions of the rotating body are known. However, calculating J is very difficult in rotating electrical machines since the rotating body (rotor) is not uniform and is usually not accessible. In addition, the rotor of the motor is normally coupled with other rotating masses of complex shape, which makes the calculation of J very difficult. In many motor control applications, however, the moment of inertia of the rotating section must be known accurately to allow advanced control and analysis.

In any rotating electrical machine system, the speed change depends on the torque imbalance in the system for a given period of time and the inertia of the system, which can be used to determine the moment of inertia.

For an electric motor where the driving electromagnetic torque is generated, the mechanical state equation can be given by

$$T_e = J\frac{d\omega}{dt} + T_m + T_{loss} \qquad (5.1)$$

where ω is the instantaneous angular velocity, T_e is the electromechanical torque, T_m is the torque required by the mechanical load, and T_{loss} is the torque required to overcome the losses in the system.

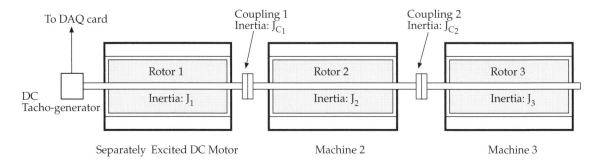

Figure 5-4
Retardation test setup used in the development stage.

As can be seen in equation 5.1, a direct computation of the inertia is possible if the rest of the parameters, which can be determined by experiment, are known.

The retardation test is used to determine J, which is very time-consuming and difficult to measure unless you use a computer-based measuring method. Fig. 5-4 illustrates the experimental setup that was used to perform a retardation test during the development stage.

In the retardation test, we assume that a separately excited dc motor is driving a mechanical load (containing two additional electrical machines connected on the same shaft), and the moment of inertia of the complete rotating part will be determined.

To achieve this, first the electrical supply to the motor's main power winding is disconnected while the machine is rotating; then the instantaneous shaft speed is measured via a dc tacho-generator and the analog input of a DAQ card.

When the power is turned off, the driving motor torque T_e disappears and the system decelerates, and, therefore, equation 5.1 can be simplified as

$$J\omega \frac{d\omega}{dt} = -P_{\text{loss}} \qquad (5.2)$$

As can be seen from equation 5.2, by plotting the speed versus time graph during the deceleration period, for a known value of P_{loss} (in watts) in the system under test, the inertia J can be calculated. However, be aware that the J in this equation represents the total inertia of the rotating bodies ($J_1 + J_2 + J_3 + J_{C1} + J_{C2} + J_{\text{tacho}}$), as shown in Fig. 5-4. The inertia of the tacho-generator is negligible, hence usually ignored.

It should be noted that, since the derivative is involved in the estimations of J, the initial estimation of the term $d\omega/dt$ may be very much in error. Therefore, for a correct result, the initial values of the estimated parameters should be discarded from the computed data, as will be explained in the following section.

5.2.1 Virtual Instrument Panel

The VI entitled `Retardation Test.vi` given in this section illustrates the method that can be implemented to determine the inertia of any rotating mass if the speed is measured and the total loss (which is usually equal to the friction and windage loss in a rotating system) is known.

Fig. 5-5 shows the two front panels that are available to simulate the experiment. Fig. 5-6 provides brief user guides for the two front panels. To perform the simulation study on the retardation method, the user should follow the flashing lights to execute the specific computation sections of the VI.

However, a data file is required in this simulation, which should contain the power loss characteristics of the rotating machine that is under investigation. This file is provided in the same folder as the VI, which is linked to the VI by a path name, `a:\dcloss`.

The first graph of the VI indicates a simulated deceleration speed curve that is normally obtained in rotating electrical machines. The consecutive graphs display the estimated values from this speed-time graph. The `Calculated Inertia J.vi` reads the file that contains a set of sample data about the losses in the machines (a dc machine in this test) and calculates the moment of inertia of the rotating mass.

The procedure for the simulation is summarized below:

- While the electrical machine is running at a steady speed, turn off the main power switch and, at the same time, initiate the data logging of the instantaneous speed n in rpm (revolutions per minute) until the dc machine stops rotating. Be aware that the stopping time depends on how large the rotating body is, and may be on the order of minutes in practice.

- After collection of the speed-time data, the angular speed in radians per second is calculated by

$$\omega = \frac{2\pi n}{60} \tag{5.3}$$

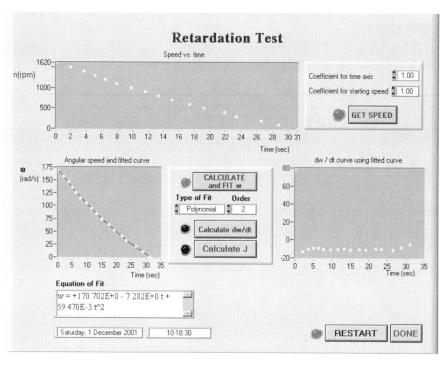
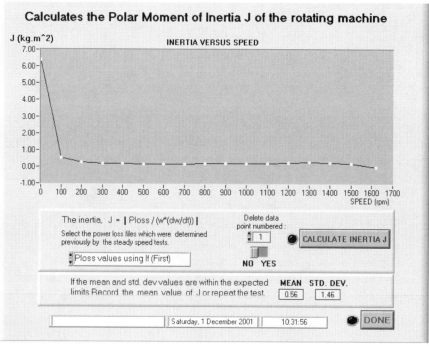

Figure 5-5
The front panels of the retardation test, Retardation Test.vi and Calculated Inertia J.vi.

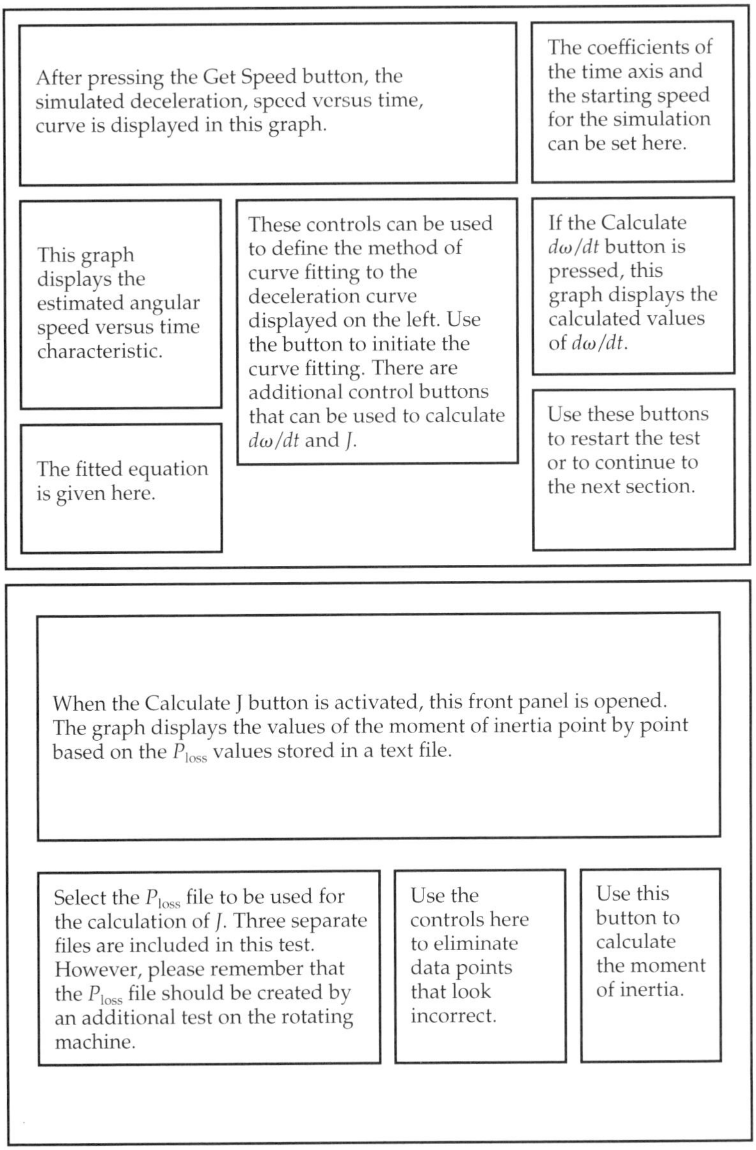

Chapter 5 • Electric Machines Laboratory

After pressing the Get Speed button, the simulated deceleration, speed versus time, curve is displayed in this graph.

The coefficients of the time axis and the starting speed for the simulation can be set here.

This graph displays the estimated angular speed versus time characteristic.

These controls can be used to define the method of curve fitting to the deceleration curve displayed on the left. Use the button to initiate the curve fitting. There are additional control buttons that can be used to calculate $d\omega/dt$ and J.

If the Calculate $d\omega/dt$ button is pressed, this graph displays the calculated values of $d\omega/dt$.

Use these buttons to restart the test or to continue to the next section.

The fitted equation is given here.

When the Calculate J button is activated, this front panel is opened. The graph displays the values of the moment of inertia point by point based on the P_{loss} values stored in a text file.

Select the P_{loss} file to be used for the calculation of J. Three separate files are included in this test. However, please remember that the P_{loss} file should be created by an additional test on the rotating machine.

Use the controls here to eliminate data points that look incorrect.

Use this button to calculate the moment of inertia.

Figure 5-6
The brief user guide of `Retardation Test.vi` and `Calculated Inertia J.vi` given in Fig. 5-5.

- The estimation of the derivative $d\omega/dt$ for the measured speed and time intervals is performed. Note that $d\omega/dt$ is the slope taken at various points of the fitted speed-versus-time graph, and the accuracy of the estimation very much depends on the frequency of sampling, specifically around the initial and final speeds. However, this is improved in this VI simply by fitting a curve to the captured speed-time data, which can generate an infinite number of data points.
- The final stage of the test is the estimation of the moment of inertia J by using equation 5.2, which requires the values of P_{loss}:

$$J = \left| P_{loss} / \left(\omega \frac{d\omega}{dt} \right) \right| \tag{5.4}$$

- Note that P_{loss} data may vary considerably depending on the speed during the retardation test. The data file included in this VI contains three sets of measured data obtained at different field current values for the conventional brushed dc machine in our Electrical Machines and Drives Laboratory.
- As stated before, to avoid the error in the initial estimated point of J, which is due to the derivative $d\omega/dt$, set the control to YES on the second front panel and enter the data points to be excluded in the computations. This adjustment is necessary in order to avoid errors in the estimation of the derivatives due to the initial and final boundary conditions. Note that the first point on the graph starts from the right-hand side.

5.2.2 Recommended Laboratory Hardware

Note that the data acquisition sub-VI has not been included in the VI provided here. Instead, only some simulated speed results are provided to study the retardation test without a real measurement.

If the real measurement is desired, the test setup given in Fig. 5-4 should be constructed and should include a separately excited dc machine equipped with a tacho-generator and driven by a prime mover (an asynchronous motor, Machine 2 in the figure). It should be emphasized that any motor can be used in this test to determine the polar moment of inertia. However, the system losses have to be either known or measured as shown in the subsequent section.

The specifications of the instrumentation used in the development stage of the real test are given next.

Computer:
Pentium PC, 32 MB RAM, and >1 GB hard disk

Data acquisition system:
National Instruments AT-MIO-16E-10 data acquisition card with 8 differential A/Ds, 12-bit resolution, 100 kHz sampling frequency, 2 Analog Outputs (for 12-bit D/A conversion), 8 Digital I/O

Signal conditioning device:
The real-time speed is measured by using a dc tacho-generator attached to the shaft of the machine. This has a moment of inertia that is negligibly small compared with the actual motor and the prime mover.

5.3 Losses in DC Motors

In this experiment, we will obtain real data and investigate the methods for isolating the components of power loss that occur in rotating machines, using the laboratory dc machine as an example. Note that a dc machine is selected here since it represents a general case.

Rotating electrical machines can lose power due to friction and windage losses, iron losses, copper losses, and stray load losses. All these losses are converted to heat, causing an increase in temperature within the machine, which places a limit on the acceptable operating range of the machine and reduces the power output available.

Here are definitions of the various losses:

- **Friction and windage loss** is due to bearings, brushgear, and commutators or slip rings, which are inseparable in an assembled machine, and is a function of rotational speed.

- **Iron losses** are due to eddy current and hysteresis losses in the magnetic core, which depend on the magnitude of the flux and the frequency of flux variation.

- **Copper losses** are due to the resistance of a machine winding, non-linear components (such as the graphite brushes), and the frequency for alternating currents (due to skin effect that increases the effective resistance).

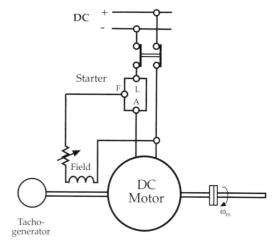

Figure 5-7
The circuit diagram used to perform the experiment.

- **Stray load losses** are due to second-order effects such as eddy current losses within the conductors themselves, extra iron losses due to waveform ripple caused by slotting, and stray fields at the ends of the machine.

The structure of the VI is explained in Section 5.3.1. It should be noted here that although this VI is designed as a teaching and learning tool, it can easily be modified to be used by machine manufacturers for testing purposes. The circuit block diagram used in this test is given in Fig. 5-7.

5.3.1 Virtual Instrument Panel

In this experiment, the following sequence of tests will be carried out:

- Determining the resistance of the armature winding as a function of current
- Running a steady-speed test
- Obtaining the polar moment of inertia of the machine set (which was studied in the previous experiment)
- Measuring the power losses as a function of speed
- Determining the no-load iron losses as a function of speed, with field current as a parameter
- Measuring the torque versus speed characteristics

The VI provided here reads a set of data files that are obtained from the real machines (see the dc machine ratings given in Chapter 1, Table 1-1). Therefore, the current version of the VI may be considered a simulation utilizing the real data set.

Fourteen data files are provided in the same folder of the VI. A number of controls are also provided to input data manually. However, a data acquisition VI can be added to read the real experimental data if the hardware is available.

Let's look at the details of each test separately now. To initiate the experiment, press the Run Continuous and Start buttons consecutively (or the Exit button to terminate the program).

After pressing the Start button, a front panel with a list of options indicating the parts of the experiment will appear as shown in Fig. 5-8. Then an option can be selected either by double-clicking on the option or by pressing the Select button.

Part 1, Armature Resistance Test

The front panel of the sub-VI is given in Fig. 5-9. The aim of this test is to determine the armature resistance R_a as a function of the armature current I_a under normal operating temperature. As expected, this resistance should be nonlinear due to the nonlinear components in the resistance circuit, such as brushes.

The test is initiated after the shaft of the machine is clamped to prevent it from moving. Then the armature voltage V_a and the armature current are measured. In case of an error in a pair of data (V_a, I_a), use the controls to recapture the data and replace it with an earlier set of data.

The two meters provided on the front panel indicate the measured dc voltage and the dc current readings at the end of the warm-up time. Then the armature resistance is calculated and plotted against the armature current on the graph, where $R_a = V_a / I_a$.

After capturing a sufficient number of data points, the equations of fitted curves can be generated by pressing the Fit Curves button. The operator is allowed to choose the appropriate type of fit by selecting Linear, Exponential, or Polynomial using the controls located on the top left-hand corner of each graph. If the polynomial fit is selected, you must provide the order of the polynomial as well.

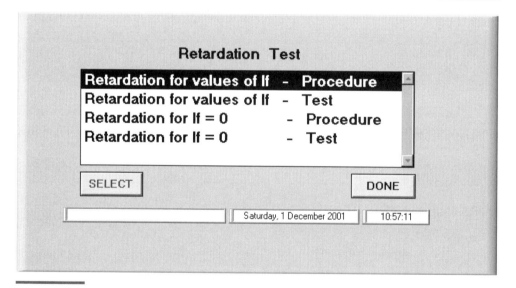

Figure 5-8
The options front panel of `Losses in DC Motors.vi`.

The coefficients of the fitted equations should be recorded for further calculations; they will be used to determine the value of the resistance for any given armature current.

After this test has been completed, the measured data points can be saved for future reference or for retrieving the data at a later stage. To return to the previous front panel, press the Done button.

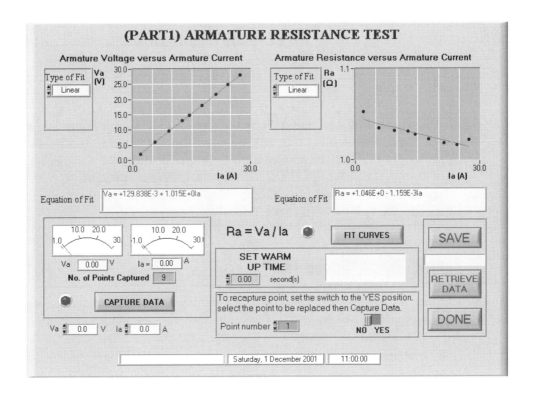

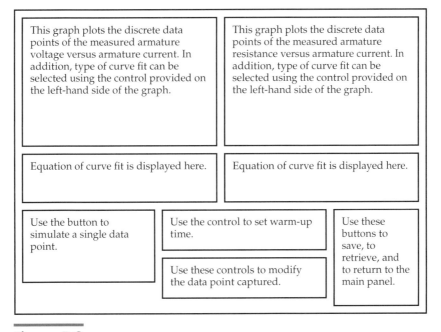

Figure 5-9
The front panel and brief user guide of the Armature Resistance Test option.

Note: In the current version of the VI, a simulated data set can be generated by the controls situated at the bottom left-hand corner of the panel, or a set of data measured from the real machine set can be retrieved by pressing the Retrieve Data button, or real data can be captured by adding a data acquisition VI and using the real machine set. In addition, please refer to the theory sections of the options for further details about the experimental procedures.

Part 2, Steady Speed Test

The aim of this test is to determine the power losses of the dc machine under steady-state conditions and with no mechanical load. Three front panels will be available to users to perform the data capturing and processing functions (Fig. 5-10, Fig. 5-11, and Fig. 5-12).

Let's consider equation 5.1 and multiply both sides by ω, which will result in an equation of motion as a function of power.

$$J\omega \frac{d\omega}{dt} = P_e - P_{loss} - P_m \qquad (5.5)$$

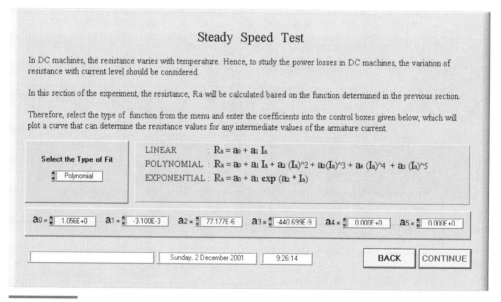

Figure 5-10
The front panel of Part 2, Steady Speed Test option.

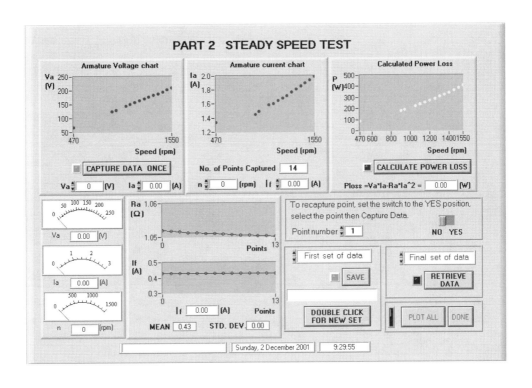

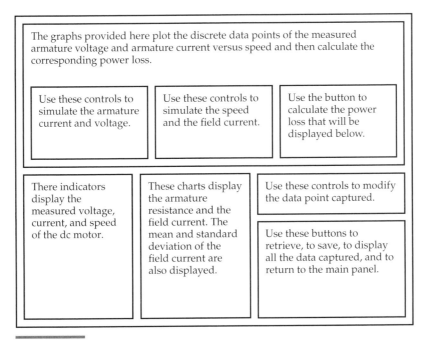

Figure 5-11
The front panel and brief user guide of the Steady Speed Test.

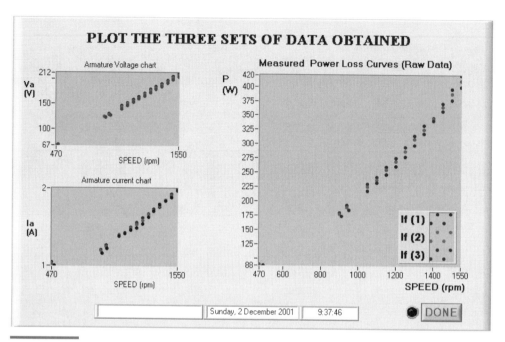

Figure 5-12
The front panel after pressing the Plot All button on the Steady State Test panel.

Under a steady-state condition and without any mechanical load, it can be seen that the input power of the motor should balance the power losses in the machine. Hence,

$$P_e = P_{loss}$$
$$V_a I_a = I_a^2 R_a + f_1(I_f, \omega) + f_2(\omega) \tag{5.6}$$

where $I_a^2 R_a$ represents the armature copper losses, the function f_1 is the no-load iron losses, f_2 is the friction and windage losses, and I_f is the field current of the motor.

When you select the Steady Speed Test option, the front panel shown in Fig. 5-10 will be displayed. The controls in this panel are used to select the type of fit and to enter its coefficients for the resistance function determined in Part 1. When a value of resistance is required in the following parts, this function is utilized to calculate its value.

Fig. 5-11 is the main front panel of the Steady State Test panel where the data points of the armature voltage and the armature current are captured and displayed as a function of operating speed. A graph of calculated power loss characteristic is also provided here. In addition, this panel provides two indicators—standard deviation and mean value of the field currents—which should be used to make sure that the value of the field current during the tests remains constant within an acceptable limit. The Save button can be used to store the measured data set into separate data files.

The power loss test is repeated for three different values of the field currents. After completing these three tests and upon pressing the Plot All button, the front panel shown in Fig. 5-12 is displayed. Here you can view the final test results and determine the fitted equations for the power losses at steady-state speed. The coefficients of the power loss curves will be used later to determine the polar moment of inertia. To return to the main menu, press the Done button.

Part 3, Retardation Test (Polar moment of inertia of the machine set)

There are two subsections in the sub-VI, where both tests are performed without any mechanical load, that is, $P_m = 0$.

In the first section, the polar moment of inertia of the system is determined, without the armature supply ($P_e = 0$), but with the field current (using one of the field current settings performed in the earlier steady-speed test). Therefore the equation of motion in equation 5.5 can be given as

$$J\omega \frac{d\omega}{dt} = -P_{loss} \tag{5.7}$$

where P_{loss} are the values found in the steady-speed tests.

The test procedure and the front panel for this section are similar to the test described previously in Section 5.2.1. The major difference, however, is in the values of the power losses. Unlike the previous VI, when the Calculate J button is pressed, the front panel shown in Fig. 5-13 is displayed, where the user can input the power loss coefficients determined in Fig. 5-12. Following this panel, the calculation of J is performed in the same manner as explained in `Calculated Inertia J.vi` of Fig. 5-5. Record the mean value of J to be used in the following section. However, remember that the boundary data points have to be removed from the characteristic of J. Note that the first point on the graph starts from the right-hand side.

In the second section, the windage and friction losses are separated by setting the field current to zero and repeating the retardation test as before. The

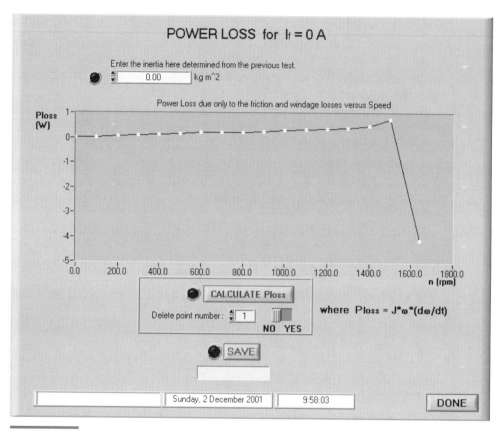

Figure 5-13
The front panel of Power Loss for the zero field current.

aim here is to determine the losses as a function of speed using the value of J determined earlier.

Note that in the retardation test here, the slowdown period of the machine set will be longer than the previous cases, since the opposing torque is due to the windage and friction only.

In addition, note that the Calculate J button in Fig. 5-5 is replaced with the Calculate Ploss button. The subsequent front panel is shown in Fig. 5-13, where the estimated value of J is entered and the windage and friction power loss is calculated as a function of speed. The data calculated here can be saved using the Save button.

$$P_{\text{loss(windage+friction)}} = J\omega \frac{d\omega}{dt} \qquad (5.8)$$

Chapter 5 • Electric Machines Laboratory

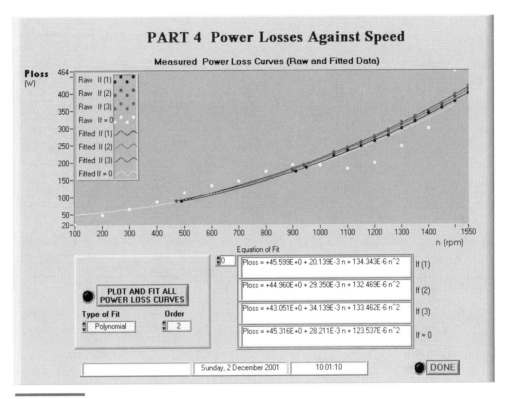

Figure 5-14
The front panel of Part 4, Power Losses Against Speed.

Part 4, Power Losses Against Speed Graphs

In this part of the experiment, no practical test is done. Instead, all the power loss characteristics are plotted on a single graph, and corresponding equations are displayed for further analysis, including the zero field current case (Fig. 5-14). One of the conclusions from these graphs is that the power loss curve for the case $I_f = 0$ is (should be) lower than the others since it represents the friction and windage losses only.

The coefficients of the power loss equations have to be recorded to calculate the no-load iron losses that will be studied in Part 5.

Part 5, No-Load Iron Losses versus Speed

Similar to Part 4, there is no practical measurement in this test. The basis of the calculation of the no-load iron losses is expressed in equation 5.9, where

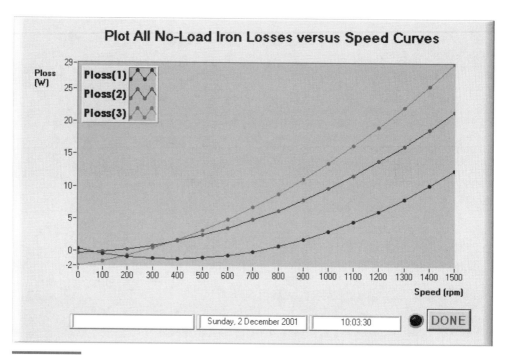

Figure 5-15
The front panel showing all no-load iron losses versus speed curves.

the power loss functions with and without a field current are determined earlier.

$$P_{\text{no-load iron loss}} = P_{\text{loss},I_f \neq 0} - P_{\text{loss},I_f=0} \quad (5.9)$$

There are three front panels available in this part: the front panel to enter the coefficients of the power loss curves for the states $I_f \neq 0$ and for $I_f = 0$, the panel that displays the no-load iron losses with a Save option, and the panel where all the no-load iron losses for the three field current values are displayed (Fig. 5-15).

Part 6, Torque/Speed Characteristic

Note that the real-time measurement of the torque-speed characteristic depends on a torque transducer that should be available on the setup of the system.

These characteristics are considered as the output characteristics of a motor, which provide very valuable information about the performance of the complete drive system where the motor drives a mechanical load. This can be

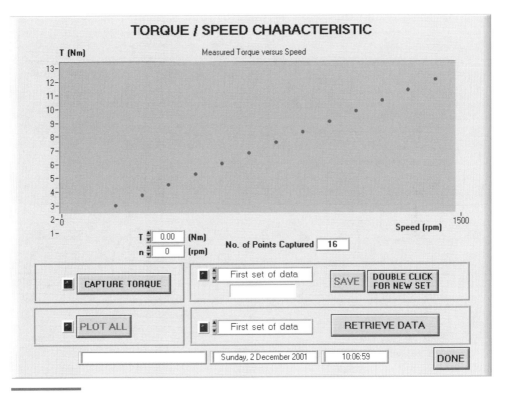

Figure 5-16
The front panel of Torque versus Speed sub-VI.

simulated in the laboratory by using the synchronous machine as a generator, which is wired to a star-connected resistance load.

The sub-VI provided here presents a solution to estimate the torque as a function of the rotational speed, which is similar to the sub-VI in Part 2. First, the field current of the motor is set to one of the values used in the steady-speed tests, and the armature voltage and the armature current readings are obtained for various operating speeds, which are multiplied to obtain the input power. Then using the corresponding power loss characteristics and the armature resistance functions determined earlier, the output power and torque are calculated (Fig. 5-16), where all known sources of loss in the system are eliminated from the calculations before producing the required torque/speed curves.

$$T_{\text{out}} = \frac{P_{\text{out}}}{\omega} = \frac{P_{\text{in}} - I_a^2 R_a - P_{\text{loss}}}{2\pi(n/60)} \qquad (5.10)$$

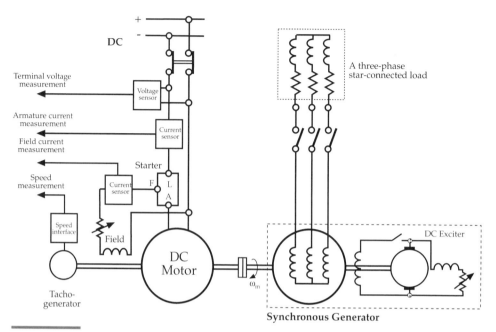

Figure 5-17
The test setup recommended in this experiment.

5.3.2 Laboratory Hardware

Fig. 5-17 illustrates the schematic of the test setup used in the development stage of the VI presented here. Note that the real-time measurement system should include a single point data acquisition sub-VI. The details of the instrumentation recommended in this test are similar to the previous experiment given in Section 5.2.2.

5.4 Electromechanical Device Experiment

In Electrical Engineering, collecting and processing data from experimental devices is usually a time-consuming practice. In addition, the collection of data needs to be accurate to enable detailed analyses to be made of the devices' characteristics.

In conventional experiments, the processing of data (such as force, current, and voltage) usually includes very lengthy calculations and the corresponding production of numerous graphs to analyze the characteristics of the device. Hence, the procedure is very expensive in terms of both human and financial resources.

The LabVIEW-based experiment presented here provides various benefits over conventional tests and facilitates an effective experimenting and learning process. However, some traditional practices, such as wiring, taking notes, and so on, which have significant benefits, are retained.

This experiment reports the design and implementation of an alternative way to derive the electromechanical energy conversion devices' characteristics by using a computer-aided setup. The devices can be excited by either ac or dc supply, and their voltages, currents, and force can be measured in real time.

The components of the test system are illustrated in Fig. 5-18, which contains the custom-built control and signal conditioning circuits, and provide a complete isolation and also allow the user to perform ac or dc tests on the device under test. As provided earlier in Fig. 4-9, the device under test can be a simple or a complex magnetic circuit with a variable air-gap. The details of the circuit diagrams are provided in the Appendix.

Table 5-2 indicates the distinct levels of the software components of the experiment. This experiment basically investigates the property of forces exerted by the magnetic field within the device for different air-gaps under ac or dc excitations. The theory and the graphs behind the conversion of electrical energy into useful mechanical energy via magnetic circuit are displayed by using LabVIEW's graphical user interface.

The main part of the experimental setup consists of four principal units: the electromechanical energy conversion device, the custom-built measurement box, the PC/data acquisition card, and the application VI.

As stated in Table 5-2, the experiment comprises four main sections. In the data-gathering stage of the experiment, the user is required to connect up the hardware outputs to the PC via the plug-in board interface channels (Fig. 5-18). In ac data acquisition, only three analog channels (voltage, current, and force) are used while in dc data acquisition three analog input channels are employed in conjunction with one digital output channel (drive signal to turn on the transistor).

Part 1, Data Acquisition

In the menu for Part 1 of the experiment, the user has the choice of entering three separate sections: Remove DC Offsets, Data Acquisition for AC, and

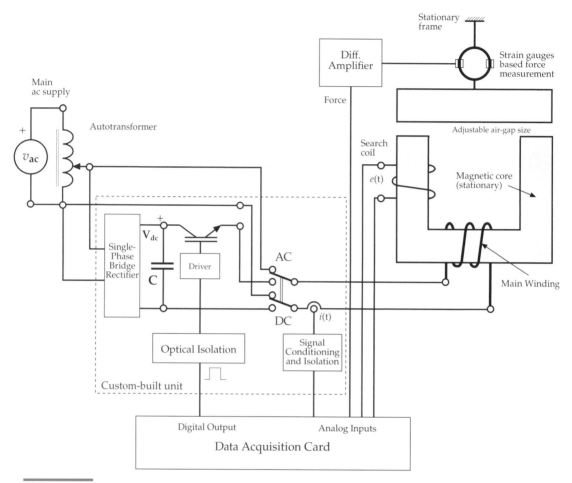

Figure 5-18
The principal components of the test system.

Data Acquisition for DC, which are associated with the collection and saving of data when both ac and dc inputs are applied to the hardware.

The practical sensors do exhibit a degree of dc offset, which has to be removed for the error-free measurement. Therefore, the option Remove DC Offset has been provided, which is used to determine the level of dc offset in the current and the force inputs and to calibrate the analog inputs before the signals are applied.

In the ac data collection section, the procedure entails the data acquisition of induced voltages via the search coil, currents, and forces at specified supply voltage levels and at air-gap sizes, and the saving of these data. The

Table 5-2 *The software levels of the experiment.*

Screen Levels	Remarks
Log-in Screen	Contains a title, an input text box, and two Boolean controls (Start, Exit).
Main Menu	Accepts an input string value corresponding to the operator's name, and contains a list that holds four subsections of the experiment.
Introduction Stage	It is a while loop that has as an (constant) output to the screen with a text dialog and a termination button. This screen also contains a list to link to the sections of the experiment.
Parts 1, 2, 3, and 4 Data Acquisition AC Characteristics DC Characteristics	The theory and the procedure link and the major experimental modules are included.

procedures for this test are clearly stated in the front panels of the subject sub-VI. Note that the settings are device-dependent, hence they have to be specified for a given magnetic circuit. The values of currents and voltages and forces for different air-gaps have to be saved for subsequent parts in this experiment to retrieve and process. However, before saving make sure that the data is of a satisfactory level by magnifying the signals and observing any irregularities. To alter a data set, use the Back button and amend the current setup for the test and proceed again.

The sample front panels given in Fig. 5-19 illustrate the procedure and the data acquisition panels for the dc data acquisition tests. The main difference is the method of collecting the data, which requires a digital signal output. As opposed to the previous test, to initiate the data collection from the device, the Capture control is utilized. After preparing the setup (as specified in the procedure panel) and pressing the Capture control, the digital pulse is applied to the switching transistor (Fig. 5-18), which turns on and provides a step voltage to the main winding of the magnetic circuit. At the same time the analog input signals are measured by using the Sampling Rate and the Scan Rate specified on the front panel controls. Note that these settings have to be amended when the air-gap size is changed. The need for adjustments arises because the saturation times differ for current in each air-gap size. Once the user is satisfied with the data for a given air-gap, the data can be saved, and the process can be repeated for other air-gap sizes.

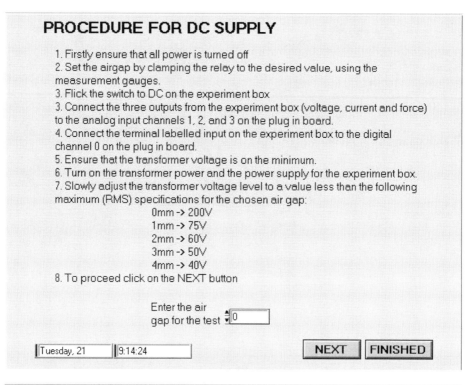

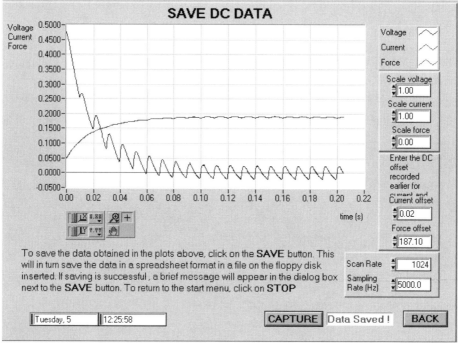

Figure 5-19
The sample front panels: Procedure and data capture panels for the Data Acquisition for DC test.

Part 2, AC Characteristics of the Magnetic Circuit

This is where processing and analysis of all the ac data obtained in Part 1 takes place. The panel for the menu of Part 2 provides some background information and a list of options: View input data for AC, View and Analyze Simulation Hysteresis Loop, View and Analyze Real Data Hysteresis Loop, and Calculate Total Power Losses and Eddy Current Losses.

View Input Data for AC

Upon selecting the first option, a set of three plots are displayed: the induced voltage, the current, the estimated flux linkage, and the measured force. The user can also select the data sets for each air-gap size to be displayed on the associated graphs. Remember that the influence of the dc-offset errors is very critical in the estimation of the flux linkage, since the integration is involved (see equation 4.18 in Chapter 4).

View and Analyze Simulation Hysteresis Loop

The second option, View and Analyze Simulation Hysteresis Loop, aims to give the user an opportunity to partake in a hands-on exercise to observe the effects of the parameters changes on the Hysteresis Loop (*BH* curve). Furthermore, the analysis of the hysteresis loss and the critical points of the *BH* curve can be studied from this and the other related panels. The front panels of this option are shown in Fig. 5-20 and Fig. 5-21.

After determining the amplitude and the frequency of the voltage and current waveforms and corresponding *BH* characteristic, the user can select two options from the panel in Fig. 5-20: Calculate Hysteresis Area (Loss) and Identify Critical Points.

Note here that the *BH* characteristic can easily be derived from the flux linkage and the current characteristics if the mean length and the cross-sectional area of the core and the number of turns of the excited winding are known (see Section 5.1).

Upon selection of the first option, the panel in Fig. 5-21a will appear, where the panel is programmed to calculate the hysteresis loss. Since the energy is proportional to the power, using Faraday's Law it can easily be shown that the hysteresis loss is proportional to the area enclosed by the *BH* loop. This is primarily due to the hysteresis effect, the phenomenon whereby the flux density

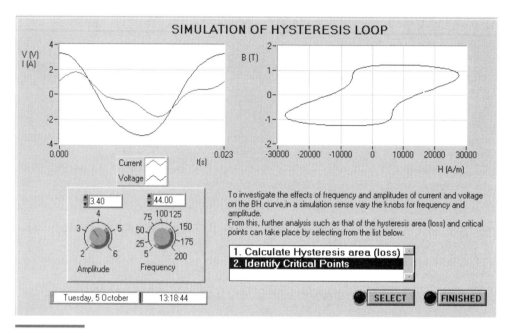

Figure 5-20
The front panel of the View and Analyze Simulation Hysteresis Loop option.

lags the field intensity throughout the magnetization cycle. Hence we can give the following equation.

$$P_{\text{Loss(hysteresis)}} = V_{\text{core}} f(\text{Area of } BH \text{ loop}) \quad (5.11)$$

where V_{core} is the volume of the core and f is the frequency of the ac supply that drives the magnetization cycle. Fig. 5-22 illustrates the trapezoidal method used to estimate the area of the BH loop.

Let's consider one element of area that is shown in Fig. 5-22. If the area of this element is calculated using the coordinates of the points indicated in the figure, we can obtain an approximated area that is a trapezoid. The total area under a curve is found by summing all the elements. Referring to the same figure, the total area of the loop can be given by

$$\text{Area of } BH \text{ loop} = 2(\text{Area}_2 - \text{Area}_1) \quad (5.12)$$

Note that an alternative method for the estimation of the areas is to fit curves to determine the corresponding functions. Then these functions can be integrated directly, which can provide a more accurate result. However, for simplicity, this is not considered here.

The second option in Fig. 5-20 opens the front panel given in Fig. 5-21b, where the user can select from the list to view various distinctive points of

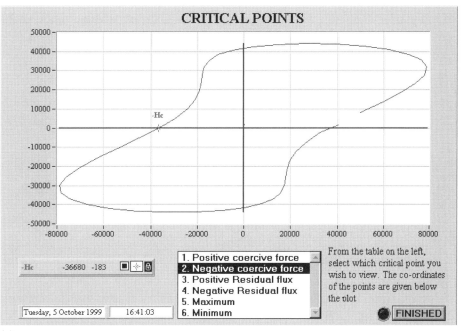

(a)

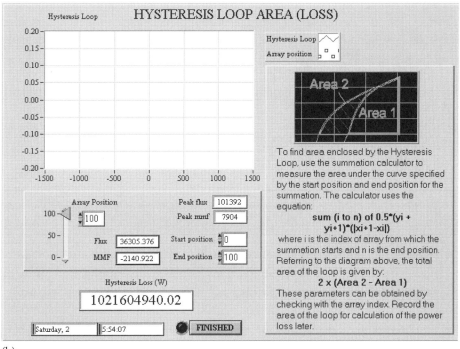

(b)

Figure 5-21
The front panels of the View and Analyze Simulation Hysteresis Loop option: (a) Critical Points Analysis panel, and (b) Hysteresis Loop Area (Loss) panel, which are accessible via the panel shown in Fig. 5-20.

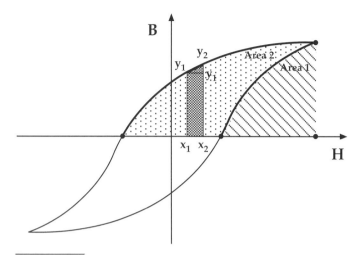

Figure 5-22
The *BH* characteristic and the trapezoidal rule.

the *BH* loop: Positive Coercive Force, Negative Coercive Force, Positive Residual Flux Density, Negative Residual Flux Density, Maximum Flux Density, and Minimum Flux Density. Note that the coordinates of the critical points (relative to the *BH* curve) are revealed in the indicator box located below the *BH* graph.

> **Note:** The methodology behind pinpointing the various critical points in LabVIEW is based on traversal techniques. It is known that residual flux densities and coercive forces theoretically should lie on their respective axes. This implies that to identify the points, one should simply search for the crossover points, that is, the axis intercepts. However, a problem arises if the saved data set does not include exact intercept points. In this case, we search for the points closest to zero. The same method is used for the maximum and minimum points, which is done searching through the H array and looking for the index with the minimum and maximum value.

View and Analyze Real Data Hysteresis Loop

This segment of the experiment is similar to the previous test, except that instead of examining the simulated waveforms, the real-time data acquired in

Part 1 is studied. In addition, an additional option is available, View all *BH* Curves.

First, the user should define the air-gap and analyze the corresponding *BH* characteristic by selecting two options (Calculate Hysteresis Area and Identify Critical Points, respectively). The results have to be recorded to be used in the next section, where the calculation of the eddy current losses will be carried out.

Calculate Total Power Losses and Eddy Current Losses

The final analysis segment available under Part 2 is the examination of the total power losses to separate the component of the eddy current losses.

The eddy current losses occur in the magnetic core when the flux varies in time. Due to the changing flux, induced currents flow in the core causing power losses since the core material has some resistance. Because the total core losses can be given as a sum of the hysteresis and the eddy current losses, if the total core loss is known, the eddy current loss can be calculated.

$$\begin{aligned} P_{\text{loss(eddy)}} &= P_{\text{loss(core)}} - P_{\text{loss(hysteresis)}} \\ &= V_{\text{rms}} I_{\text{rms}} \cos \theta - P_{\text{loss(hysteresis)}} \end{aligned} \quad (5.13)$$

where V_{rms} is the emf across the main winding, which can be estimated using the turns ratio of the windings in the core.

The panel used to calculate the total core loss is shown in Fig. 5-23. The graph area of the panel displays the waveforms of the emf and the current of the excited winding, which should be used to determine the phase angle θ in degrees. Then the user should enter the value of θ into the control provided and press the Calculate the Total Loss button, which should be recorded to calculate the eddy current losses. This process should be repeated for all the air-gap sizes.

Part 3, DC Characteristics of Magnetic Circuits

This part of the experiment processes and analyzes all the dc data obtained in Part 1. The analyses that will take place are flux linkage–current characteristics, inductance–current characteristics, and estimated forces that are associated with different currents and air-gaps.

Two options are provided in the main front panel of this part: View DC Data and Theoretical Force Calculation.

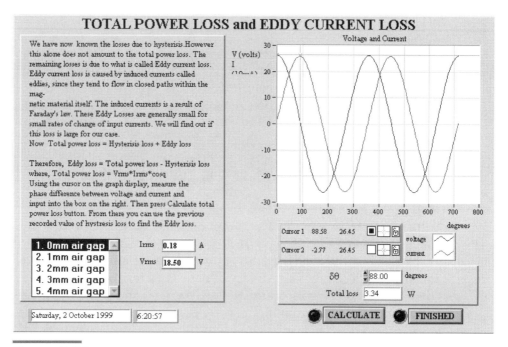

Figure 5-23
The front panel to calculate total power losses.

View DC Data

This first option is devoted to viewing the data obtained in Part 1 under dc input supply. The front panel of this segment is shown in Fig. 5-24. This panel displays three plots that are captured earlier—voltage, current, and force—and provides two options to link to the other panels for viewing flux linkage versus current and inductance versus current characteristics.

Theoretical Force Calculation

After selecting this option, the panel shown in Fig. 5-25 is displayed, where the derivative of the inductance with respect to the air-gap is estimated, then used to determine the force theoretically.

$$F = \frac{1}{2} I_{\text{rms}}^2 \frac{dL}{dx} \tag{5.14}$$

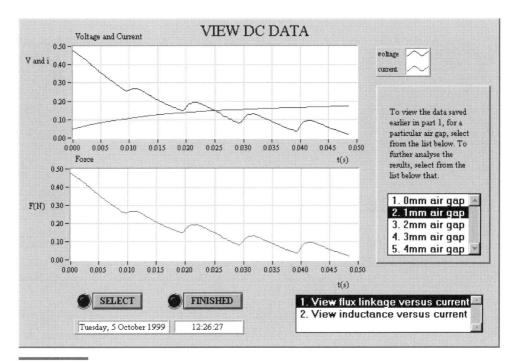

Figure 5-24
The front panel of the View DC Data option.

where F is the force in newtons, L is the inductance of the winding circuit, and x is the distance (air-gap).

5.4.1 Laboratory Hardware

The laboratory hardware used in this experiment was given earlier in the block diagram in Fig. 5-18. As indicated in the figure, a custom-built measurement and interface unit is required to perform the experiment. This unit provides a complete isolation and accommodates various components and circuits. The details of the unit are given in the Appendix, which includes the printed circuit board layout and the circuit diagrams.

Note that three quantities—the excitation current (ac or dc), the voltage of the search coil, and the force—are measured. All other circuit parameters are either entered using the controls on the front panels or specified in the block diagrams of the relevant sub-VIs. Therefore, if the device under test has

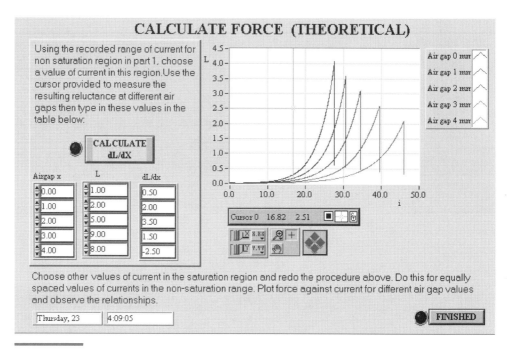

Figure 5-25
The front panel to aid the calculation of the force.

different parameters, some minor modifications may be required for the correct measurements.

Fig. 5-26a illustrates the structure of the real magnetic device. Note that the device has a variable air-gap allowing the user to carry out the tests at different air-gap sizes. The air-gap of the device is manually adjusted and kept unchanged during each test.

The specifications of some of the other hardware used in the experiment are a computer, a data acquisition card, an ac supply (240 V, 50 Hz), and a single-phase autotransformer (240 V, 8 A, 50 Hz).

The dimensions of the device are shown in Fig. 5-26b. Note that due to the complex shape of the original magnetic circuit, some approximations are made and only the equivalent dimensions are given in the figure. The main coil current of the magnetic circuit is 300 mA max. This level of current is used to determine the maximum dc input voltage, when the dc test is performed. Note that the dc voltage level will vary for a given air-gap size. Four air-gap gauges (1 mm, 2 mm, 3 mm, 4 mm) are used to set the gap in the device.

Chapter 5 • Electric Machines Laboratory

(a)

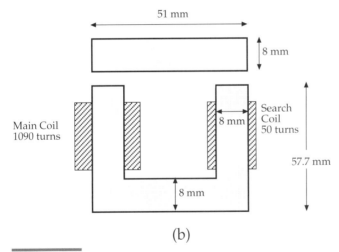

(b)

Figure 5-26
(a) The photo of the device under test, and (b) the equivalent magnetic circuit.

5.5 Tests for AC Circuits

In Chapter 4, we gave basic definitions and provided various VIs so you could understand and visualize the operation of ac circuits. Since sinusoidal (or ac) signals constitute the most important class of signals in the power systems, covering a wide spectrum of applications from households to industries, it is important to observe such signals in real time.

Furthermore, ac circuit elements can be represented in terms of their impedance, which may be conceptualized as a frequency-dependent resistance. The only difference between the analysis of ac and resistive circuits lies in the use of complex algebra instead of real algebra. Note that the sign of the imaginary part of the impedance determines the nature of the load.

Impedance measurements can be made using Wheatstone bridge methods or special instruments. However, such methods may be time-consuming, difficult, expensive, or may have limitations (such as frequency, voltage, and current levels).

Due to the availability of the data acquisition system featured in this book and some custom-written VIs, we will study an alternative measurement method here. This method can provide a quick and medium accuracy impedance measurement, where the components of the impedance can be separated into two basic components: resistance and inductance or capacitance.

This experiment should be considered as a review of the ac circuits and all associated definitions, which were studied earlier. The experiment has three major tests on single-phase and three-phase circuits, where the principal definitions will be routinely visited and parameters will be calculated, which are summarized as follows:

- Voltage and current relationships for resistors, inductors, and capacitors
- Computation of rms values, powers, and power factors for periodic real waveforms and study of phasors

The software levels of the experiment are summarized in Table 5-3. Note that no theory is provided in the VIs because of the simplicity of the setups.

Table 5-3 *The software levels of the experiment.*

Screen Levels	Remarks
Log-in Screen	Contains a title, an input text box, and two Boolean controls (Start, Exit).
Main Menu	Contains a list that holds three sub-sections of the experiment.

Parts 1, 2, and 3
   Single-Phase AC Circuit
   Three-Phase AC Circuit and Harmonics
   Power Factor Correction
	The major experimental modules are included.

If you need help, refer to Chapter 3 and the hardware explanations provided here.

The principal sections of the experiment are

- Single-Phase AC Circuit and Definitions
- Three-Phase AC Circuit and Harmonics
- Power Factor Compensation (Correction)

5.5.1 Single-Phase AC Circuit Test

The ac circuits made up of resistors, inductors, and capacitors constitute reasonable models for many practical devices, such as transformers and electric motors. Therefore, the voltage-current relationship for a practical inductive load will be tested in this section. The front panel of Single Phase.vi used in this experiment is shown in Fig. 5-27.

Note that when an inductive load is tested, the calculated reactance will be positive (or, alternatively, the phase angle is positive). However, in the voltage-current phasor, the current should lag the voltage.

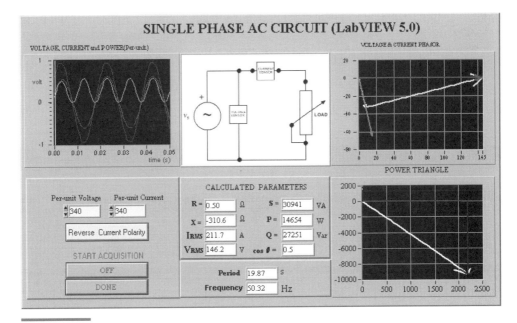

Figure 5-27
The front panel of the VI for the Single-Phase AC Circuit Tests.

The power triangle of the inductive load is also demonstrated by the `Single Phase.vi`. In addition, the components of the complex power and the rms values of the voltage and the current of the load are calculated. An additional control provided on the panel allows the user to change the polarity of the current sensor.

The principal objective in this test is to connect an inductive load to an ac supply, observe and record the front panel information, and then verify the results analytically.

> **Note:** Unlike the ac circuit elements mentioned in the previous chapters, there are no ideal components in practice. A practical capacitor can be modeled by an ideal capacitor in parallel with a resistor (largely capacitive). This resistance represents the losses in the capacitor. Similarly, a practical inductor can be modeled by an ideal inductor in series with a resistor (largely inductive). Further, a practical resistance can be modeled by an ideal resistor and an inductor connected in series (largely resistive).

Hardware for the Single-Phase Circuit Test

The essential parts of the real-time system (Fig. 5-28) used in this test consist of the following principal units: a computer, a data acquisition system, a rheo-

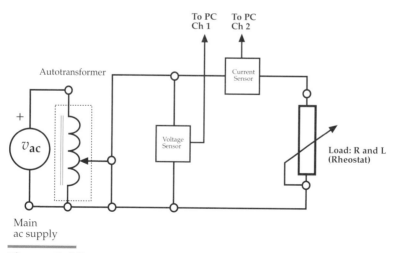

Figure 5-28
A sample wiring diagram for the single-phase ac test.

stat (50 Ω with an unknown inductance, 5 A), an ac supply (240 V, 50 Hz), an autotransformer (240 V, 8 A, 50 Hz), and two signal conditioning devices (one current and one voltage transducer).

5.5.2 Three-Phase AC Circuit Test

Similar to the previous test, the objective of this test is to study the ac circuit parameters—but this time in a three-phase ac circuit containing a star-connected balanced impedance load. All the circuit parameters, including the active, reactive, and apparent powers are computed from the measured voltage and the currents.

The front panel of ThreePhase.vi is shown in Fig. 5-29. In the front panel, three graph areas display the measured voltages and currents, estimated complex powers, and the associated phasor diagrams that explicitly indicate the amplitude and phase of the sinusoidal signals. In addition, two links are made

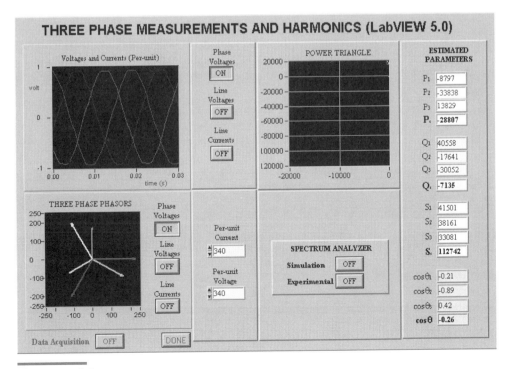

Figure 5-29
The front panel of the VI for the Three-Phase AC Circuit Tests.

available to study the spectrum analysis either using the simulated waveforms or the measured waveforms.

The key objectives in this test are to study the ac circuit parameters in a three-phase circuit, to observe and record the front panel information, and to verify the results analytically.

Hardware for Three-Phase AC Circuit Test

As shown in Fig. 5-30, the phase voltages and the line currents of the star-connected load are measured using suitable signal conditioning devices. The key devices used in this experiment are a computer, a data acquisition system, three rheostats to form a three-phase load (each 50 Ω with an unknown inductance, 5 A), three-phase ac supply (415 V, 50 Hz), a three-phase autotransformer (415 V, 15 A, 50 Hz, output: 0–470 V), and six signal conditioning devices (three current and three voltage transducers).

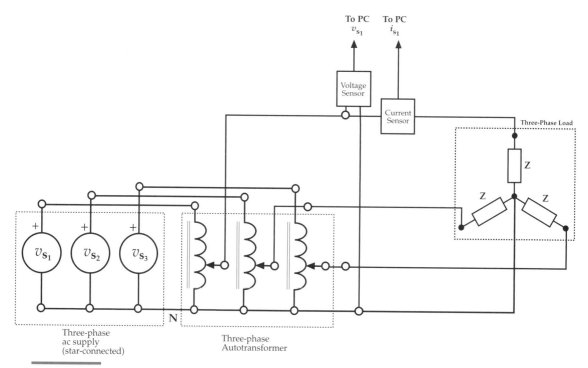

Figure 5-30
A sample wiring diagram for the three-phase ac test.

> **Note:** The circuit provided here is a four-wire system. If the star-point is not accessible either from the supply side or from the load side, an artificial star-point (fourth-wire) can be created by using three identical high-value resistors star-connected across the line terminals. If the load is delta-connected, however, the voltage and the current sensors have to be relocated to measure the phase quantities.

5.5.3 Power Factor Correction Test

The front panel of PFC.vi used in this experiment is shown in Fig. 5-31.

The wiring diagram for this test is given in Fig. 5-32, which is similar to the single-phase ac circuit test in Fig. 5-28. Therefore, an identical test should be carried out here first without the capacitor connected and then with the capacitor.

Unlike the previous circuit, however, there are two current sensors in this test to determine the capacitor's current. The front panel of PFC.vi illustrates three power triangle graphs, which should be studied to observe the improvements in the power factor of the supply.

Hardware for Power Factor Correction Test

The specifications of the hardware used are a computer, a data acquisition system, a rheostat (50 Ω with an unknown inductance, 5 A), a capacitor (4 μF, 1000 V), an ac supply (240 V, 50 Hz), an autotransformer (240 V, 8 A, 50 Hz), and two signal conditioning devices (one current and one voltage transducer).

5.6 Transformer Test

Transformers play a major role in the power system network and are an indispensable component of electrical and electronic engineering, covering a wide spectrum of applications. The most common functions of transformers are

- Changing the voltage and current levels: the distribution transformers (step-up, step-down)
- Impedance matching: to match a load to a line for improved power transfer

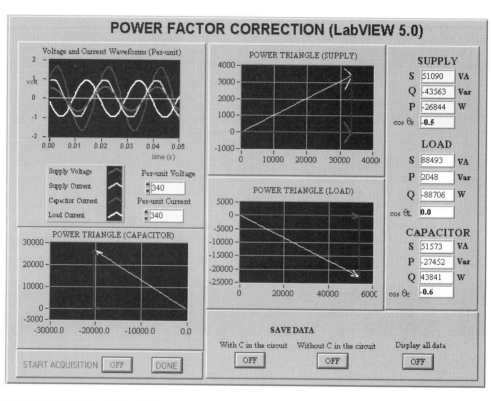

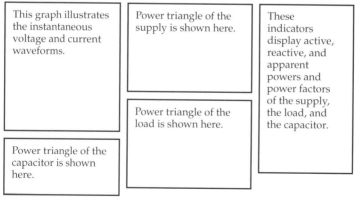

Figure 5-31
The front panel and brief user guide of PFC.vi.

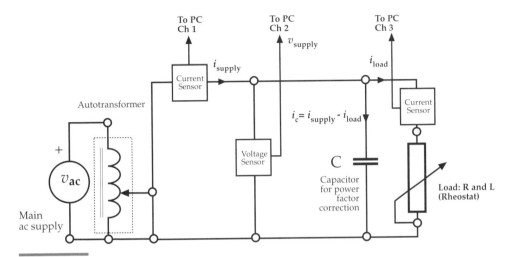

Figure 5-32
A sample wiring diagram for the power factor correction test.

- Electrical isolation: to eliminate electromagnetic noise, to block dc signals, for safety

A transformer consists essentially of a ferromagnetic core built up of insulated laminated iron plates and two distinct sets of windings suitably located with respect to one another, which are named primary and secondary. These two (or more) windings are coupled by a mutual magnetic field that is created by an ac supply connected to the primary side. The secondary side is directly connected to the load.

It should be noted here that, as in other types of electrical machines, the functions of the primary and secondary windings are interchangeable, providing step-up or step-down voltage or current conversion.

The transformer we will consider in this study is a real, small-scale step-down distribution transformer. Unlike ideal transformers, such a practical transformer will have

- Leakage flux (primary and secondary leakage fluxes)
- Winding resistances (primary and secondary windings resistances, resistive losses)
- Finite core permeability (requires a finite mmf for its magnetization)
- Loss in magnetic core (hysteresis and eddy current losses)

The equivalent circuit of a real transformer has to take all these nonideal measures into account. Hence, impedance, voltage regulation, exciting current, losses, and efficiency can be studied, which are each in some respect direct measures of the departure of the transformer from the ideal case.

Practically every one of these parameters affects the operation of the transformer or the system to which it is connected. For example, short-circuit phenomena in systems are affected by the impedance of the transformer, or high efficiency at light loads becomes a dominant consideration in the distribution transformers. Furthermore, a careful selection of the appropriate approximations greatly facilitates the solutions of various transformer problems.

This study provides two basic equivalent circuits (exact and approximate) that can be used to analyze a single-phase as well as a three-phase transformer (using the per-phase equivalent). However, if needed, further simplified analysis can be carried out, such as

- The transformer, as a series reactance: the value of the short-circuit current can be calculated to size the circuit breakers and relay systems
- The transformer, as a series impedance: to calculate the voltage regulation and for successful operation in parallel
- Transformer, as a shunt impedance: to study core losses and rush of magnetizing current and the harmonics in the magnetizing current

Practical Transformer

A practical transformer is equivalent to an ideal transformer plus external impedances that represent the imperfections of an actual transformer. Referring the secondary side to the primary gives the complete equivalent circuit of a transformer under steady-state ac conditions shown in Fig. 5-33.

Since the mutual flux magnetic circuit of an iron-cored transformer is of high permeance and low loss, the magnetizing current I_0 is usually 2 to 4% of the rated primary current. Consequently, moving the magnetizing branch in the equivalent circuit representation of the loaded transformer to the input terminals causes little error. This simplifies computation of currents, because both the exciting branch impedance and the load branch impedance are directly connected across the supply voltages, and the winding resistances and leakage reactances can be lumped together. The new equivalent circuit is known as Approximate Equivalent Circuit and is shown in Fig. 5-34.

The parameters of the equivalent circuit model for the actual transformer can be used to predict the behavior of the transformer. These parameters can

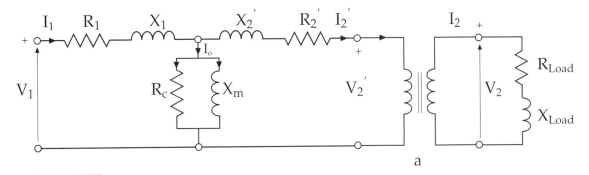

Figure 5-33
Exact equivalent circuit of a transformer (per phase), secondary side referred to primary side.

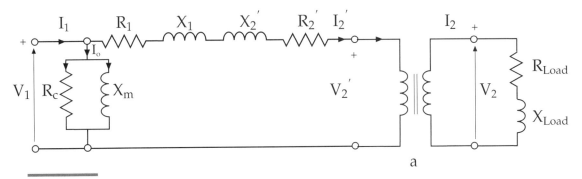

Figure 5-34
Approximate Equivalent Circuit of the transformer.

be directly and easily determined by performing two tests: Open-Circuit Test (No-Load) and Short-Circuit Test. In this study, however, an additional test, Full-Load Test, will be carried out to observe the behavior of the transformer.

Open-Circuit Test

Normally, the rated voltage at the rated frequency is applied to either the high voltage or low voltage side of the transformer (Fig. 5-35). The secondary side of the transformer is left open. The resultant equivalent circuit is shown in the figure.

In this test, the phase voltage, current, and power at the terminals of the primary winding have to be known. Then, if R_1 and X_1 are known (see Short-

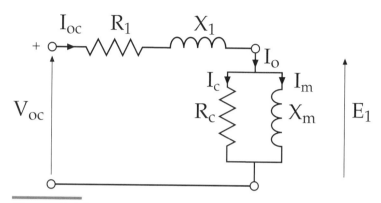

Figure 5-35
Equivalent Circuit under open-circuit conditions.

Circuit Test) the parameters R_c and X_m can be determined using the circuit in Fig. 5-35.

$$R_c = \frac{E_1^2}{P_{core}} \tag{5.15}$$

$$X_m = \frac{E_1}{I_m} \tag{5.16}$$

where

$$E_1 = V_{oc}\angle 0° - I_{oc}\angle \phi_{oc}(R_1 + jX_1) \tag{5.17}$$

$$\phi_{oc} = \cos^{-1}\frac{P_{oc}}{V_{oc}I_{oc}} \tag{5.18}$$

$$P_{core} = P_{oc} - I_{oc}^2 R_1 \tag{5.19}$$

$$I_m = \sqrt{I_{oc}^2 - I_c^2} \tag{5.20}$$

$$I_c = \frac{P_{core}}{E_1} \tag{5.21}$$

Note that with LabVIEW's ability to measure real-time data through the use of the data acquisition system, there is no longer a need to use a wattmeter to measure the input power of the transformer under open-circuit conditions, P_{oc}. Therefore, the real waveforms of primary voltage and current are used to calculate the input power.

$$P_{oc} = V_{oc}I_{oc}\cos\phi_{oc} \tag{5.22}$$

where ϕ_{oc} is the phase angle between the voltage and current waveforms.

Short-Circuit Test

In this test, one of the windings is short-circuited across its terminals, and reduced voltage is applied to the other winding. The choice of the winding to be short-circuited is usually determined by the available measuring equipment. Usually the rated current flows in the short-circuited winding, and the voltage, current, and power at the primary winding are measured. In medium and high power transformers, the magnetizing current I_0 is usually 2 to 4% of the rated primary current.

When one of the windings is short-circuited, only a small voltage will be required. Consequently, the flux density in the core will be small in this test, so the core loss and magnetizing current will be smaller. Therefore, the simplified equivalent circuit (see Fig. 5-36) is used for computations in the short-circuit test.

Since this test requires a low voltage supply, an autotransformer has to be used to ensure that a large voltage is not accidentally applied to the transformer. After the voltage, current, and active power are measured, the equivalent circuit parameters can be calculated using the equations given next.

Equivalent impedance, equivalent resistance, and equivalent leakage reactance are

$$Z_e = \frac{V_{sc}}{I_{sc}} \qquad (5.23)$$

$$R_e = \frac{P_{sc}}{I_{sc}^2} \qquad (5.24)$$

$$X_e = \sqrt{Z_e^2 - R_e^2} \qquad (5.25)$$

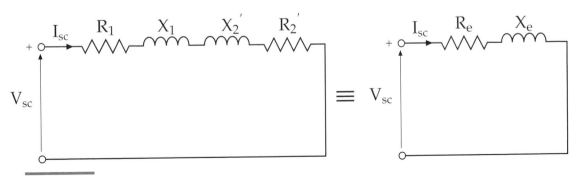

Figure 5-36
Equivalent Circuit under short-circuit conditions.

Hence for given R_1 (which can be measured directly) and a (turns ratio), R_2 can be found from

$$R_e = R_1 + R'_2 = R_1 + a^2 R_2 \tag{5.26}$$

Finally, it is usually assumed that the leakage reactance is divided equally between the primary and the secondary. Therefore,

$$X_1 = X'_2 = a^2 X_2 = \frac{1}{2} X_e \tag{5.27}$$

Full-Load Test

In this test, the transformer is loaded using a suitable variable resistance bank. This load resistance bank connected across the secondary winding is gradually reduced until the rated primary current is drawn from the supply. The secondary voltage is measured under load, and then the load is thrown off, and the secondary voltage is measured once more.

In this test, the voltage regulation is defined as the change in secondary terminal voltage that occurs when the load current is reduced from rated, or full load, value to zero, which depends on the power factor of the load.

The voltage regulation can then be calculated using

$$\text{Voltage regulation (VR)} = \frac{|V_2|_{oc} - |V_2|_{load}}{|V_2|_{load}} \tag{5.28}$$

The efficiency of the transformer is calculated using the output and the input active powers of the transformer.

$$\text{Efficiency \%} = \frac{V_2 I_2 \cos \phi_2}{V_1 I_1 \cos \phi_1} 100\% \tag{5.29}$$

5.6.1 Virtual Instrument Panel

Table 5-4 indicates the distinct levels of the software components of the transformer experiment, which fundamentally obtains the equivalent circuit parameters of the transformer with a series of real-time tests, and performs a steady-state analysis. The hierarchical structure of this experiment is similar to the common structure described in the other experiments.

Table 5-4 *The software levels of the experiment.*

Screen Levels	Remarks
Log-in Screen	Contains a title, an input text box, and two Boolean controls (Start, Exit).
Main Menu	It holds a list of five subsections that link to the sections of the experiment. The main menu is a while loop that has as an output to the screen with a text dialog and a termination button.

Parts 1 to 5:
   Introduction
   Experimental Tests
   Analysis
   Full-Load Test
   Simulation with Phasors
	The theory and the procedure link and the major experimental and analysis modules are included.

Note that the detailed descriptions of the front panel items are accessible via the Help menu of each VI. To perform the single-phase transformer tests, the voltage and current sensors need to be assigned to specific channels of the data acquisition board.

In the custom-written VI provided here, the following channel assignments are used, which is crucial for the correct measurement. However, note that an alternative arrangement can be made if the programming diagram is modified and the correct measurements gains are provided.

Primary voltage	to	Channel 0
Primary current	to	Channel 1
Secondary voltage	to	Channel 2
Secondary current	to	Channel 3

In the LabVIEW-based test VIs, menus are programmed using while loops, case loops, and sequence structures together with controls and indicators. The program starts with an initial screen where a specific logo of the educational institution is shown. After the program starts, a Log-in menu is displayed, where students can register their details. These details are used in the consecutive panels to relate to the operator.

The following menu is about the principal sections of the transformer test as indicated in Table 5-4. The main menu gives the students a choice of which practical test they are to perform and gives them an opportunity to navigate between the different study modules.

Most sections of the custom-written VI contain an experimental and a theory section. The theory section is noninteractive and contains basic information that aims to reinforce the practical tests and analysis. Selecting the experimental section, however, displays a customized front panel that contains graphs, numeric controls, Boolean switches, and other interactive elements.

There are five main sub-VIs implemented in the transformer test: `TTIntro.vi`, `Experimental Tests.vi`, `Analysis Menu.vi`, `Load Test Menu.vi`, and `Simulation with Phasors.vi`. Navigation through the menu is explained and the front panel images of the major sub-VIs are given in the following paragraphs.

Experimental Tests: Open-Circuit and Short-Circuit

The VIs designed to perform the open- and short-circuit tests are similar in terms of the data capturing and presentation requirements. The front panels of these sub-VIs are given in Fig. 5-37 and Fig. 5-38. As can be seen, once the circuit is connected and the VI is executed, the front panel can display the real-time signal of the waveforms.

From the waveform, the phase difference is measured. However, due to the fact that saturation occurs in the transformer, the magnetizing current has a nonsinusoidal shape, hence not enabling the phase difference to be measured easily.

In this section, to simplify the measurement, an ideal sine waveform of the current is generated using two definitions for the power factors of the nonsinusoidal current waveforms.

$$\text{Power factor} = \frac{P_{\text{average}}}{V_{\text{rms}} \cdot I_{\text{rms}}} \qquad (5.30)$$

In equation 5.30, P_{average} is calculated by averaging the value of instantaneous power (which is obtained by multiplying the real voltage and current waveforms). Calculated Power Factor is then used to generate a hypothetical phase angle to generate an ideal sine wave current. The test results indicate that the previous method provides an accurate measurement of power factor of a nonsinusoidal waveform.

The Short-Circuit Test executes similarly to the Open-Circuit Test. However, some difficulties may be encountered in this test due to the low level of voltage measurement (see the sample results in Section 5.6.3 for the measured voltage). To obtain accurate measurements, the gain settings of the

Chapter 5 • Electric Machines Laboratory

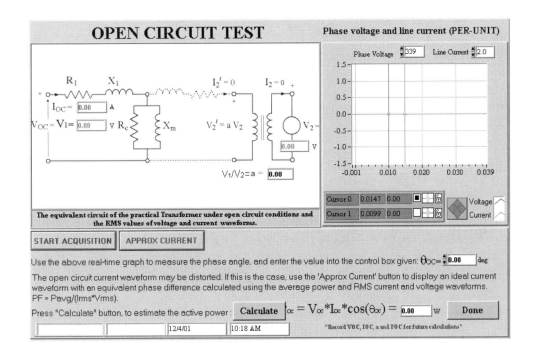

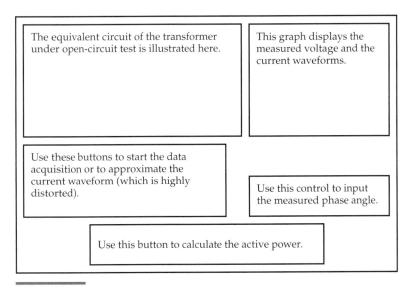

Figure 5-37
The front panel and brief user guide of the Open-Circuit Test.

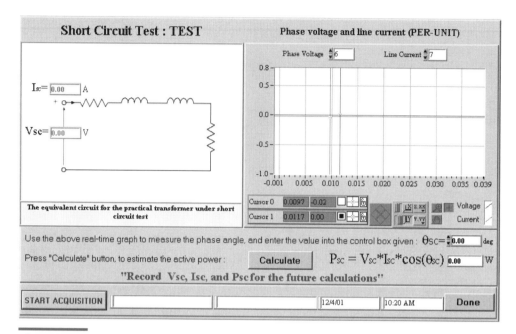

Figure 5-38
The front panel for the Short-Circuit Test.

DAQ card can be adjusted or a low-voltage, full-scale voltage sensing device can be accommodated. In addition, a higher sampling frequency and some form of digital filtering may be necessary to improve the quality of the measured voltage.

Analysis

Two VIs are provided in this section to help students perform the necessary analysis that will be based on the test results obtained from open- and short-circuit measurements. The measured parameters from these previous tests are the inputs to the sub-VIs in the Analysis option. The Analysis option performs the necessary calculations, and the corresponding equivalent circuit parameters are obtained.

A sample front panel is given in Fig. 5-39, where the electrical equivalent circuit for the Open-Circuit Test is displayed. The user can input the mea-

Chapter 5 • Electric Machines Laboratory

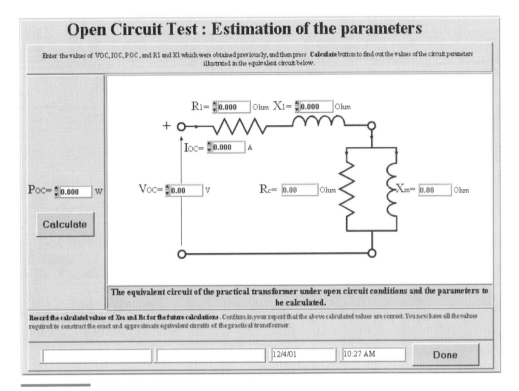

Figure 5-39
A sample front panel for parameter estimation: Open-Circuit Parameters.

sured voltage, current, and power values using the controls provided on the panel, and corresponding circuit calculations are performed and displayed on the indicators.

Full-Load Test

In this section, students can conduct the Full-Load Test and calculate the efficiency and voltage regulation. The phasors corresponding to the testing condition are also illustrated here.

The operation of the first sub-VI is simple. First, the circuit is wired and energized. Then the sub-VI displays the waveforms and draws the corresponding phasor diagrams in real time (Fig. 5-40), and the load impedance under the test condition is estimated and displayed on the front panel. Students can

LabVIEW for Electric Circuits, Machines, Drives, and Laboratories

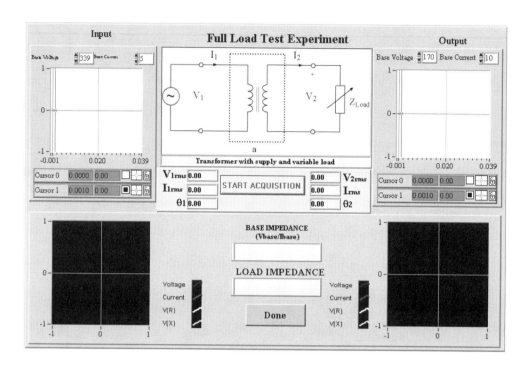

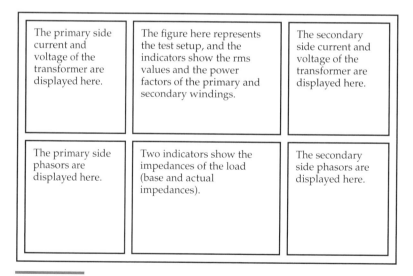

Figure 5-40
The front panel and brief user guide of the Full-Load Test.

perform an efficiency calculation manually using the rms voltage, rms current, and phase angles on the primary and the secondary side of the transformer.

Simulation with Phasors

This sub-VI (Fig. 5-41) is purely theoretical with no hardware requirements. The aim of this section is to allow students to investigate the different aspects of the transformer by changing the input and circuit parameters. Two circuit options, Approximate and Exact Equivalent circuits, are provided.

5.6.2 Self-Study Questions

1. Open-Circuit Test
   a. Why is the magnetizing current waveform not sinusoidal?
   b. What loss is being measured in this part of the experiment?
2. Short-Circuit Test
   a. Why is the input voltage waveform slightly distorted?
   b. Why is the shunt branch ignored in calculations involving the Short-Circuit Test?
   c. What loss is being measured in the Short-Circuit Test?
3. Analysis
   a. What have you noticed about the values of R_c and X_m as the voltage increases?
   b. Verify your results for the parameters of the practical transformer. *Hint:* See the theory sections for the open-circuit and short-circuit tests.
4. Full-Load Test
   a. Calculate the voltage regulation and efficiency of the transformer under this test.
5. Simulation with Phasors
   a. Use the equivalent circuit parameters of the transformer under test and verify that there isn't significant error in using the Approximate and Exact Equivalent circuits. Note that the simulation study can be performed by using a variety of loads and comparing resultant currents, voltages, and efficiencies.
   b. For given transformer parameters obtain two separate graphs for Efficiency versus Input Voltage and Power Factor versus Efficiency.

LabVIEW for Electric Circuits, Machines, Drives, and Laboratories

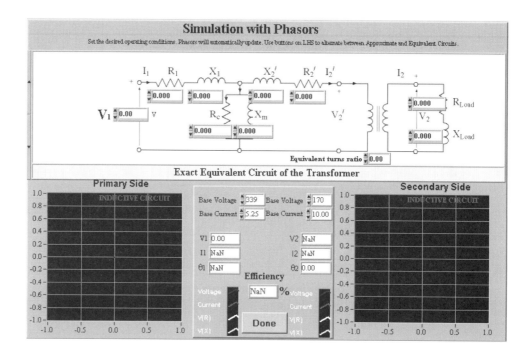

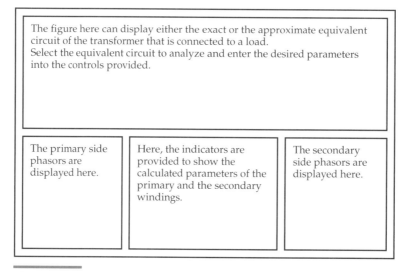

Figure 5-41
The front panel and brief user guide of the Simulation of Phasors sub-VI.

5.6.3 Sample Results

The nameplate of a transformer provides all the essential information operating under loads. The nameplate data of the single-phase transformer used in this study was 240/120 V, 1 kVA.

The transformer test results given in Tables 5-5, 5-6, and 5-7 are measured using the real test system available in our laboratory. We expect that these results can provide some guidance for students in terms of the profile of the waveforms and the quantities measured and estimated using the VIs in this section.

Table 5-5 *Comparison of Open-Circuit Test results.*

Voltage V_1	Using the VI Provided			Using Conventional Instruments and by Manual Calculations		
	Core Loss	R_c	X_m	Core Loss	R_c	X_m
240 V	16.69 W	3.487 kΩ	216.82 Ω	32.2 W	1.79 kΩ	268.7 Ω
200 V	10.97 W	3.668 kΩ	458.99 Ω	21.8 W	1.84 kΩ	549.3 Ω
160 V	10.24 W	2.504 kΩ	1.577 kΩ	13.4 W	1.91 kΩ	1.61 kΩ

Table 5-6 *Comparison of Short-Circuit Test results.*

VI Results	Conventional
Core Loss	Core Loss
10.17 W	10.57 W

Table 5-7 *Comparison of Full-Load Test results.*

VI Results		
No Load V_2 119.56 V	Full Load V_2 118.47 V	Full Load I_1 4.09 A
Measured Voltage Regulation = 0.92%		Calculated VR (at PF = 1.0) = 1.08%
Conventional Measurements		
No Load V_2 120.8 V	Full Load V_2 119 V	Full Load I_1 3.74 A
Measured Voltage Regulation = 1.51%		Calculated VR (at PF = 1.0) = 1.05%

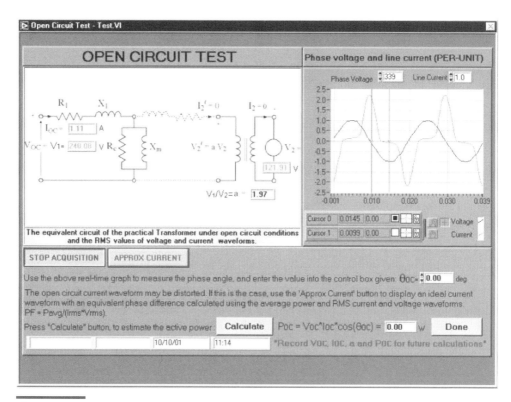

Figure 5-42
Open-Circuit Test at 240 V.

The tables provide comparative results, which are based on the estimated parameters of the transformer under test: using the VI here and the conventional measuring instruments (ammeter, voltmeter, wattmeter). The turns ratio of the transformer was estimated as 1.97 using the VI and 1.98 using conventional instruments.

As can be observed from the tables, there are discrepancies between the results of the VI and the conventional tests. You can conclude that the method presented here provides more accurate results since it considers true rms measurements.

In the front panel illustrated in Fig. 5-42, note that there is 0.0147 − 0.0099 = 0.0048 time interval (using the cursor positions), which corresponds to a lag of 86.4 degrees (for a frequency of 50 Hz and $T = 0.02$ s). Consequently the power loss is estimated as 16.69 W. The front panels in Fig. 5-43 and Fig. 5-44 illustrate the other test results.

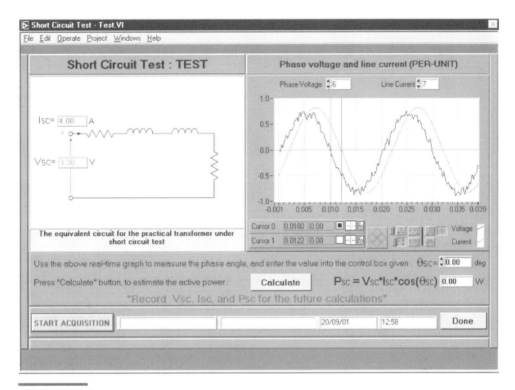

Figure 5-43
Short-Circuit Test results.

5.7 Asynchronous (Induction) Motor Test

After Nicola Tesla revealed an elementary form of a multiple-phase asynchronous (induction) motor in 1891, Dobrowolsky described a machine with a cage rotor. Since then various improvements to the earlier asynchronous motor have been made, and today it is the most popular motor. In fact, the majority of the motors used in industrial systems are asynchronous motors (usually three-phase).

Although the construction of asynchronous motors is simple and their equivalent circuits and operations are well understood, it is still considered the most difficult topic to learn and teach. Therefore, a number of VIs developed in this book aim to provide a bridge between theory and real asynchronous machines. Specifically, some powerful visual aids and flexible analysis tools are provided to encourage learning. I recommend that students study

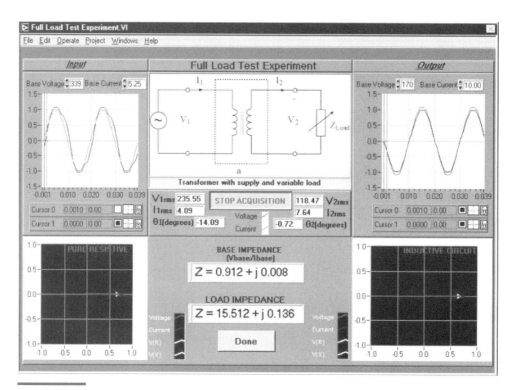

Figure 5-44
Full-Load Test results.

Section 7.1 first before initiating this experiment, where the rotating field concept in three-phase ac machines is explained. The dynamic simulation of the asynchronous motor is also provided in this book and should be investigated after completing this experiment.

After performing this experiment, students should be able to

- understand how the equivalent circuit parameters of a three-phase induction motor are determined.

- understand the differences between the exact and the approximate equivalent circuits.

- study the behavior of the motor and the motor characteristics under the sinusoidal steady-state and various operating conditions, including the torque-speed characteristics at different supply frequencies, supply voltages, and rotor resistances.

5.7.1 Theory

A three-phase asynchronous motor has two main sections: a stationary part (stator) and a rotating part (rotor). The stator winding of a three-phase asynchronous motor consists of three multi-turn phase windings. The windings are evenly spaced around the cylindrical surface of the stator and may be connected either in star (generally) or in delta.

The rotor conductors may be connected in two different ways. In the so-called slip-ring or wound rotor machine, the rotor windings are connected in star as a rule and are terminated on three slip rings mounted on the rotor shaft. Brushes contacting the slip rings enable closing the rotor circuits through externally controllable resistances for the purpose of speed and torque control. In the squirrel cage rotor, however, all rotor conductors are connected together at both ends of the rotor by means of two end rings. In this type of machine, the rotor windings are not accessible, and, therefore, no external resistances can be inserted into the rotor circuit for the torque or speed control.

If the motor is connected to a three-phase supply, and if the rotor is free to move, the torque due to the forces acting on the rotor conductors will cause the rotor to rotate in the direction the flux density wave (rotating field) is moving.

If the shaft of the motor is not subject to any external mechanical load, the rotor speed increases to near synchronous speed, such that the corresponding rotor-induced voltage and rotor current is just sufficient for providing the torque necessary for overcoming the retarding torque due to friction, windage, and rotor iron losses. The rotor could assume a speed equal to n_s only in the absence of the above no load losses.

Remember that a practical asynchronous motor always rotates at a lower speed than synchronous speed. If it rotated at synchronous speed, there would be no change in flux linkage, no induced voltage and current in the rotor, and hence no torque would be generated.

However, any mechanical load applied to the shaft causes the rotor speed to fall, and the rotor-induced voltage and current to rise to a value sufficient to meet the torque requirement of the load.

To indicate the deviation between the synchronous speed and the operating speed (n) of the motor, a term called *slip* (s) is used, which is a per-unit value.

$$s = \frac{n_s - n}{n_s} \quad \text{or} \quad s = \frac{\omega_s - \omega}{\omega_s} \tag{5.31}$$

where ω_s and ω are the synchronous speed and the actual speed of the motor in rad/s, respectively. Note that $\omega = 2\pi n/60$.

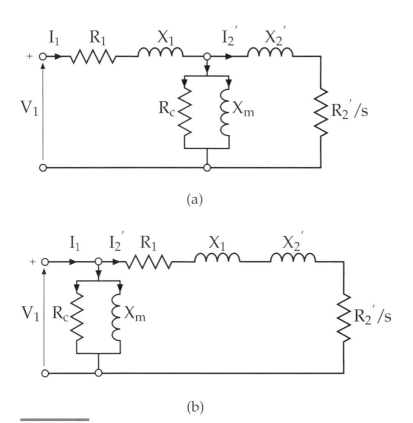

Figure 5-45
(a) The exact and (b) the approximate equivalent circuits of the asynchronous motors given per-phase.

Full-load condition (rated operating condition) in the motor is reached when the stator and rotor currents reach their rated values as stated on the nameplates. Furthermore, the speed of the motor will be equal to the rated speed that is less than the synchronous speed.

Whether the motor is connected in star or in delta, or has the windings on the rotor or not, the exact equivalent circuit per-phase of the motor is given as in Fig. 5-45a. The circuit in Fig. 5-45b is an approximate representation of the per-phase equivalent circuit and can be used in the motors at medium or higher power range, which simplifies the analysis without significant error in the estimated parameters.

In Fig. 5-45, V_1 is the phase voltage, I_1 is the phase current, R_1 is the resistance of the stator winding, X_1 is the stator leakage reactance, R_c is the core loss resistance, X_m is the magnetizing reactance, I_2' is the rotor current referred to the stator side, X'_2 is the rotor leakage reactance referred to the stator, and R'_2 is the rotor resistance referred to the stator. The referred values can be given using the actual values of the parameters and the turns ratio a between the stator and the rotor windings as

$$R'_2 = a^2 R_2, \qquad X'_2 = a^2 X_2, \qquad I'_2 = \frac{I_2}{a} \qquad (5.32)$$

In practice, the circuit parameters of the exact equivalent circuit are deducible from no-load (light-running or light-load) and blocked-rotor tests, which will be performed in this experiment. Then the results will be utilized to study the performance of the motor under numerous operating conditions using either the exact or the approximate equivalent circuits.

In the following paragraphs, some background information will be provided about the principal tests, the equivalent circuits, and the output parameters of the asynchronous motors.

Measuring the Equivalent Stator Winding Resistance R_1

In three-phase star-connected balanced circuits, such as asynchronous motors, the dc resistances of the windings are measured between two stator terminals with the third terminal open-circuited. Then the average value of these resistances is calculated as

$$R_{ave} = \frac{R_{AB} + R_{BC} + R_{CA}}{3} \qquad (5.33)$$

where R_{AB}, R_{BC}, and R_{CA} are the dc resistances across the three-phase terminals A, B, and C.

To take the frequency of the supply into account, the skin effect can be included in the calculations, and the per-phase winding resistance is estimated by

$$R_1 = \frac{R_{ave}(1 + 0.05)}{2} \qquad (5.34)$$

Blocked-Rotor Test

The blocked-rotor test on the induction machines is analogous to the short-circuit test on the transformers, which was studied earlier. If its rotor is blocked, it behaves as a transformer. In this test, the rotor is blocked or clamped so that it cannot rotate, and a reduced voltage is applied to the stator

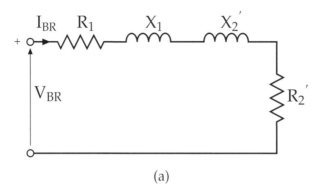

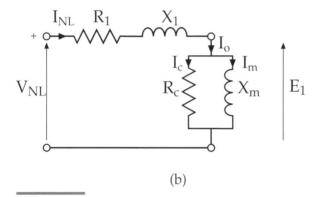

Figure 5-46
The equivalent circuits for the (a) Block-Rotor and (b) No-Load tests.

terminals of the motor. The value of the voltage is chosen to ensure that the rated current flows in the stator winding.

It is assumed that at the reduced voltage, the magnetizing branch current is negligibly small, therefore, the shunt branch ($R_c // X_m$) in the equivalent circuit can be neglected. Hence, the rotor of the machine is blocked $s = 1$. Then the equivalent circuit becomes as in Fig. 5-46a.

In the Blocked-Rotor Test, the line current (I_{BR}) and the phase voltage (V_{BR}) are measured, and the input active power (P_{BR}) is calculated. Note that the current and the voltage values are the per-phase rms values.

After taking these measurements, the equivalent circuit parameters for Fig. 5-46a can be calculated using the following equations.

$$Z_e = \frac{V_{BR}}{I_{BR}} \tag{5.35}$$

$$R_e = \frac{P_{BR}}{I_{BR}^2} \quad (5.36)$$

$$X_e = \sqrt{(Z_e^2 - R_e^2)} \quad (5.37)$$

where Z_e is the equivalent impedance, R_e is the equivalent resistance, and X_e is the equivalent leakage reactance.

For a given value of R_1 and the equivalent turns ratio a, R_2 and R_2' can be determined as

$$R_e = R_1 + R_2' = R_1 + a^2 R_2 \quad (5.38)$$

It is usually assumed that the equivalent leakage reactance is divided equally between the primary (stator) and the secondary windings (rotor). Therefore,

$$X_1 = X_2' = a^2 X_2 = \frac{1}{2} X_e \quad (5.39)$$

No-Load (Light-Running) Test

The No-Load Test on the asynchronous machine is analogous to the open-circuit test on the transformers. In this test, the rated voltage is supplied to the stator of the motor without any mechanical load attached to the shaft. At this operating condition, the motor runs very close to the synchronous speed, which implies slip s is very small and, therefore, the term R_2'/s becomes very large. Since this term is very large, the current of the rotor winding is ignored, and the equivalent circuit can be given as in Fig. 5-46b.

Similar to the Blocked-Rotor Test, the line current (I_{NL}) and the phase voltage (V_{NL}) are measured in real time and the input per-phase active power (P_{NL}) is estimated. In the equivalent circuit, the voltage E_1 can be given as a function of the measured voltage and current as

$$\overline{E}_1 = V_{NL} \angle 0° - I_{NL} \angle \theta (R_1 + jX_1) \quad (5.40)$$

Note that $\overline{E}_1$ is a complex quantity. In equation 5.40, the angle of V_{NL} is equal to 0 degrees, and the angle of the line current I_{NL} is given by θ.

$$\theta = \arccos \frac{P_{NL}}{V_{NL} I_{NL}} \quad (5.41)$$

Hence, the parameters in the equivalent circuit in Fig. 5-46b can be calculated easily.

$$R_c = \frac{E_1^2}{P_c} \quad (5.42)$$

$$X_m = \frac{E_1}{I_m} \tag{5.43}$$

where

$$P_c = P_{NL} - I_{NL}^2 R_1 \tag{5.44}$$

$$I_m = \sqrt{(I_{NL}^2 - I_C^2)} \tag{5.45}$$

$$I_c = \frac{P_c}{E_1} \tag{5.46}$$

The equivalent turns ratio is a $\sim V_{NL}/E_2$, where E_2 is the rotor-induced voltage measured when the rotor is open-circuited.

Performance Characteristics

Once the circuit parameters are determined from the tests previously described, the operating characteristics of the motor can be obtained easily. However, to understand asynchronous motors and to obtain performance characteristics, the relationship between the input electric and the output mechanical power must be known, which can be illustrated by the power flow diagram in Fig. 5-47. A number of parameters can be estimated, including stator current, power factor, input power, and rotor current.

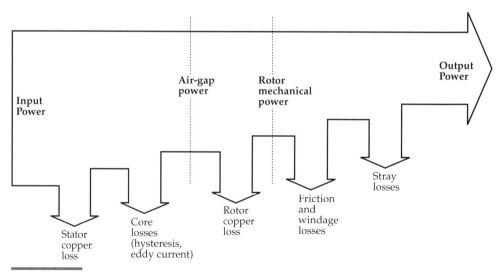

Figure 5-47
The power flow diagram of the asynchronous motor.

If we consider the exact equivalent circuit given in Fig. 5-45a, the input current (phase current) can be estimated easily by using the total impedance seen across the phase terminals.

$$\bar{Z}_1 = R_1 + jX_1 + (R_c // jX_m) // (R_2'/s + jX_2') \tag{5.47}$$

$$\bar{I}_1 = \frac{V_1 \angle 0°}{\bar{Z}_1} \tag{5.48}$$

Then the rms phase current can be calculated as

$$I_1 = \frac{V_1}{|Z_1|} \tag{5.49}$$

where V_1 and I_1 are the rms values of the phase voltage and current, respectively, and $|Z_1|$ is the magnitude of the total impedance given above.

The input power factor can be calculated using the phase angle between the voltage and the stator current, which then can be used to estimate the input active power.

$$P_{in} = V_1 I_1 \cos\theta \tag{5.50}$$

Similarly, referring to Fig. 5-44a, the rotor current can be derived as

$$\bar{V}_1 = \bar{I}_1(R_1 + jX_1) + \bar{E}_1 \tag{5.51}$$

$$\bar{E}_1 = \bar{I}_2'(R_2'/s + jX_2') \tag{5.52}$$

$$\bar{I}_2 = a\frac{\bar{V}_1 - \bar{I}_1(R_1 + jX_1)}{(R_2'/s + jX_2')} \tag{5.53}$$

As can be visualized from Fig. 5-47, the power flow diagram, the mechanical power on the rotor develops the torque on the shaft. Therefore the shaft torque can be derived as

$$T_m = \frac{3I_2'^2 R_2'}{s\omega_s} \tag{5.54}$$

Note that in the estimation of the total torque in equation 5.54, all three phases of the motor are considered.

Then the output power of the motor and efficiency of the motor can be obtained easily as

$$P_{out} = P_m - P_{rotor\,losses}$$
$$= \omega T_m - (P_{rotor\,core} + P_{windage+friction}) \tag{5.55}$$

$$\eta\% = \frac{P_{out}}{P_{in}} 100 = \frac{\omega T_m - (P_{rotor\,core} + P_{windage+friction})}{3 V_1 I_1 \cos\theta} 100 \tag{5.56}$$

It should be noted that the performance characteristics of a motor can be extended further by varying one of the circuit parameters and observing its influence on others.

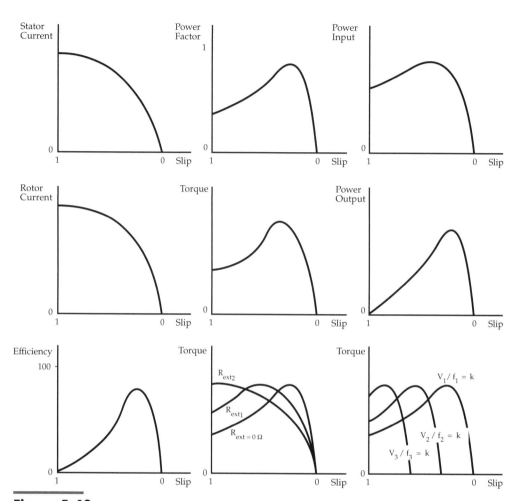

Figure 5-48
Typical performance characteristics of asynchronous motors.

Fig. 5-48 illustrates the typical performance characteristics that can be obtained solving the given equations. These characteristics can be used to verify the profile of the analysis results, which will be carried out in the VI explained in the next section.

5.7.2 Virtual Instrument Panel

The structure of the VI developed in this experiment is similar to the previous experiment, the Transformer Test. Although further preparations are necessary to initiate the LabVIEW-based tests, the overall experiment is self-explanatory. There are a number of subsections in this experiment, and each section includes a theory section that highlights some of the basic knowledge about each segment.

This experiment consists of five major sections:

- Measuring Stator Resistance
- Blocked-Rotor Test
- No-Load Test
- Performance Characteristics

Measuring Stator Resistance

This segment contains a brief guide on how to measure the phase resistance of a three-phase star-connected asynchronous motor as explained in the previous section.

Blocked-Rotor Test

When this VI is executed via the main menu, it produces three options: Theory, Test, and Parameters. The Theory option provides an instant reference about the Test section. The Test VI is basically identical to the test performed in the Transformer Experiment for the Short-Circuit Test, where two graphs are displayed: real time voltage and current waveforms of one of the phase windings of the motor. Then the user determines the phase angle and calculates the per-phase active power under the blocked-rotor condition. The measured rms phase voltage, the current, and the estimated power are recorded for further use.

Upon pressing the Done button and selecting the Parameters option, the equivalent circuit for the blocked-rotor test is displayed where the user can enter the recorded values and calculate the unknown circuit parameters as indicated in Fig. 5-46a.

No-Load Test

This test is performed based on the procedure as described in its theory section, as also explained in Section 5.7.1. Again, the rms values of the applied phase voltage, current, and the corresponding active power are recorded.

Similar to the previous test, the real test should follow the Parameters section, where all other remaining circuit parameters of the exact equivalent circuit are calculated.

Performance Characteristics

A great deal of information regarding the steady-state performance of the induction motor can be obtained from the equivalent circuit of the asynchronous motor, which is determined by the two tests described previously.

After the equivalent circuit parameters of the motor are determined, it is possible to analyze the motor's behavior under the various operating conditions and at sinusoidal steady-state. Note that the dynamic behavior of the asynchronous motor will be studied in Chapter 7, where various other electrical machine simulations are presented. The performance characteristics of the motor can be studied using the measured equivalent circuit parameters or using alternative motor parameters.

The front panel of the main VI is given in Fig. 5-49, which displays two reference circuit options (exact equivalent circuit and approximate equivalent circuit, which can be selected from the picture ring), and a drop-down menu that includes five study modules for the performance characteristics. The performance characteristics are as follows.

- View the performance characteristics based on the current circuit parameters.
- View the performance characteristics for the different values of the external rotor resistances, R_{1ext}, R_{2ext}, R_{3ext} (Fig. 5-50).
- View the performance characteristics for the different values of the stator phase voltages, V_1, V_2, V_3.
- View the performance characteristics for the different values of the supply frequencies, f_1, f_2, f_3.
- View the performance characteristics for the different values of V/f ratios, $V_1/f_1 = V_2/f_2 = V_3/f_3 = V_{rated}/f_{rated}$ = constant (Fig. 5-51).

Remember that all the parameters mentioned in the real measurements and in the circuit diagrams are per-phase quantities. However, the performance characteristics are based on the three-phase quantities.

Chapter 5 • Electric Machines Laboratory

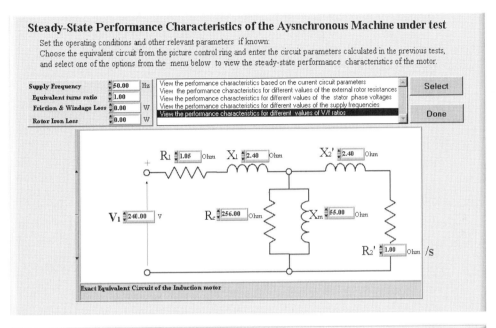

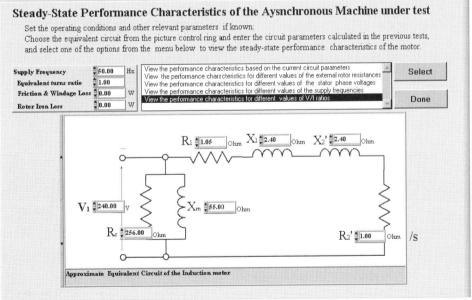

Figure 5-49
The front panels to select the exact or approximate equivalent circuits and to set circuit parameters.

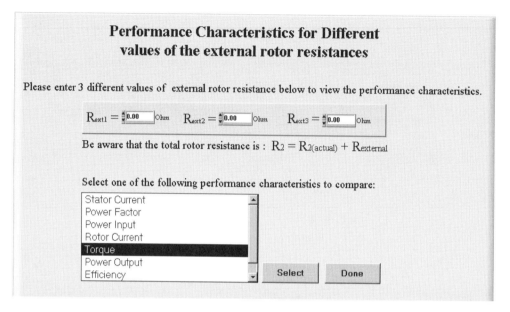
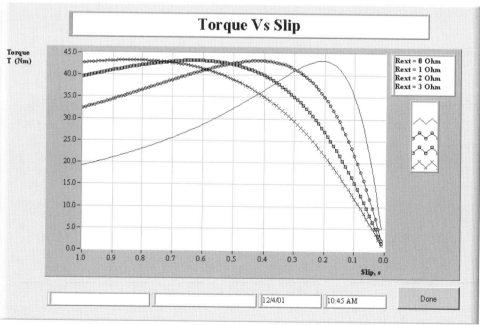

Figure 5-50
Sample front panels to study the performance characteristics for the different values of the external rotor resistances and the torque versus slip curves.

Chapter 5 • Electric Machines Laboratory

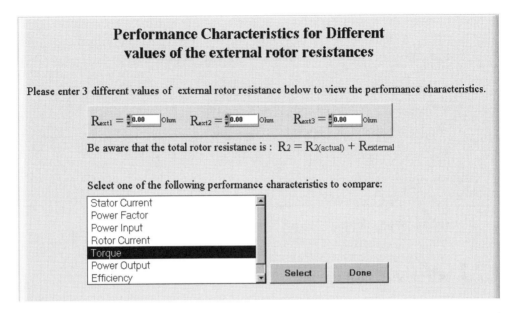

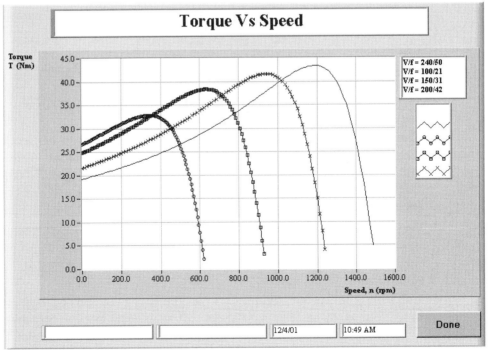

Figure 5-51
Sample front panels to study the performance characteristics for the different values of V/f ratios.

After the selection of one of the options, another front panel is displayed, which includes a set of custom controls and a menu containing various characteristic curves options that are generated automatically:

- Stator Current versus Slip (or Speed)
- Power Factor versus Slip (or Speed)
- Power Input versus Slip (or Speed)
- Rotor Current versus Slip (or Speed)
- Torque versus Slip (or Speed)
- Power Output versus Slip (or Speed)
- Efficiency versus Slip (or Speed)

5.7.3 Self-Study Questions

Open and run the custom-written VI, and investigate the following questions. After you run the experiment, determine the equivalent circuit parameters of the motor.

1. Estimate the total (three-phase) apparent and reactive powers under the no-load and the block-rotor conditions.
2. What are the maximum values of the voltage and the current under the no-load and the block-rotor conditions?
3. The breakdown torque is the maximum torque the motor can generate. Develop an expression for the breakdown and for the starting torque, and highlight these torque values on a sample torque-speed characteristic of your own choice. You may use the approximate equivalent circuit for the calculations.
4. For the three-phase induction motor, derive analytical expressions for the input power, the stator current, the rotor current, the output power, the electromagnetic torque, and the efficiency and the power factor, all as a function of the speed.
5. Induction motors are frequently used in industry to start up high inertia loads and as variable speed drives. After you observe the torque-speed characteristics under the various conditions in the Performance Characteristics section, identify which characteristic(s) are most suitable for these industrial applications. Print the most suitable torque versus slip (speed) graph(s).

6. Induction motors draw very large currents during the start-up process, which can be 4–5 times greater than their rated currents. The starting current decreases in time when the speed of the motor increases. Observe the current-speed characteristics under various settings and determine an operating condition, which can limit the starting current to twice the rated current of the test motor.
7. For the different values of the external rotor resistances, plot the torque-speed characteristics and determine the operating speeds on the graph for a selected value of the constant load torque.
8. For the identical operating condition as in question 7, what is the ideal starting and running characteristic of this wound-rotor machine?
Hint: Observe the intersections of the torque-speed curves.

5.7.4 Laboratory Hardware

Note that this experiment does not depend on the size of the motor under test. The only requirement is to measure the phase voltage and current of a three-phase machine considering a complete isolation under two principal test conditions: Blocked-Rotor and No-Load.

The wiring diagram used and recommended for the asynchronous motor experiment is shown in Fig. 5-52. After setting up the test system, the analog input channels of the DAQ card should be assigned via the suitable signal conditioning devices: voltage and current transducers. The channel assignment is required to ensure the correct measurements in the associated VIs.

- Channel 1 **to** the Phase 1 Voltage
- Channel 2 **to** the Line 1 Current.

The no-load and the blocked-rotor tests on the laboratory wound-rotor (slip-ring) asynchronous motor are performed using the circuit diagrams in Fig. 5-52. Before each test, the external resistances in the rotor circuit are set to zero and Phase 1 voltage and Line 1 current waveforms are measured and processed as explained in the previous section.

The per-phase active power is calculated from the rms values of the current, the voltage, and the phase angle that are obtained from the measured waveforms.

In the Blocked-Rotor Test, a reduction in the applied voltage is achieved by inserting resistances in series with the stator terminals of the motor as shown.

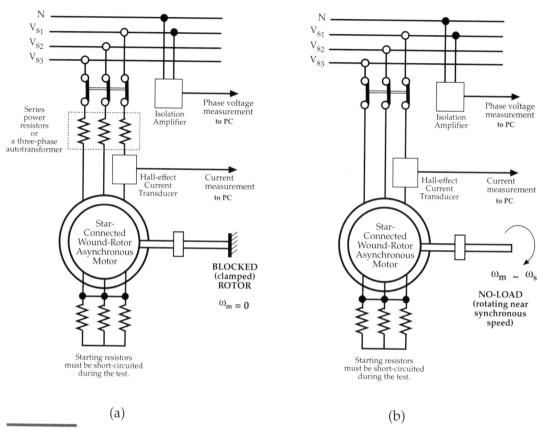

Figure 5-52
Wiring diagrams for the (a) Blocked Rotor and (b) No-Load tests.

However, if available, these resistors can be replaced with a three-phase autotransformer.

The specifications of the laboratory equipment, all necessary auxiliary devices, and data acquisition devices are given in the Appendix as a reference.

5.8 Synchronization Observer

Synchronous generators are the main source of ac power generation in the world and are usually linked and operated in parallel. The number of parallel generators can be as high as hundreds, interconnected by hundreds of kilometers of transmission lines.

The principal purpose of this interconnection is the continuity of the service and the economy in plant investment and operating cost. For example, since the power requirements of a large system vary during the day, generators are successively connected to the system or temporarily disconnected from the system to provide the demanded power.

A synchronous generator can be connected to an infinite bus (large power grid; the bus that ideally has a constant voltage and a constant frequency) by driving it at synchronous speed and adjusting its field current so that its terminal voltage is equal to that of the bus.

However, in the real system more conditions should be met before the generator is connected to the infinite bus, such as phase sequence and phase difference.

If the frequency of the incoming generator is not exactly equal to that of the bus, the phase relation between its voltage and the bus voltage will vary at a rate equal to the difference between the frequencies of the two voltages.

Therefore, if the synchronization action is not performed correctly, very large equalizing currents may flow in the system, causing serious system disturbances.

An early form of synchronization tool comprised the use of three incandescent lamps connected across the open main switch and is still used occasionally. Today, electronic solutions for synchronization are adapted. However, none of the devices and methods illustrate the concepts behind the synchronization action.

After performing this experiment, students should be able to

- study the requirements for the synchronization of a three-phase synchronous generator.
- understand the concept of synchronization by observing the real-time voltage waveforms; three-phase voltage phasors (voltage vectors); and other parameters, such as frequencies, phase angles, and phase differences.

The conditions for synchronization follow.

1. Phase sequence of the generator must be same as that of the bus.

    Phase 1 of the generator should be connected to Phase 1 of the bus.
    Phase 2 of the generator should be connected to Phase 2 of the bus.
    Phase 3 of the generator should be connected to Phase 3 of the bus.

    This condition depends on the initial wiring of the generator. Once the wiring is done, there is no need to check this condition again. Either a special device or a three-phase induction motor can be used to check the phase sequence. In case of identical phase sequence, the motor should rotate in the same direction both when it is connected to the bus side and to the generator side.

2. The generator frequency must be equal to the bus frequency.
   In a balanced three-phase system, this condition can be observed by measuring the frequency of, say, Phase 1 voltages on both sides, generator and bus.

3. The generator voltages must be equal to the corresponding bus voltages.
$$V_{g1} = V_{b1}, \quad V_{g2} = V_{b2}, \quad V_{g3} = V_{b3}$$

4. The generator and bus voltages must be in phase.
$$\phi_{g1} = \phi_{b1}, \quad \phi_{g2} = \phi_{b2}, \quad \phi_{g3} = \phi_{b3}$$

Fig. 5-53 illustrates three typical phasor states for the three-phase bus voltages and three-phase generator voltages. As can be seen, the synchronization conditions have not been met in any of these operating conditions. It should be noted here that, in practice, the conditions for synchronization do not need to be met exactly. Hence, small voltage, phase, and frequency differences are acceptable.

5.8.1 Virtual Instrument Panel and Laboratory Hardware

The objective of this experiment is to study the synchronization concept by observing the actual three-phase voltages of both the generator and the infinite bus. To perform the test, prepare the test setup shown in Fig. 5-54. Ap-

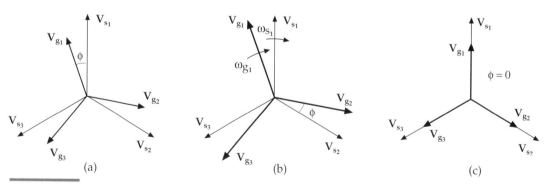

Figure 5-53
Typical phasor status during the synchronization process:
(a) Equal rotational speed (equal frequency) but different voltage magnitudes and with phase difference
(b) Equal voltage magnitudes but different frequency and with phase difference
(c) Equal frequency, no phase difference but different voltage magnitudes

Chapter 5 • Electric Machines Laboratory

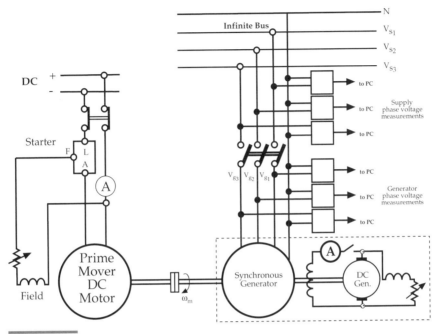

Figure 5-54
The wiring diagram for the real-time synchronization test.

propriate manipulation of the speed of the prime mover (a dc motor as used in this experiment) should be available. In total, six parameters—three-phase bus voltages and three-phase generator voltages—are measured.

Remember that in a practical system, the measured voltages (generator and bus voltages) can be as high as thousands of volts. Make sure that the high voltages are attenuated to an appropriate level before connecting to a DAQ card. The channel assignments used in this test are

Channels 1, 2, and 3 to the bus phase voltages 1, 2, and 3, respectively
Channels 4, 5, and 6 to the generator phase voltages 1, 2, and 3, respectively

The front panel of the synchronization test, Synchronization.vi, is shown in Fig. 5-55.

Procedures for the Synchronization Test

- Wire the experimental setup.
- Ensure that the three-phase switch, in the open position, is connected to the circuit between the generator terminals and the infinite bus

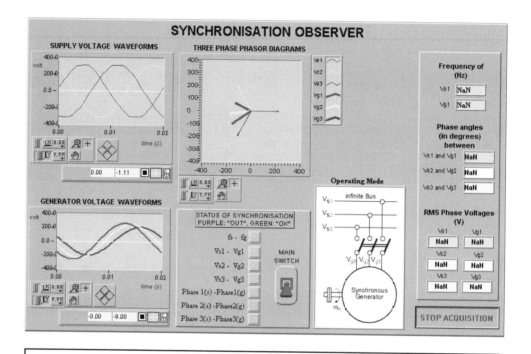

Figure 5-55
The front panel and brief user guide of the synchronization test.

terminals, and the voltage sensors are assigned to the channels as indicated in Fig. 5-54.

- Open `Synchronization.vi`, and press the Continuous Run button to run.
- Press the Start Acquisition button to initiate the data acquisition.
- Run the prime mover (a dc motor here) and adjust the speed so that the generator frequency is close to the infinite bus frequency.
- Adjust the excitation of the synchronous generator so that the generator voltage is equal to the bus voltage.
- Observe the phase angle between the phase voltage of the generator and the bus by using the phasor graph (analogous to a synchronoscope in practice). As illustrated in Fig. 5-55, the phasor graph has two sets of three-phase voltage vectors covering the entire range from 0° to 360°.
- During the synchronization process, the three-phase phasors rotate as the phase angle between the generator and the bus voltages varies. If the generator frequency is slightly higher or lower that the bus frequency, the direction of the phasor rotation will be different (clockwise or counterclockwise).
- At the instant of synchronization, when all four conditions listed above are met, the corresponding LED lights on the front panel turn green and the switch on the picture ring changes its position indicating the correct instant.
- After the synchronization conditions are obtained, the VI can be stopped and the front panel can be printed to prove the synchronization condition.

> **Note:** Although the synchronization status of the switch on the front panel can be used to control the main contactor, it is not suggested due to the potential risks in the system. These risks are potential wiring mistakes, wrong scaling factors, or imperfect or faulty signal conditioning devices. Remember that in a closed-loop system (fully automated synchronization) an additional signal conditioning device (containing an opto-coupler and a high voltage static relay) will be required to interface the main contactor to the PC.

5.9 Synchronous Machine Test

The main aim of this experiment is to present an interactive learning tool, using LabVIEW, to guide students through the theory of three-phase synchronous machines. Although some background information (used in the VI) will be given next, additional reading may be required to understand the operation and behavior of the synchronous machines in depth.

The synchronous machine is a doubly excited system: ac in the stator and dc in the rotor. Note that the stator and rotor functions of the synchronous machine are interchangeable. In the slots of the stationary section, the stator, there are uniformly distributed three-phase ac windings, which are either excited by alternating currents (motoring mode) or produce three-phase ac power (generating mode).

The rotor winding, however, is excited by a dc supply. The type of rotor depends on the speed required for application. Round rotors are required in applications involving high speeds, whereas low-speed applications have large salient pole-type rotors. The dc rotor winding produces a magnetic field, which acts as the catalyst in the energy conversion process. Furthermore, there is also a third winding, which is cage-embedded in the pole shoes of the rotor field. This winding provides a motor-starting mechanism in conjunction with the stator's ac windings and a stabilizing influence by damping any mechanical oscillations of the rotor.

Synchronous machines are the primary sources of power generation in power systems and are usually driven by a turbine. A power generation station normally contains several ac generators to supply the total load. During the light load on the station, only a few generators are operated to supply the demand. Therefore, when the load demand increases, other generators are connected to the main bus in order to cope with the increased load.

If the rotor winding is excited by a dc source and rotated by a prime mover, a rotating field is created in the air-gap, which produces a rotating flux that induces voltages in the stator windings. The induced voltages are equal in magnitude, but they are phase-shifted by 120 electrical degrees as described in Chapter 3.

With a synchronous machine operating in motor mode with a constant mechanical load, changes in the field current change the armature (stator) current and its operating power factor. Hence three modes of field excitation can be defined.

- **Normal field excitation:** Armature current is minimum and the operating power factor is unity (as a resistive type of load).

- **Underexcitation:** Defined reference to the normal excitation. Armature current increases and the operating power factor is lagging (as an inductive load).
- **Overexcitation:** Armature current again increases, but the operating power factor is leading (as a capacitive load).

Equivalent Circuit and Phasors

A synchronous generator can be considered to consist of three components in series: an ac generating source, a resistor representing copper and iron losses, and an inductor representing the inductance of the winding and magnetic leakage. Moreover, ac load is also connected in series with these components.

The terminal voltage of an armature (ac stator) winding of a round rotor synchronous generator can be given as

$$V_t = E_f - I_a R_a - jX_s I_a \qquad (5.57)$$

where V_t is the terminal voltage and R_a is the resistance of the armature winding per phase, E_f is induced voltage, I_a is the armature current, and X_s is known as the synchronous reactance (a fictitious reactance).

$$X_s = X_l + X_a \qquad (5.58)$$

Here, X_l is the leakage reactance (allows for flux leakage paths in air), and X_a is the armature reaction reactance (accounts the fundamental field of armature reaction). It is usual to define the synchronous impedance in the equivalent circuit. Hence,

$$V_t = E_f - I_a(R_a - jX_s) = E_f - I_a Z_s \qquad (5.59)$$

Equation 5.59 provides the basis for synchronous machine analysis, which will be widely used in this study. Note that the resistance R_a is frequently omitted from calculations.

Fig. 5-56 illustrates an equivalent circuit that satisfies equation 5.59 and provides three distinct phasor diagrams.

A number of tests can be used to determine the equivalent circuit parameters of the synchronous machine R_a and X_s. However, since the saturation of the iron sections of the magnetic circuits of the machine affects the armature reaction, the value of X_s depends on the field excitation and the load conditions. Therefore, the value of the synchronous reactance cannot be determined precisely but can be approximated for saturated conditions. Let's look at these tests briefly.

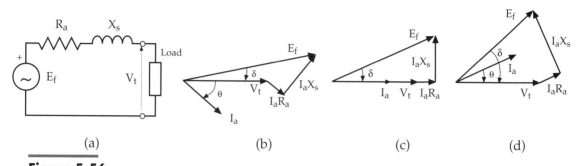

Figure 5-56
(a) Per-phase equivalent circuit of a round rotor synchronous generator, (b) phasor diagrams for lagging power factor, (c) unity power factor, and (d) leading power factor.

Open-Circuit Test

In this test, the synchronous machine is driven at synchronous speed. The open-circuit terminal voltage $V_t = V_o (= E_o)$ is measured as a function of the field current.

The open-circuit characteristic (OCC) is shown in Fig. 5-57. Due to the fact that the terminals are open, the curve illustrates the variation of the excitation voltage E_o (i.e., E_f at open circuit) with the field current I_f. The reluctance of

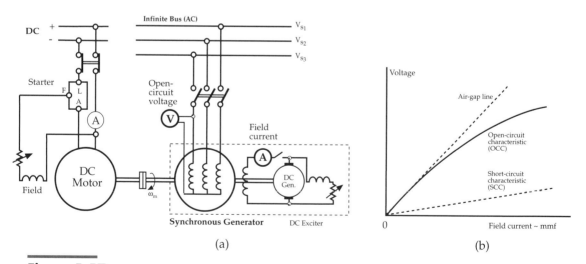

Figure 5-57
(a) Circuit diagram for the open-circuit test, and (b) the characteristic curve obtained.

the magnetic circuit of the machine may be regarded as consisting of two main parts: the reluctance of the iron path (which changes as a function of field current) and the reluctance of the air-gap (which remains constant). For small values of field current, the reluctance of the air-gap is large compared with that of the iron path. Therefore, for low-field current, the OCC approximates a straight line (known as the air-gap line).

Short-Circuit Test

The short-circuit characteristic (SCC) provides the relationship between the short-circuit armature current and the field current. The test is performed while all three phases of the machine are short-circuited and the machine is running at constant synchronous speed (Fig. 5-58).

The field current I_f is varied and the average of the three armature currents is calculated (I_{sc}). Under short-circuit conditions, the machine is not subject to saturation, and, therefore, the SCC is a straight line.

Since the terminal voltage is zero, the internal voltage is used to drive the short-circuit current through the machine's synchronous impedance. The air-gap voltage is used in driving the short-circuit current through the armature resistance and armature leakage reactance of the machine.

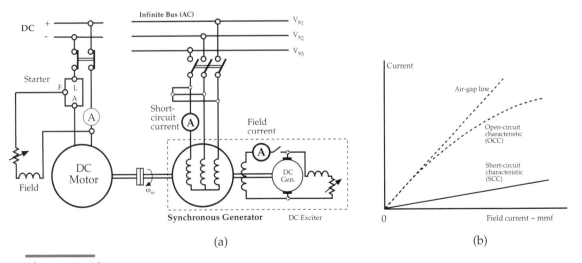

Figure 5-58
(a) Circuit diagram for the short-circuit test, and (b) the characteristic curve obtained.

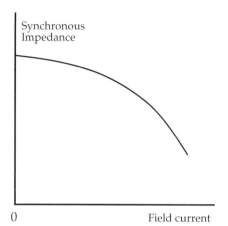

Figure 5-59
Synchronous impedance characteristic obtained from OCC and SCC.

Synchronous Impedance

Synchronous impedance can be defined as the ratio of open-circuit voltage (or induced emf) to short-circuit current (or mmf) corresponding to the same field excitation. Under short-circuit conditions, the machine is not subject to saturation and so for any given field current the air-gap line must give the internal voltage E_f. It must be noted that the unsaturated value of the synchronous impedance corresponds to a low value of field current, as the air-gap line and OCC are the same for low field current I_f (where there are no saturation effects).

Zero Power Factor Characteristic and Potier Triangle

The Zero Power Factor Test is performed to determine leakage reactance component of the synchronous reactance of a synchronous machine. In this test, the synchronous generator is loaded using an inductive load or an underexcited synchronous motor (as in this case) (Fig. 5-60).

While the armature winding current is maintained at full load, the terminal voltage V_t is measured as the field mmf (or field current) is varied. The Zero Power Factor Characteristic (ZPFC) is plotted as a function of terminal voltage versus field current (Fig. 5-61). This characteristic is similar to OCC but displaced by distance that is defined by the Potier triangle. Since the armature winding current remains constant in the ZPFC, both the leakage reactance and the armature reaction reactance must be constant for any field excitation. This implies the OCC and ZPFC are parallel.

Chapter 5 • Electric Machines Laboratory

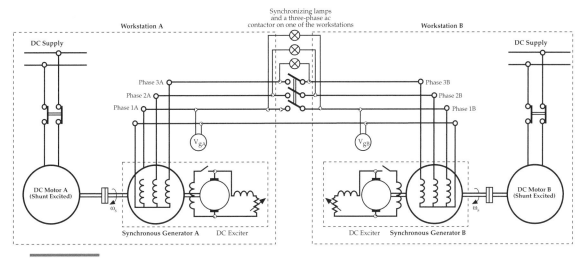

Figure 5-60
Circuit diagram for synchronization and Zero Power Factor Test.

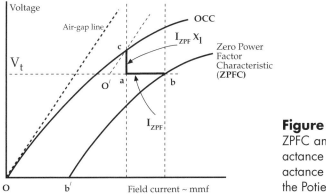

Figure 5-61
ZPFC and determining leakage reactance and armature reaction reactance utilizing OCC and ZPFC, the Potier triangle.

The armature winding current I_a lags the terminal voltage V_t by 90 degrees so that the armature leakage reactance drop jI_aX_l is in phase with V_t. If the resistance (R_a) is neglected, the phasor diagram becomes

$$V_t = E_f - I_a X_L \qquad (5.60)$$

The leakage reactance and armature reaction reactance can both be determined by constructing the Potier triangle.

The lamps shown in Fig. 5-60 represent the synchronizing lamps. To indicate the correct equalization of frequency and phase sequence of the incoming generator, the dark lamp synchronization connection is used in this test.

In the ZPFC test, the individual field currents are adjusted until the voltages of each machine are equal. The lamps should now turn on and off simultaneously and at a slow rate. After achieving the synchronization conditions (all lamps dark), the machines can be synchronized by closing the contactor as shown in Fig. 5-60.

Slip Test

The Slip Test measures the ratio of direct and quadrature reactance in a salient pole synchronous machine. The synchronous machine is considered as passive impedance, the ratio of whose maximum and minimum values is measured by a voltage/current test.

For the slip test, the synchronous machine is driven at a speed slightly less than the synchronous speed (about a slip less than 5%) while the field winding is open-circuited (Fig. 5-62a). Then a three-phase balanced reduced voltage is applied to the stator windings of the machine. Thus the only mmf present in the machine is that due to stator current, rotating in space at synchronous

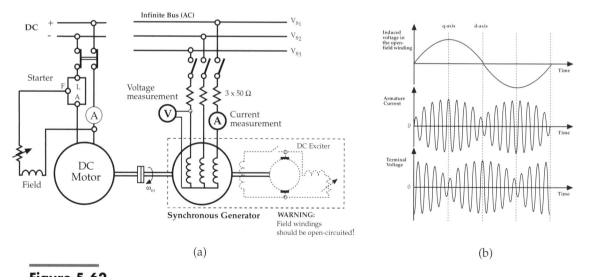

Figure 5-62
(a) Circuit diagram for the Slip Test, and (b) typical voltage and current waveforms.

speed. The unenergized rotor structure drifts through the mmf wave with a relative velocity equal to the slip speed of the rotor.

Under this operating condition, the changing reluctance presented to the mmf wave can be observed by monitoring the current and voltage in the stator windings. Typical waveforms of the armature current, terminal voltage, and the voltage across the field winding are shown in Fig. 5-62b.

Using these waveforms, approximate values of direct-axis synchronous reactance and quadrature-axis synchronous reactance can be given as

$$X_{ds} = \frac{V_{max}}{I_{min}} \tag{5.61}$$

$$X_{qs} = \frac{V_{min}}{I_{max}} \tag{5.62}$$

Alternatively, if the voltage is low enough to ensure unsaturated operation, the ratio X_{qs}/X_{ds} can be given by

$$\text{(Voltage peak to trough ratio)(Current peak to trough ratio)} = \frac{X_{ds}}{X_{qs}} \tag{5.63}$$

However, since the field circuit is unexcited during the slip test, the magnetic conditions are not the same as with the normal operating condition. Therefore, the best results for obtaining X_q can be achieved by using the ratio X_{qs}/X_{ds} from the slip test and using the value of X_d from the open-circuit and short-circuit tests, which is virtually the same as the synchronous reactance X_s for the round rotor machine for the unsaturated case.

$$X_d = X_s = \sqrt{Z_s^2 - R_a^2} \tag{5.64}$$

Hence X_q can be calculated using the ratio X_{qs}/X_{ds}.

$$X_q = \frac{X_{qs}}{X_{ds}} X_d = \frac{V_{min}}{I_{max}} \frac{I_{min}}{V_{max}} X_d \text{ per unit} \tag{5.65}$$

Power and Phasor Analysis

The power flow in a salient pole synchronous machine is a general case. Hence the analysis of such a machine can provide a much broader view on the operation of the synchronous machines.

The power and phasor analysis requires testing the machine to find the terminal voltage, armature current, and excitation voltage phasors under particular operating conditions. Fig. 5-63 shows the phasor diagram for the salient pole synchronous machine, acting as a generator.

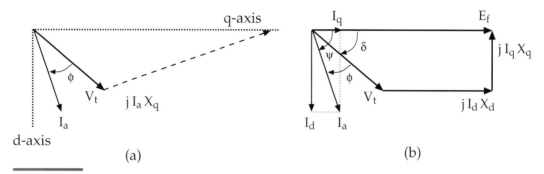

Figure 5-63
Phasor diagrams for a salient pole synchronous machine, which ignores the armature resistance.

In Fig. 5-63, the armature current I_a is lagging the excitation voltage E_f by an angle ψ, known as the power factor angle. If the angle between I_a and V_t (ϕ) is known, the angle between V_t and E_f must be known to obtain currents I_d and I_q (Fig. 5-63b).

In a practical machine, however, usually only V_t, I_a, and ϕ are available. To find δ (and hence break up I_a into I_d and I_q) a fictitious phasor of jI_aX_q is added to V_t. This terminates on the q-axis (Fig. 5-63a).

The equations representing the phasor diagrams are given next.

$$E_f = V_t + I_aR_a + I_djX_d + I_qjX_q \tag{5.66}$$
$$I_a = I_d + I_q \tag{5.67}$$
$$\psi = \phi \pm \delta \tag{5.68}$$
$$I_d = I_a \sin \psi = I_a \sin(\phi \pm \delta) \tag{5.69}$$
$$I_q = I_a \cos \psi = I_a \cos(\phi \pm \delta) \tag{5.70}$$
$$V_t \sin \delta = I_qX_q = I_aX_q \cos(\phi \pm \delta) \tag{5.71}$$
$$\tan \delta = \frac{I_aX_q \cos \phi}{V_t \pm I_aX_q \sin \phi} \tag{5.72}$$

The real per-phase power in a salient pole synchronous machine is composed of two components ($P = I_dV_d + I_qV_q$). Hence, using equations 5.69, 5.70, and 5.71 gives

$$P = \frac{|V_t||E_f|}{X_d} \sin \delta + \frac{|V_t|^2(X_d - X_q)}{2X_dX_q} \sin 2\delta = P_f + P_r \tag{5.73}$$

The first term P_f represents the power due to the excitation voltage E_f. The second term P_r represents the effects of salient poles and produces the reluctance torque, which is independent of field excitation. Note that if $X_d = X_q$, P_r is zero. Hence, under this assumption, the power equation can represent the round rotor synchronous machine.

The reactive power per phase is also dependent on terminal voltage and excitation voltage, X_d and X_q.

$$Q = \frac{|V_t||E_f|}{X_d} \cos \delta - |V_t|^2 \left| \frac{\sin 2\delta}{X_q} + \frac{\cos 2\delta}{X_d} \right| \qquad (5.74)$$

The power angle characteristic, shown in Fig. 5-64, illustrates the excitation component P_f and the reluctance component P_r of the real power. It can be seen that if the machine has salient poles, the maximum resultant power occurs at $\delta < 90°$, making the curve steeper in the region of positive slope.

Note that the power angle characteristic is different for different values of field excitation and constant terminal voltage. If the field excitation is reduced to zero, the machine can still develop power (or torque) because of the saliency of the rotor structure.

In a salient pole synchronous machine, if the field excitation is varied over the normal operating range, saliency effects on power or torque are negligible. Power and torque (due to saliency) become important when the machine is running with low excitation.

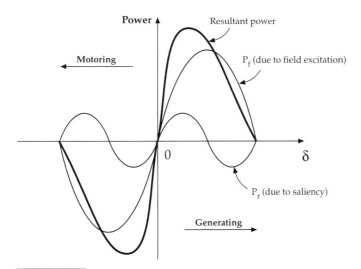

Figure 5-64
Power angle characteristic for a salient pole synchronous machine.

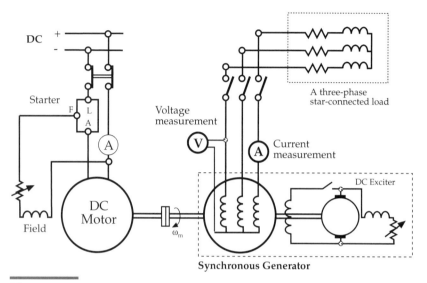

Figure 5-65
Circuit diagram for the Power and Phasor Test.

Fig. 5-65 illustrates a wiring diagram that can be used to perform the Power and Phasor Test, as will be described in the next section.

5.9.1 Virtual Instrument Panel

Table 5-8 indicates the distinct levels of the software components of the synchronous machine experiment, which fundamentally obtains the equivalent circuit of the machine with a series of real-time tests and performs a steady-state power analysis.

Note that the hierarchical structure of this experiment is similar to the common structure described for other experiments at the beginning of this chapter. In addition, you should save the data obtained during the tests on a floppy disk.

Once the program has started, a main menu appears, listing the eight tasks given in Table 5-8 to be carried out. Students are required to work through each option in the menu sequentially. However, if needed, each option can be repeated independently. Each option is selected by either double-clicking on the option or selecting it and pressing the Select button.

In the following paragraphs, the operation details of each option are explained. However, note that for correct measurements in each VI the gains of

Table 5-8 *The software levels of the experiment.*

Screen Levels	Remarks
Log-in Screen	Contains a title, an input text box, and two Boolean controls (Start, Exit).
Main Menu	It holds a list of eight subsections that link to the sections of the experiment. The main menu is a while loop that has an output to the screen with a text dialog and a termination button.

Parts 1 to 8:
  Measuring Stator Resistance
  Open-Circuit Test
  Short-Circuit Test
  Synchronous Impedance
  Synchronization
  Zero Power Factor Test
  Slip Test
  Power and Phasor Analysis
	The theory and the procedure link and the major experimental and analysis modules are included.

the voltage and current measurement units may require adjustments, which depends on the measuring interface circuits.

Measuring Stator Resistance

This option essentially outlines the procedure of measuring and calculating the stator resistance of the synchronous machine, which is very simple and does not contain any sub-VI.

Open-Circuit Test

When this option is executed, a menu with two options is produced: Theory and Test and Characteristic.

The Test and Characteristic option is a two-in-one VI in that it plots the OCC as the experiment is conducted. The front panel shows the OCC as well as indicating the phase voltages and the field current at each moment data is acquired from the machine (Fig. 5-66). Some additional features have been incorporated such as a Save option, which saves the values in a file, allowing the

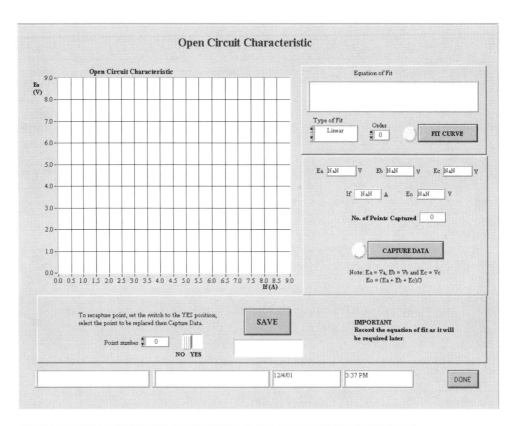

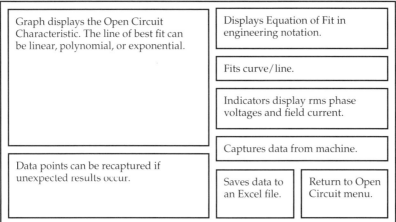

Figure 5-66
The front panel and brief user guide of Open-Circuit Test and Characteristic (LabVIEW 5.0).

student to print a copy for the final report. Moreover, data points can be re-captured if unexpected results occur.

In this test, the data is acquired by using a data acquisition sub-VI. Voltage and current gains have also been added to compensate for the attenuation, introduced by the transducers in the internal interface.

The channel assignments used in the test are

Phase A voltage	to	Channel 0
Phase B voltage	to	Channel 1
Phase C voltage	to	Channel 2
Field Current	to	Channel 3

The other main features of the test accessible via its front panel are the line/curve of best-fit function and the save function (which is currently defined to save to the A drive).

Students are required to record the coefficients of the fitted curves to be used later.

Short-Circuit Test

This VI has many similarities to the Open Circuit Test VI described previously. The program structure is identical in that the menu provides two options. Furthermore, the Test and Characteristic front panel has identical features, but this time three-phase currents are measured rather than phase voltages (Fig. 5-67). The VI module calculates the average phase currents for a specified field current and consequently plots the SCC. Again, students are required to record the coefficients of the fitted curves to be used later.

Synchronous Impedance

This part of the experiment explains the significance of saturation effects on the characteristics of the synchronous machine. The data is acquired as in the previous practical sections and is used to plot a graph of synchronous impedance versus field current.

Students are required to enter the coefficients of the OCC and the SCC using the controls available on the front panel (Fig. 5-68). Then by varying the field current control, the curve/line can be plotted. In this sub-VI the value

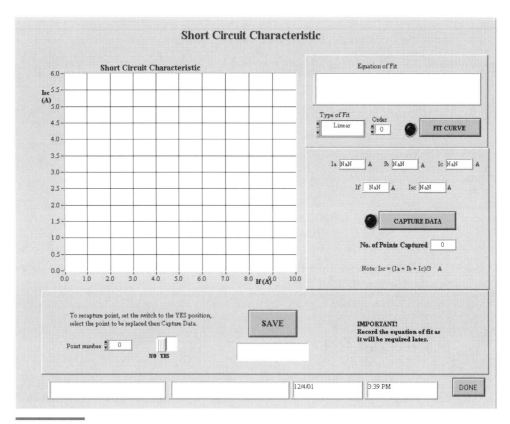

Figure 5-67
Short-Circuit Test and Characteristic front panel.

of Z_{su} (for 3 A) is used to calculate the unsaturated direct-axis synchronous reactance. However, note that the current level depends on the ratings of the machine under test.

Synchronization

In order to prepare for the Zero Power Factor Test, which requires two synchronous machines and precedes this section, the machines must be synchronized. This module simply provides some theoretical explanation and outlines the method with which synchronization can be achieved.

Chapter 5 • Electric Machines Laboratory

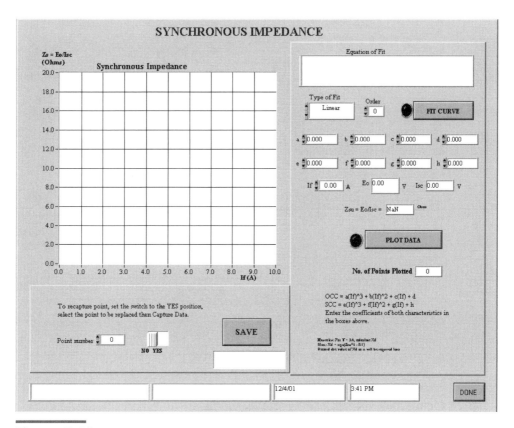

Figure 5-68
Synchronous Impedance Characteristic front panel.

Zero Power Factor Test

The Zero Power Factor Test is a significant section of this practical test, which has three options: Theory, Test and Characteristic, and Calculation of Armature Reaction and Leakage Reactance.

The Test and Characteristic section performs the relevant data acquisition and also provides the ZPFC. The block diagram includes many of the data structures and sub-VIs used in the VIs for plotting the OCC and SCC.

The channel assignments used in this sub-VI are

Phase A voltage	to	Channel 0
Field Current	to	Channel 1
Phase A current	to	Channel 2

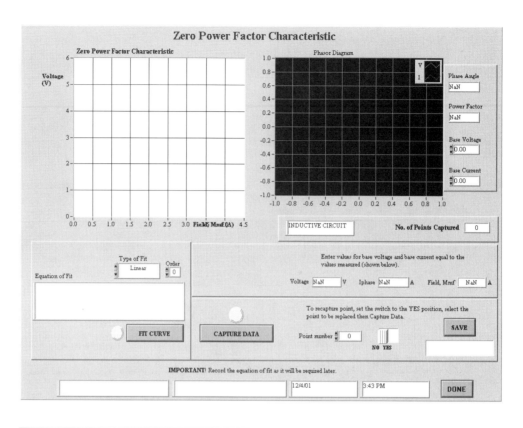

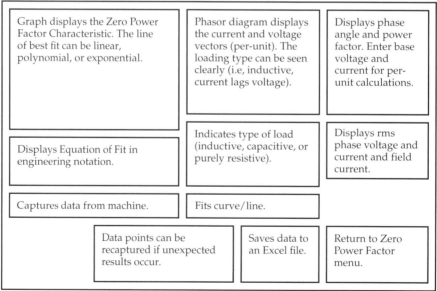

Figure 5-69
The front panel and brief user guide of the Zero Power Factor Test and Characteristic.

The sub-VI has additional functionality in that it calculates the power factor and phase angle between the voltage and current waveforms, which is used to display the phasor diagram that is visible on the front panel (Fig. 5-69).

The final option provided in this module is a step-by-step demonstration of the calculation of armature reaction X_a and leakage reactance, X_l.

This VI incorporates a formula node, which enables the implementation of equations given specific inputs, and then outputs the solution. To perform the calculations in this module, the students are required to input the coefficients of both the OCC equation and the ZPFC equation, which were obtained previously. The program then plots these two functions for further use in the construction of the Potier triangle (used in the calculation of X_l). Then the successive steps should be followed and the in-built cursor function should be used to measure the values of the required parameters.

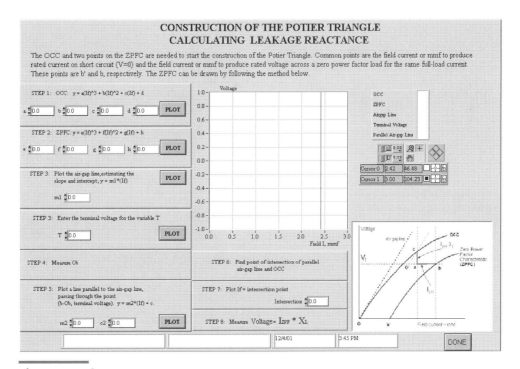

Figure 5-70
Calculation Leakage Reactance front panel.

Slip Test

This section consists of a theoretical explanation module and a test module. The theoretical explanation briefly outlines what is to be done during the practical and the relevant equations. The experiment is performed in the 'Test and Calculations' module. In this part, the waveforms of current and voltage (versus time) are acquired, using the data acquisition capability. The maximum and minimum voltage and current values can be obtained, using the two cursors provided with both graphs. The front panel displays voltage and current peak-to-trough ratios and the resultant ratio X_d/X_q (Fig. 5-71). In order to find X_q, the student needs to enter the value of X_d, calculated earlier. It was found that the best result for X_q is obtained by finding the ratio

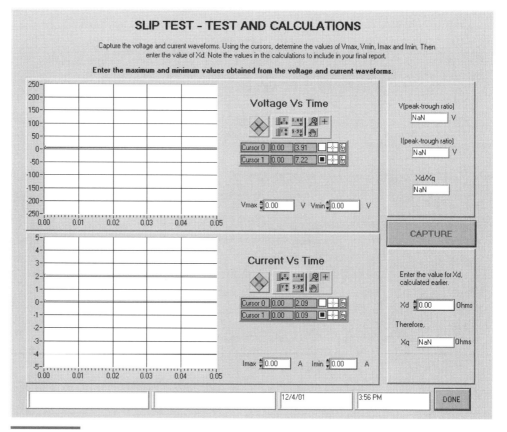

Figure 5-71
Slip Test and Calculations front panel.

X_d/X_q, then using the value of X_d from the open-circuit and short-circuit tests.

Power and Phasor Analysis

This section consists of three modules: Theory, Test, and Analysis. The purpose of the test module is to obtain the phase angle (phi) between terminal voltage and armature current (Fig. 5-72). This experiment is dependent on the operating conditions.

Once the phase angle has been found, the power angle characteristic and the phasor diagram can be obtained. The analysis section illustrates the effect of machine parameters (such as voltage, current, phi, and reactance) on the power angle characteristic. The phasor diagram is drawn as the values are entered (Fig. 5-73). It must be noted that the value of δ will be different for each operating condition. Therefore, students will be required to find the value of δ, which completes the phasor diagram.

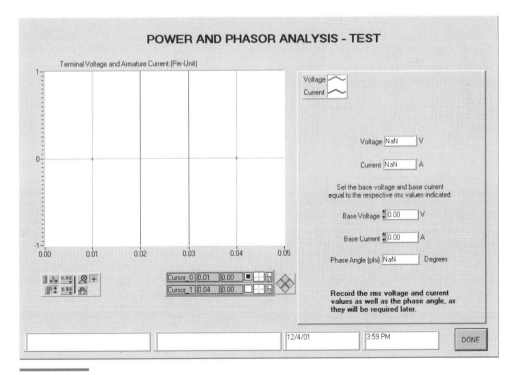

Figure 5-72
Power and Phasor Analysis Test front panel.

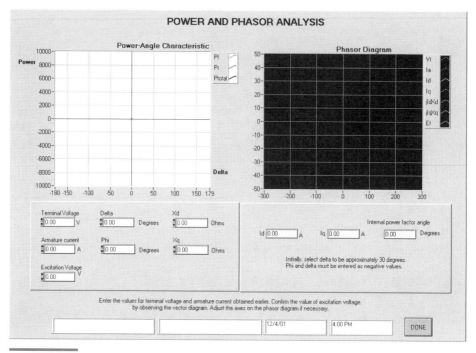

Figure 5-73
Power and Phasor Analysis Simulation front panel.

The phasor diagram is drawn using sub-VIs that draw vectors. The block diagram is relatively complicated, as the starting point of each vector must be specified (and is not necessarily zero).

5.9.2 Self-Study Questions

Obtain the following characteristics and constants of a laboratory salient pole synchronous machine:

1. Stator Resistance
2. Open-Circuit Characteristic
3. Short-Circuit Characteristic
4. Synchronous Impedance Characteristic
5. Zero Power Factor Characteristic
6. The unsaturated direct-axis synchronous reactance (X_d)
7. The unsaturated quadrature-axis synchronous reactance (X_q)
8. The armature reaction and leakage reactance

Study power transfer within a synchronous machine by power and phasor analysis.

5.9.3 Sample Results

The characteristics and parameters estimated in this section very much depend on the specifications of the actual machine(s) and test setup. The primary purpose for using the software in the laboratory is to ensure that the correct scaling constants are used in the data acquisition stages.

The following results are obtained from a real salient pole synchronous machine set. Our aim is to provide some guidance for students and potential developers.

Open-Circuit Test Results (Fig. 5-74)

Open-Circuit Voltage (V)	77.7335	144.1406	194.6892	231.5764	250.8014	267.4778
Field Current (A)	0.5078	0.9857	1.4731	2.0005	2.389	2.8547

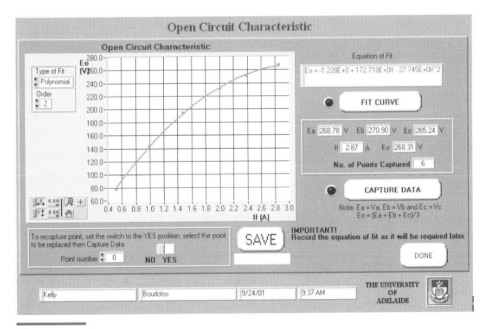

Figure 5-74
Open-Circuit Characteristic test result.

Short-Circuit Test Results (Fig. 5-75)

Short-Circuit Current (A)	2.1685	4.3437	6.3175	9.4481	10.8052
Field Current (A)	0.2533	0.4961	0.7159	1.0647	1.2146

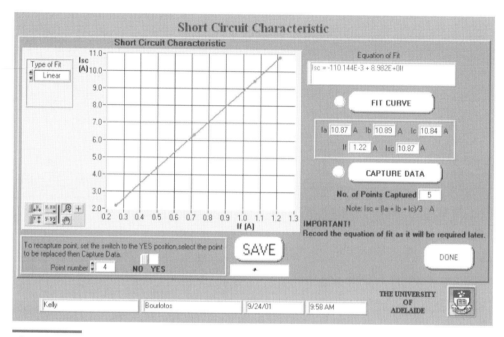

Figure 5-75
Short-Circuit Characteristic test result.

Synchronous Impedance (Fig. 5-76)

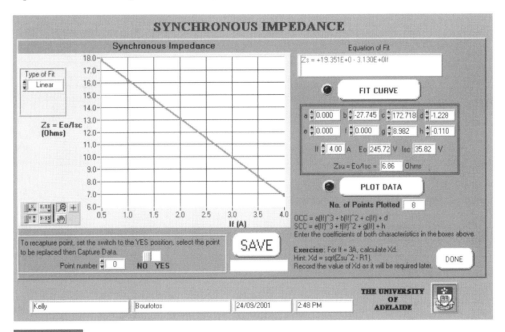

Figure 5-76
Synchronous Impedance Characteristic test result.

Zero Power Factor Test (Fig. 5-77 and Fig. 5-78)

Voltage (V)							
243.3499	228.4135	211.2717	199.5928	184.174	172.7469	153.8505	142.1851
Field Current (A)							
3.062	2.706	2.4141	2.233	2.0612	1.9308	1.7551	1.6623

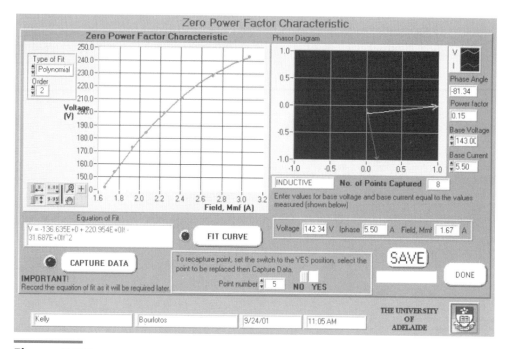

Figure 5-77
Zero Power Factor Characteristic test result.

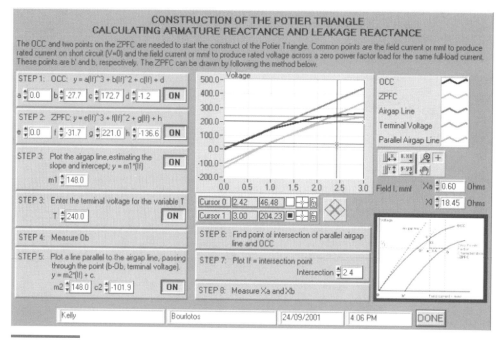

Figure 5-78
Potier triangle test result.

Slip Test (Fig. 5-79)

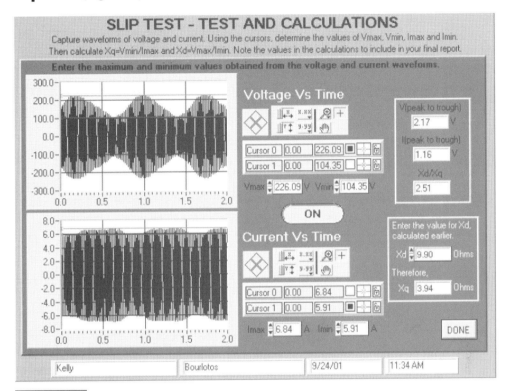

Figure 5-79
Slip Test and Calculations.

Power and Phasor Analysis (Fig. 5-80 and Fig. 5-81)

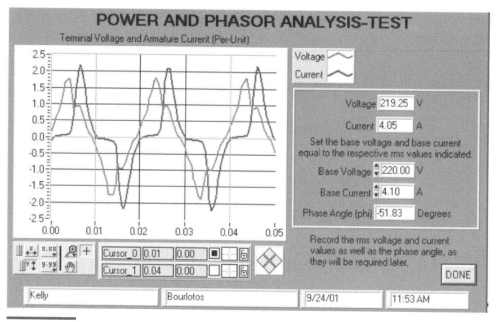

Figure 5-80
Power and Phasor Analysis test results.

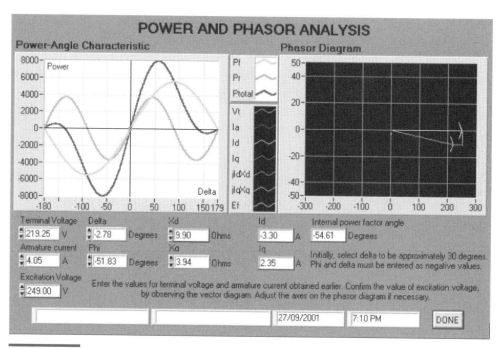

Figure 5-81
Power and Phasor Analysis simulation results.

The operating conditions for which this test was carried out are as follows: secondary side of a three-phase autotransformer was used as a load while its voltage setting was 120 V, which loaded the synchronous machine with a line current of 4 A and phase voltage of 240 V.

5.10 References

Bourlotos, K., and I. Hur Lao. "Final Year Project Report on Transformers and Synchronous Machines." Department of Electrical and Electronic Engineering. University of Adelaide, 2001.

Dubey, G. K. *Fundamentals of Electrical Drives.* New Delhi: Narosa, 1995.

Ertugrul, N. "Electric Power Applications Lecture Notes." Department of Electrical and Electronic Engineering. University of Adelaide, 1997.

Ertugrul, N. "Towards Virtual Laboratories: A Survey of LabVIEW-Based Teaching/Learning Tools and Future Trends." *International Journal of Engineering Education* 16(2). Special Issue on LabVIEW Applications in Engineering Education (2000).

Gray, C. B. *Electrical Machines and Drive Systems.* New York: Longman Scientific and Technical (Wiley), 1989.

Gross, Hans (Ed.). *Electrical Feed Drives for Machine Tools.* Berlin: Siemens Aktiengesellschaft (Wiley), 1983.

Padde, S., and A. Santoso. "Final Year Project Report on Electromechanical Energy Conversion Device Experiment." Department of Electrical and Electronic Engineering. University of Adelaide, 1999.

Parker, A. M. "Characteristics and Losses of Machines." Level 3 Energy Conversion Practical Handout. Department of Electrical and Electronic Engineering. University of Adelaide, 1989.

Say, M. G. *Alternating Current Machines.* New York: Longman Scientific and Technical (Wiley), 1986.

Van, L. H. "Final Year Project Report on Force Measurement and Electromechanical Energy Conversion Device Experiment." Department of Electrical and Electronic Engineering. University of Adelaide, 1998.

Wildi, T. *Electrical Machines, Drives, and Power Systems.* Englewood Cliffs, NJ: Prentice Hall, 1991.

Introduction to Power Electronics Circuits

6

Power Electronics has won universal acceptance due to its ties with almost all engineering subjects and is rapidly expanding as an enabling technology supporting many sectors of our modern industrial society, such as switched mode power supplies in computers and multi-level high-power inverters in large motor drives.

The control of electric power is frequently desired in Power Electronics for regulation of one or multiple parameters in industrial systems, such as the speed, position, and torque of an electric motor; the temperature of an oven; the rate of an electrochemical process; or the intensity of lighting. Power Electronic converters control and convert electric energy by all methods of switching, which involve various switching devices, circuit design and analysis, control systems theory, heat transfer, electric machines, and so on.

This chapter specifically focuses on certain power electronics converter topologies to give an insight into the operation of these circuits by providing a highly flexible simulation environment in LabVIEW, which closely integrates the real operation and the theoretical studies. A number of fundamental circuits and converters are simulated. Although not all the questions are addressed here, the principal objective is to provide basic simulation ideas that easily can be adapted to other more complex circuit topologies.

As will be demonstrated in the relevant VIs, the powerful graphical user interfaces of the VIs will allow students to study the circuit operation in great depth. Each VI allows the user to set a number of component values and observe all the available parameters both in graphs and in indicators. An additional dimension provided to the user is the knowledge of the active circuit (current carrying circuit), which is delivered by using multiple circuit drawings, each highlighted in red. Additional controls are also available to the user, such as start, reset, pause, quit, change per-unit value, and speed of the simulation. The simulation methodologies used here are incorporated into the motor drives that are presented later in Chapter 7.

The first two sections of this chapter introduce the operation of two switching devices, diodes (uncontrolled) and thyristors (controlled turn on). Section 6.3 demonstrates one of the most commonly used switching techniques, PWM. The last four sections present basic converter topologies utilized in Power Electronics.

6.1 Diode Conduction

The diode is the most fundamental nonlinear electronic element, and it forms the basis of nearly all solid-state devices. This section aims to introduce the operation of a diode and discuss the current-voltage characteristics of the diode through the examination and visual demonstration of the diode equation, test results, and other circuit parameters.

The circuit considered in this section consists of a diode, a voltage source (ac or dc), and a resistor in series as illustrated in Fig. 6-1. The current-voltage characteristic curve of a typical diode, illustrated in the same figure, has three distinct regions: the forward bias region ($v_d > 0$), the reverse-bias region ($v_d < 0$), and the breakdown region ($v_d < V_z$).

Theoretically, the diode saturation current can be determined using the Schockley diode equation:

$$i_d = I_s(e^{v_d/nV_T} - 1) \qquad (6.1)$$

where I_s is the reverse saturation current, n is a constant between 1 and 2, depending on the material physical structure and the operating region, V_T is the thermal voltage given by

$$V_T = \frac{kT}{q} \qquad (6.2)$$

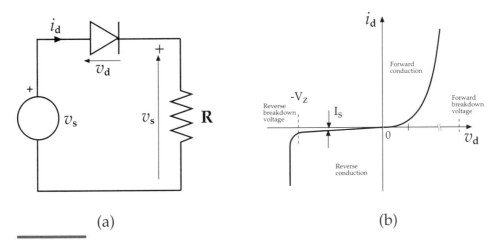

Figure 6-1
(a) Circuit for the diode simulation, and (b) the diode current-voltage relationship.

where k is the Boltzmann constant (1.38×10^{-23} joules/kelvin), T is the absolute temperature in Kelvins, and q is the charge of the electron (1.6×10^{-19} coulombs). At room temperature (20°C), $V_T = 25$ mV.

The voltage source in the circuit can be either ac or dc, and the resistor is variable. The principal operation of the simulated circuit is that as the source voltage is varied, the diode current i_d and the diode voltage v_d are determined. Writing the voltage equation around the loop given in Fig. 6-1a yields

$$v_d = v_s - i_d R \tag{6.3}$$

This equation is known as a load line and can be plotted on the same axis as a plot of the diode characteristic. The intersection of these two plots is called the operating point of the circuit and yields the diode current and the voltage across the diode.

In the VI presented here, the load line is generated using two simple formulas.

$$i_{d(N)} = \frac{v_{s(N)}}{R}\bigg|_{v_d=0} \quad N = 1, 2, 3 \ldots \tag{6.4}$$

$$v_{d(N)} = v_{s(N)}|_{i_d=0} \quad N = 1, 2, 3 \ldots \tag{6.5}$$

where $i_{d(N)}$ and $v_{d(N)}$ are the current and voltage values of the diode at sample point N, which provide discrete values on the y-axis and x-axis, respectively.

The resistance of a diode at a particular operating point is called the dc or static resistance of the diode, which may seem independent of the shape of the characteristic. However, this is not true for an ac source, since the varying input moves the operating point up and down the characteristic curve and defines a specific change in the values of current and voltage, which will be illustrated as a moving dot point in the simulated graphs in the LabVIEW VI `Diodesim.vi`.

6.1.1 Virtual Instrument Panel

The developed front panel for the diode test is given in Fig. 6-2. The figure also describes the functions of each control, indicator, and graph. The principal features of the VI are that the user is able to control the supply voltage, the resistance, and the model used to simulate the diode. The diode can be modeled either using theoretical formulas or from values entered by the user into a look-up table. The look-up table contains some default values, but the results of an experiment have also been included on the drop-down menu. The experimental values are stored as a text file in the same directory as the simulation, formatted as a tab-delimited text file with two columns (voltage and current).

There are two graphs on the front panel. The first graph displays the diode characteristic, the operating point, and the load line. In addition, depending on the setting of a switch, the graph can display either the first or the third quadrant of the voltage-current characteristics in order to distinguish the forward and reverse characteristics of the diode. Two indicators are also included to display the present values of the diode voltage and diode current. The second graph exhibits the source voltage and the load current (resistor current or diode current).

The VI can also generate two error messages in two distinct operating conditions: when forward maximum diode current is exceeded and when maximum diode voltage in the reverse-biased mode is exceeded.

6.1.2 Self-Study Questions

1. Obtain a manufacturer's data sheet for a diode and test the simulation with the parameters given. Also include the ratings of the diode to test the error messages.

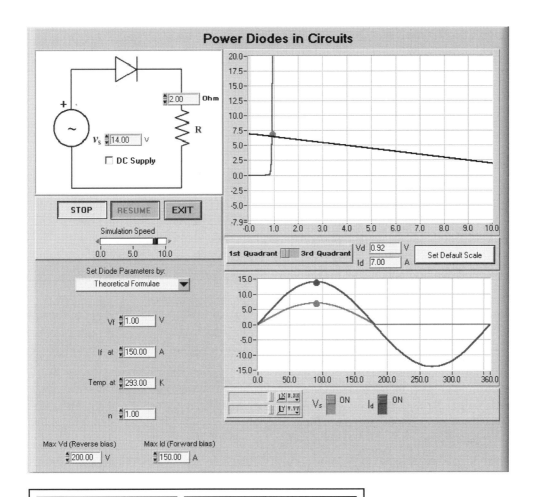

Figure 6-2
The front panel and brief user guide of Diodesim.vi.

2. Use equation 6.1 to create a look-up table to generate a diode characteristic that is close to the experimental data provided for the International Rectifiers' diode 1N5404 (3 A, 400 V).
3. As mentioned above, the static resistance of the diode changes when the diode current changes (as in the ac source), and it is fixed for a specific operating condition. The small signal resistance (or incremental resistance) of the diode can be given by $r_d = nV_T/i_d$. Select an operating condition and verify the small signal resistance calculated by this equation. Consider the operation at room temperature ($n = 1$).

6.2 SCR Conduction

A thyristor (SCR) has a p-n-p-n structure power electronic switch. Therefore, an equivalent circuit can be given by a two-transistor model (a pnp and an npn), which is interconnected to form a regenerative feedback pair (Fig. 6-3a). As can be seen in such an equivalent circuit, the collector of the npn transistor provides the base drive for the pnp transistor. Similarly, the collector of the pnp transistor and gate current supply the base drive for the npn transistor.

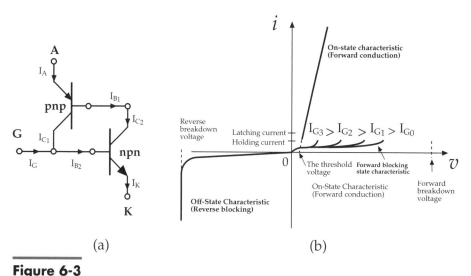

Figure 6-3
(a) Two-transistor model of SCR, and (b) SCR's voltage characteristic curves.

Thus, a regenerative situation exists when the positive feedback gain exceeds unity.

In order to trigger (fire or turn on), the SCR must be forward-biased (p emitter positive, n emitter negative). A positive control (gate) current I_G is applied to the p base of the npn transistor. This causes a collector current I_{C2} in the npn transistor. Then this enters the n base of the pnp transistor in the form of a negative control current I_{B1}. This causes a collector current I_{C1} in the pnp transistor, and, therefore, it increases the npn transistor's original control current. Thus the two transistors build each other up until continuous conduction is achieved. Therefore, when a thyristor is triggered, it conducts a load current I_A between the emitter electrodes, which are marked A and K.

In Fig. 6-3b, the characteristic curves of a thyristor have been drawn to show critical operating points in detail. Despite the simplicity of a diode's characteristic curve, it is more complex in a thyristor. As can be seen in the figure, there are three states in the characteristic curves: forward blocking state, reverse blocking state, and on-state. During these three states, there is a particular link between a thyristor's current and its voltage. The thyristor's off-state (reverse blocking) characteristic curve is practically identical to the diode's off-state characteristic curves. Positive voltages in a thyristor result in two characteristic curves—the on-state characteristic and the forward blocking state characteristic—depending on whether the thyristor is turned on or not, respectively.

If you examine the on-state characteristic curve closely, you can see that the extended curve (shown with dashed lines) does not end at zero; instead it intersects the voltage axis at a particular value, called the threshold voltage. This means that before a noticeable current can flow, voltage across the thyristor must increase to more than the threshold voltage.

Fig. 6-3b also shows how the gate currents affect the forward maximum voltage. It is clear that a certain gate current is necessary in order to ensure triggering of a thyristor. However, an additional condition must be added to triggering requirements: triggering pulse duration. The minimum on-state current is called the latching current I_L. In terms of triggering, if the triggering pulse duration is too short, then the latching current is not achieved, and the thyristor does not conduct.

Here are the requirements to turn on a thyristor:

- The thyristor must be forward-biased ($V_{AK} > 0$), and
- A control signal must be applied to the gate to the cathode ($V_{GK} > 0$), and
- The width of the control signal must be long enough to allow current flow greater than the latching current I_L ($i_T > I_L$).

Here are the requirements to turn off a thyristor:

- The current must be below the holding current I_H ($i_T < I_L$), and
- The voltage across the thyristor must be zero ($V_{AK} = 0$), or
- A reverse voltage must be applied for sufficient time to enable the recovery of the thyristor blocking state ($V_{AK} < 0$).

However, as can be seen in these explanations, the thyristor switch cannot be turned off easily with a dc supply, where the thyristor is always forward-biased, or within the positive half period of an ac supply. This is the main drawback of an SCR switch. Using an SCR in such applications needs an auxiliary SCR and extra arrangements to turn the main SCR off. In this study, however, the turn-off action is achieved by using a mechanical switch that can be controlled by the user. Furthermore, in practical applications, the maximum current rating of the thyristor should not be exceeded, and the maximum voltage of the thyristor must be about twice the rated operating supply voltage for safety. The voltage and current ratings of the thyristor are included in the simulation study.

Note that critical rate of rise of on-state current di/dt and critical rate of rise of off-state voltage dv/dt capabilities are also important for SCRs. However, these parameters are not studied in the VI provided here.

The principal aim of the VI is to demonstrate to students a typical use for the thyristor within a basic electric circuit, which is illustrated in Fig. 6-4. Note that the figure also illustrates the possible current conduction states of the circuit. The circuit has a resistive and inductive load, an SCR, and a mechanical switch. In addition, there is also a freewheeling diode within the circuit, which exists for protection purposes and to reduce the conduction losses of the SCR in inductive loads.

Let's perform a circuit analysis for this circuit. During the time that the gate signal is applied to the thyristor, or during the time that the thyristor is conducting, Kirchhoff's Voltage Law gives the following linear differential equation.

$$V_{dc} = Ri + L\frac{di}{dt} \qquad (6.6)$$

In equation 6.6, the state variable, the unknown, is the thyristor current. Note that the instantaneous value of this current very much depends on the initial value of the current at the previous instant, and the values of V_{dc}, R, and L.

However, when the mechanical switch is opened to stop conduction (when the freewheeling diode takes over the current conduction), or when the thy-

Chapter 6 • Introduction to Power Electronics Circuits

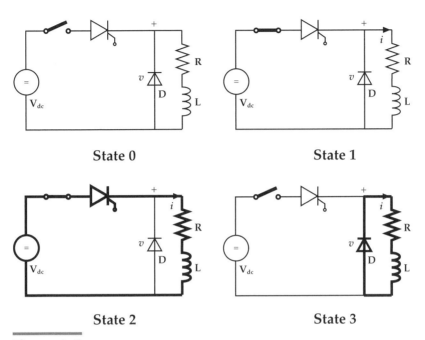

Figure 6-4
The SCR circuit and the possible circuit states.

ristor stops conducting because of its failure, applying Kirchhoff's Voltage Law provides a second linear differential equation.

$$0 = Ri + L\frac{di}{dt} \tag{6.7}$$

In the VI, differential equations 6.6 and 6.7 are solved numerically using the trapezoidal integration rule.

6.2.1 Virtual Instrument Panel

As a common feature in all LabVIEW programs, the front panel of the VI constructed here consists solely of a set of controls, to allow the user to select values for the circuit parameters, and indicators, to display the output parameters that result from running the VI. In addition, various other features are added to enhance user interactivity and understanding, such as animations.

As stated earlier, the principal use of the thyristor in electrical/electronic circuits is as a diode (i.e., it only allows current conduction in one direction) and as a turn-on switch (with no turn-off capability). The SCR circuit under discussion is simulated in the VI whose front panel is shown in Fig. 6-5.

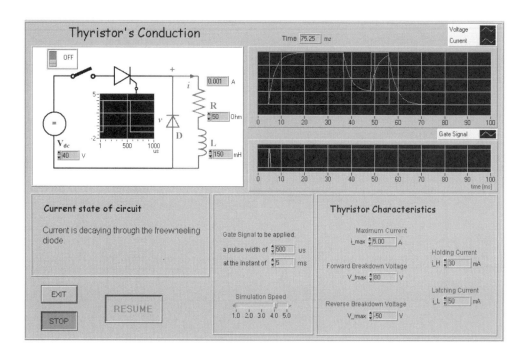

Figure 6-5
The front panel and brief user guide of Thyristor.vi.

First, to explain the features of the VI, let's revisit the turn-on conditions of the SCR. To turn on:

- The thyristor must be forward-biased, meaning the anode voltage is positive with respect to the cathode voltage.
- A positive control signal (voltage) is applied to the gate of the thyristor.
- The duration of the control signal must be long enough.

As indicated earlier, the control signal needs to be on long enough for the current to build up to a value greater than some threshold called the latching current i_L. Once the current through the thyristor is greater than i_L, it can continue to conduct indefinitely (even without the gate control signal).

Other important parameters of the thyristor studied in the VI include the following.

- Maximum current: This is the maximum current that the thyristor can carry safely.
- Holding current i_H: If the current through a conducting thyristor falls below this value, it will cease to conduct.
- Forward and reverse breakdown voltages: If the forward and reverse bias voltages are greater than these values (given in their data sheets), the thyristor will be damaged permanently.

In the VI's front panel, there are buttons for starting/stopping, pausing the simulation, and exiting the VI. The indicators are a set of graphs, of the voltage and current through the load and of the gate signal voltage. Simulation speed can be varied using the control available on the front panel. In addition, there are two sets of controls in the VI: one set of controls for the thyristor parameters and one set for the circuit parameters.

Generation of a gate trigger pulse and its phase shift for a thyristor can be achieved in many ways, such as by using the ac supply voltage as a reference. In the VI presented here, the gate pulse is generated in reference to the time step of the integration. The gate pulse(s) during the simulation occur at preset time(s) specified by the user, which can be updated during operation (after the mechanical switch is turned off and turned on consecutively).

The animation ring given in the front panel illustrates the different states that the circuit could be in, depending on the states of the mechanical switch and the thyristor and the status of the freewheeling diode. The current conduction state in the circuit is shown in red in the front panel; all possible states of the circuit are shown by heavier lines in Fig. 6-4.

State 0: Initial state, mechanical switch is off, no circulating current in the freewheeling diode

State 1: Mechanical switch is on but no current flowing in the thyristor

State 2: Mechanical switch and the thyristor are on

State 3: While the SCR was conducting, the mechanical switch was opened, hence the load current circulates (decays) through the freewheeling diode

Along with the circuit states, the area that is just under the circuit diagrams displays a textual description of the states while the integration is performed. The animation is achieved by using a picture/text ring. The visible item of this object is a single picture/text at any one time, out of many pictures/texts. Each item of the ring is associated with an (unsigned) integer value. By making the ring an indicator, wiring it to a different integer value displays a different item that corresponds to a state of the circuit.

Note that the execution of the VI and the simulation of the thyristor are logically not the same thing, due to the presence of the user-interactive buttons. Therefore, the outermost loop (which is exited by the Exit button) is made the prominent programming structure. It is the structure that keeps track of the instantaneous value of the simulated time, the instantaneous value of the circuit current, and a flag indicating whether the thyristor is burned or not. The pseudo-code for the outermost loop is shown in Table 6-1.

The pseudo-code for a one-step simulation is given in Table 6-2. Note that only the major details are shown. This sub-VI runs once for each instant of

Table 6-1 *A pseudo-code structure of the outer loop.*

```
initialize the loop variables; in particular, set the simulated time
t to zero
while (not exit) loop
   if (start)
      reset the display if t = 0
      check for the error "holding current >= latching current"
      call sub-VI "one_step_sim" // calculates the thyristor
      behavior for one Δt
      increment t
   else // stop
      reinitialize the time to zero
   end if
   if (pause)
      wait until resume
   end if
end loop
```

Table 6-2 *Pseudo-code for one-step simulation of SCR conduction.*

```
check that the thyristor is not burned
check that the current < the allowed maximum
check that the voltage < the allowed maximum // note: voltage >= 0 for this
circuit
if (the checks do not all hold) update the state and the current // decaying
current
else
   if (the switch is on)
      if (the thyristor was conducting prior to this)
         if (the current >= holding current) // there is still conduction
            update the current
         else // no more conduction
            update the state and the current
         end if
      else // the thyristor was not conducting in the previous time step
         if (the gate pulse is present)
            if (the current >= the latching current) // there is conduction
            now
               update the state and the current
            else // there is still no conduction
               if (the pulse is too short)
                  display a message
               end if
               check the resolution of the gate pulse
               update the current // it continues to rise
            end if
         else // the gate pulse is not present
            if (the current was decaying prior to this) // for whatever
            reason
               update the current and maybe the state // it is still decaying
            else // the current was not decaying prior to this
               if (the current >= the latching current) // there is
               conduction now
                  update the state and the current
               else // there is no conduction
                  update the current and the state // current will decay now
               end if
            end if
         end if
      end if
   else // the switch is off
      if (the thyristor was conducting prior to this)
         update the state and the current // the current is decaying now
      else // the thyristor was not conducting prior to this
         update the current and maybe the state // the current continues to
         decay
      end if
   end if
```

simulated time, and outputs the circuit states or errors. Note that these pseudo-codes in Tables 6-1 and 6-2 are not necessarily performed in the sequential manner presented, due to the object-oriented and event-driven nature of the LabVIEW codes. However, sequential execution will occur along wires (originating at controls and ending at indicators), and between nested structures, with the earlier objects getting their updated values before the later objects (like a flow diagram).

The variables used to store the state are directly linked to the circuit diagram and message currently being shown to the user. This means that the underlying code is reliant on the order that the pictures/text messages were added to their respective rings.

6.2.2 Study Guides

Coding for an audience means that many small details have to be attended to, to contribute to a user-friendly interface. Although the front panel presented here considers the basic parameters only, the number of options are many. Therefore, I suggest that students create various scenarios and observe the outcome of sequences of events. A basic question can be a good starting point.

1. A load of 20 Ω and 0.5 H is connected to a dc power supply of 100 V via a thyristor. The thyristor has a latching current of 50 mA. If a triggering pulse of length 50 μs is applied to the gate of the thyristor, will the thyristor remain on when the firing pulse ends? The thyristor has a 2 V on-state voltage-drop.

The simulation result should indicate that the thyristor current has failed to reach its latching current level within the given length of triggering pulse. It will not remain ON. Therefore, either the triggering pulse width must be increased or parameters of the load must be changed.

Note that the holding current is necessarily always less than the latching current. Hence this condition has to be considered in the control settings. Furthermore, this VI can also be utilized to study the dynamic behavior of an R-L load switching with a dc supply.

6.3 Three-Phase Half-Way Diode Rectifier

Rectification is the process of power flow from an ac source to a dc load. The converter topologies utilized in rectification are called rectifiers, which are connected to an ac source (single phase or multiple phase). Some types of rectifiers can be controlled so that power flow occurs from the dc terminals of the load back into the ac line (as in an active load, such as a dc machine load while it is generating), but these will not be discussed here.

In practice, three-phase rectifier circuits are used because they have less ripple than the single-phase rectifiers. Although the number of different rectifier circuits is very large, we will only investigate and simulate the three-phase half-way diode rectifier as shown in Fig. 6-6.

Commutation is the transfer of current from one device to another. As studied earlier, the diodes are naturally commutated devices since their conduction is defined by the positive voltage across the anode and cathode terminals. In rectifiers, the rectification process also occurs naturally (except in pulse width modulated rectifiers).

Note that the dc load is characterized by a resistor, an inductor, and a constant voltage source, which represents a broad-spectrum load. Most practical dc loads can be generated using this arrangement; for example, a battery load can be represented by R and E only.

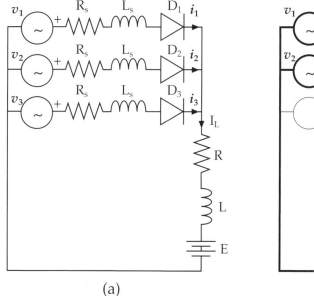 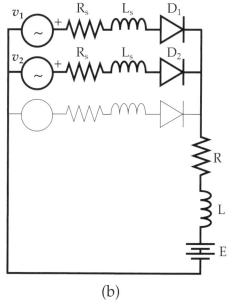

(a)   (b)

Figure 6-6
(a) Circuit diagram of the three-phase half-way diode rectifier circuit, and (b) a circuit status during commutation.

6.3.1 Fundamental Theory

There are two modes of current conduction in the rectifiers: the continuous current conduction and the discontinuous current conduction. In the discontinuous conduction mode, the outgoing diode current goes to zero before another phase diode starts conducting.

Let's now consider the rectifier topology given in Fig. 6-6a. Under continuous conduction mode, which is normally the case in multiple-phase rectifiers, due to presence of supply inductance and resistance (since transformers are usually used in the input of the rectifiers), some time is required for a current commutation to take place.

Neglecting the effect of the resistance associated with the ac supply and assuming that the diodes are ideal, circuit conditions within the commutation interval can be described by two voltage equations.

$$v_1 = V_m \cos(\omega t + \pi/3) \tag{6.8}$$

$$v_2 = V_m \cos(\omega t - \pi/3) \tag{6.9}$$

The commutation instant described here is illustrated in Fig. 6-6b. If we use the Kirchhoff's Voltage Law for the loop where the active diodes are,

$$v_1 - v_2 = L_s \frac{di_1}{dt} - L_s \frac{di_2}{dt} \tag{6.10}$$

In addition, let's assume that the load is inductive enough to give a sensible level of constant load current I_L.

$$i_1 + i_2 = I_L \tag{6.11}$$

If we use these three equations and rearrange them, the following differential equation can be obtained.

$$2L_s \frac{di_1}{dt} = -\sqrt{3} V_m \sin \omega t \tag{6.12}$$

Hence, the solution of differential equation 6.12 can provide the instantaneous current during the commutation interval.

At the end of the commutation interval, the current of the outgoing diode reaches zero and the commutation period ends. The commutation angle γ (also known as overlap angle or overlap period) is given by

$$\cos \gamma = 1 - \frac{2\omega L_s}{\sqrt{3} V_m} I_L \tag{6.13}$$

Furthermore, the variation of the output voltage within the commutation angle is

$$v_o = \frac{1}{2} V_m \cos \omega t \tag{6.14}$$

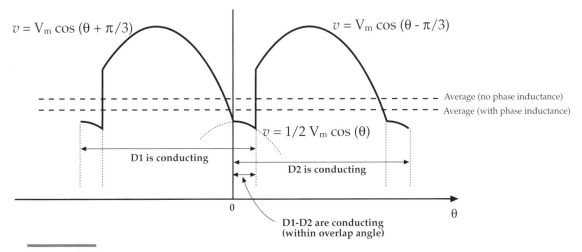

Figure 6-7
Output waveform during continuous current conduction mode.

As you examine this equation, you will see that the output voltage during commutation varies as a cosine function, and its peak value is equal to half the ac supply voltage.

The principal conclusion of the preceding analysis is that if there is a supply inductance, the average output voltage is reduced. Fig. 6-7 illustrates the output voltage of the rectifier in the presence of supply inductance.

In addition to the rms and the mean values of the output voltage, the efficiency and the ripple factor (RF) of the rectifier can be estimated, which are used to quantify the quality of the rectifier circuits.

$$\eta = \frac{V_{dc}I_{dc}}{V_{rms}I_{rms}} \tag{6.15}$$

$$RF = \frac{V_{ac}}{V_{dc}} = \frac{\sqrt{V_{rms}^2 - V_{dc}^2}}{V_{dc}} \tag{6.16}$$

Development of Current Conduction States for the VI

If Fig. 6-6a is considered, the circuit can be in six possible circuit states, which will be explained next.

State 1: When only one diode is on (say, D_3), the differential equation is given as

$$\frac{di}{dt} = \frac{R + R_s}{L + L_s}i + \frac{V - E - V_d}{L + L_s} \tag{6.17}$$

where V is the supply voltage and V_d is the voltage drop across each diode, which is assumed constant.

State 2: The D_1 in the circuit turns on when it is forward-biased. As the characteristics of each phase of the supply (R_s and L_s) and the voltage drop across each diode are identical, the change in the state (to both D_1 and D_3 on) occurs when $V_1 = V_3$. The equivalent circuit when analyzed gives two differential equations used in the simulation of the state when two diodes are on.

$$\frac{di_1}{dt} = \frac{V_1(L_s + L) - LV_3 + (R_sL - L_s(R + R_s))i_1 - L_s(E - V_d) + (R_sL + RL_s)i_3}{2LL_s + L_s^2}$$

(6.18)

$$\frac{di_3}{dt} = \frac{V_3(L_s + L) - LV_1 + (R_sL - L_s(R + R_s))i_3 - L_s(E - V_d) + (R_sL + RL_s)i_1}{2LL_s + L_s^2}$$

(6.19)

State 3: The next change in state is obtained when the current going through D_3 has been observed to decrease to zero. In this state, D_3 turns off, and only D_1 stays on.

State 4: D_1 and D_2 are both on.

State 5: Only D_2 is on.

State 6: D_2 and D_3 are both on, then return to State 1.

Any state of the circuit terminates when the voltage(s) of the diode(s) that are on in the current state equals the voltage of the next state. Note that when $L_s = 0$, the circuit will be in discontinuous conduction mode and only states 1, 3, and 5 will be present, where the ongoing diode current goes to zero before another phase diode starts conducting, and where only one diode is on at any one time.

6.3.2 Virtual Instrument Panel

The VI in this section is named `ThreePhaseHalfWayDiodeRectifier.llb`, and its front panel is shown in Fig. 6-8. The VI has a number of features. These include the ability to input values for the parameters of the circuit, to observe a number of waveforms, to animate the current flow through the circuit, and to calculate various quantities. It also offers a number of simulation controls.

Chapter 6 • Introduction to Power Electronics Circuits

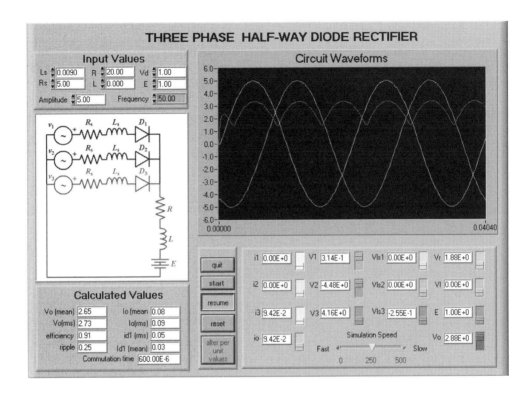

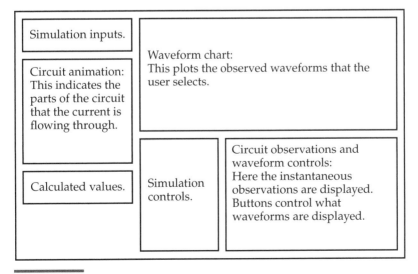

Figure 6-8
The front panel and brief user guide of `ThreePhaseHalfWayDiodeRectifier.vi`.

Simulation Controls

When the program is first started, a pop-up screen asks the user if he or she wishes to use the default values for per-unit values or use per-unit values from a presaved source.

The simulation is controlled by the six control buttons available on the front panel: Quit, Start, Pause, Reset, Alter Per Unit Value, and Simulation Speed. The Reset button can stop the simulation at the end of the current period that it is performing.

The Alter Per Unit Value button allows the user to change the per-unit value of any of the observations that can be plotted on the chart. Once pressed, a pop-up window lists all the observations' per-unit values that can be changed on a scroll-down menu. The current per-unit value will be displayed when each item is selected. When the Done button is pressed after the new settings have been made, a second pop-up screen is activated, which allows the user to save the per-unit values for later use.

The sliding bar on the front panel can control the simulation speed.

Circuit Settings

The VI allows the user to set a number of parameters in the circuit so that the implications of the values can be observed. Here is the list of these parameters that the user can set.

L_s: Phase (supply) inductance
R_s: Phase resistance
R: Load resistance
L: Load Inductance
E: Load side voltage source
V_d: Constant voltage drop that occurs across the diode when it is on. Note that an ideal diode has $V_d = 0$.
Amplitude and frequency of the supply

Circuit Animation

The current flow through the rectifier circuit is shown at every instance through the animation ring. The animation will show all black as no current is flowing through the circuit, and red when there is current in the circuit.

Waveform Chart

The waveform chart plots all values estimated during the simulation. Any grouping of waveform observations can be made. If the waveform cannot be observed on the chart (i.e., too small or too large), the per-unit value can be altered by selecting the control as discussed earlier. Note that the waveform can only be altered by the per-unit value after the point in time where the change occurred and not before.

For the rectifier circuit, the following observations can be plotted on the chart.

V_1, V_2, and V_3: Supply voltages of Phase 1, 2, and 3, respectively

V_{ls1}, V_{ls2}, V_{ls3}: Voltage drops across the supply inductances of Phase 1, 2, and 3, respectively

I_1, I_2, I_3: Currents through the diodes

I_o: Output current

V_r: Voltage drop across resistive load component

V_l: Voltage drop across inductive load component

E: Constant voltage source

V_o: Output voltage (the voltage across the load, $R + L + E$)

These waveforms can be displayed on the chart by selecting the corresponding switches. The per-unit value of a waveform is also displayed using the indicators that are located below each switch.

When the circuit observation has been selected to be observed on the chart, the per-unit value is revealed, if it is not equal to unity 1.0.

Calculated Values

The user can observe a number of calculated values. The calculated values are provided at the end of each period, excluding commutation time. Since there are three commutation instants within a period of the supply voltage, the commutation time is calculated and displayed up to three times each period.

6.3.3 Implementation Details

Per-Unit Values and Chart

The calculated values of the waveforms are placed into an array during each iteration of the period. Each element of the array is then divided by the corresponding per-unit value. After each element has been put into the per-unit form, the array is made into a bundle and acts as the input to the chart.

The chart can plot all waveforms, but they are not made visible until the user selects the switch corresponding to each waveform. The terminals of these switches are all placed on an array to allow for programming simplification. This array is then used in a for loop that determines both what waveforms are displayed and what per-unit digital displays are used.

In each iteration of the for loop, it is determined whether the corresponding switch is active. A case loop is then used to display the plot or not, which used a two-component attribute node (active plot and plot color). When the switch has not been activated, the corresponding plot color is set to transparent, else the plot color is set to the predetermined color.

As stated earlier, the animation of the current flow through the circuit is achieved through the use of a picture ring. The picture ring contains all possible states that circuits displaying current flow through the circuit. The illustration that is shown is determined by the state number that is entered into the picture ring terminal. The state number is overridden in the case where the output current is equal to zero. In this case, the default picture, where no current is flowing, is shown. This same state is shown when Pause is pressed or the simulation is stopped or reset.

Commutation Time

The number of states that the circuit goes through in one period is determined by L_s. If L_s is equal to zero, then there are only three states, otherwise there are six states in the simulation.

Commutation time is the duration of time that the two diodes are on. When $L_s = 0$, two diodes are never on concurrently, hence, within the case frame a constant (with the value 0.0) is connected to the commutation time indicator.

The more difficult case to program is the commutation time when L_s is not equal to zero. As stated before, the commutation time is updated three times every period. In each state calculation case, a constant is set. In the case of two diodes on, the constant is 0, which represents the start of the commutation time. When only one diode is on, the constant is 1, which represents the end of the commutation time.

The commutation time is likely to last more than one iteration. To obtain the correct input values for the calculation, a case statement is used, which is dependent on what the last state was and the current state. When the states are equal, the system has not experienced a change in state. Hence the value of when the commutation time started and the commutation time are just relayed on. When these states are not equal, the system is experiencing a change in state during that iteration. It is this time that is either the start or finish of the commutation time.

Therefore, when the case is false, the system has alternative case statements dependent on the constants discussed previously. When the constant is a 0, the iteration number is placed on a shift register (or in memory) to be used at a later time. When the constant is 1, the calculation of the iteration number is carried out. The current iteration number and the iteration number when the commutation period started are subtracted to give the number of iterations that took place during the commutation period. By multiplying this by the time between each sample the commutation time is obtained.

Three-Phase Supply

The three-phase sub-VI provides three outputs corresponding to the voltage of each phase. This VI primarily uses the Sine Pattern VI. Three Sine Pattern VIs are used with the start position set to 240°, 120°, and 0°. This allows the three different phases to be simulated.

As given on the front panel, the amplitude, the number of periods, and the sampling rate are all inputs to the three-phase VI. These allow the sampling rate combined with the number of periods to determine the number of points to be calculated. These calculated supply voltage values are then placed into an array, and an index array is used to select the appropriate calculation value dependent on the iteration number.

Two Equation Runge-Kutta

The Two Equation Runge-Kutta sub-VI is a manipulated version of the Runge-Kutta VI that the G-Math library provides. The sub-VI has the input of array of the formulas to be solved and also requires the initial currents observed and the period of time between calculations. The outputs are the two currents observed.

6.4 Single-Phase AC Chopper

With appropriate static switches (SCR and Gate Turn-Off [GTO] thyristors) and switching arrangements, ac power can be controlled directly. Fixed ac voltage can be converted to variable voltage of the same frequency by ac choppers, or to variable ac voltage and frequency by cycloconverters, which will be studied in Section 6.5.

In practical systems, the control of ac power may be occasional, as in soft starting to limit the starting current of a motor, or continuous, as in a lamp dimmer and a speed control of a motor (such as an induction motor or a universal motor), or semicontinuous as in temperature control by integral cycles control of an ac supply.

The ac power control normally requires two-way conduction. An obvious way to achieve single-phase two-way conduction is the triac or the anti-parallel connection of thyristors.

Fig. 6-9 shows the single-phase ac chopper including possible circuit components in a practical system. Note that, as in the previous section, the dc load is modeled by a resistor, an inductor, and a constant voltage source, which represents a broad-spectrum load. In addition, the supply inductance and the resistance are also considered in the circuit.

For three-phase applications, the single-phase switching circuit may be placed in one, two, or all three of the ac supply lines, which can make three-phase ac power control possible.

Fig. 6-10 illustrates two typical load voltage and current waveforms for a pure resistance and an inductive ($R + L$) load, respectively. Referring to Fig. 6-9, the thyristor Th1 conducts when its voltage is positive and has received a triggering signal. Likewise the reverse thyristor (Th2) will turn on when the voltage is positive (during the negative half cycle of the ac supply). If the load is a pure resistance only and neglecting the small thyristor voltage

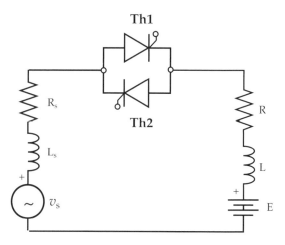

Figure 6-9
Single-phase ac chopper circuit.

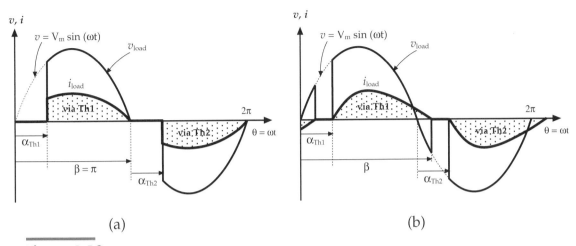

Figure 6-10
Typical voltage and current waveforms across the load in Fig. 6-9: (a) Pure resistance, no E, and (b) Inductive load, R + L, no E.

drop, the load current flows via Th1 and Th2 until the end of each positive and negative half cycle, respectively (Fig. 6-10a).

In the case of an inductive load (R + L), however, the current continues beyond the zero crossings of the supply voltage (Fig. 6-10b). You can observe that the continuation of current conduction beyond the zero crossings of the supply voltage is related to the value of inductance, which restricts the triggering angle control beyond a certain angle (β). This angle is known as the turn-off angle and is a function of the phase angle of the resistive/inductive load. Furthermore, due to the voltage source E in the circuit, a dc shift occurs on the load side, which also limits the range of triggering angle.

Let's consider the load in Fig. 6-9, which is connected in series with an ac supply. When one of the thyristors is turned on at the instant α (the triggering angle), the voltage equation can be given by

$$V_m \sin(\omega t) = Ri(t) + L\frac{di}{dt} + E \tag{6.20}$$

During the on period of the thyristors, the load current can be determined by equation 6.20. During the positive cycle of the supply voltage, the thyristor Th1 is forward-biased; therefore, it will conduct when the triggering signal is applied to the thyristor, and the ac supply voltage will be observed across the load. Note that the triggering angle (or firing delay angle) is expressed relative to the zero crossings of the ac supply.

In this study, differential equation 6.20 is solved numerically using LabVIEW. However, it is possible to solve the equation analytically for one half period as

$$i(t) = \frac{V_m}{\sqrt{R_t^2 + (\omega L_t)^2}}[\sin(\omega t - \phi) - \sin(\alpha - \phi)e^{(R_t/L_t)[(\alpha/\omega) - t]}] \quad (6.21)$$

where R_t and L_t represent the total resistance and inductance in the circuit when either thyristor is on, α is the triggering angle, and ϕ is the impedance angle ($= \arctan(\omega L/R)$).

As shown in Fig. 6-10, each thyristor remains on a time period that can be defined by the triggering angle and the turn-off angle. Hence, in general, the current conduction angle is given by

$$\gamma = (\beta - \alpha) \quad (6.22)$$

For a known function of load current, the rms values of the current for the thyristor can be estimated and used to size the component.

6.4.1 Virtual Instrument Panel

Similar to the previous VIs, a number of input values are available in this VI, where various calculations are performed and relevant waveforms are plotted. The front panel of the VI `Single Phase AC Chopper.vi` is given in Fig. 6-11. When the VI is first started, a pop-up screen asks the user if he or she wishes to use the default values for per-unit values or use per-unit values from a presaved source.

The VI is controlled by the standard control buttons described previously: Quit, Start, Pause, Reset, Alter Per Unit Value, and Simulation Speed. Moreover, the current flow through the ac chopper is shown at every instance through the circuit animation.

Simulation Inputs

Here is the list showing the parameters that the user sets as input values.

L_s: Supply inductance
R_s: Supply resistance
R: Resistance of the load
L: Inductance of the load

Chapter 6 • Introduction to Power Electronics Circuits

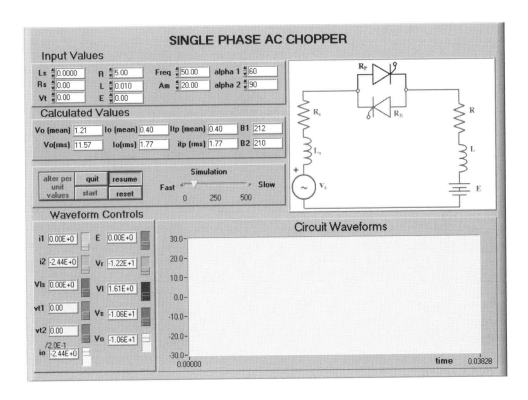

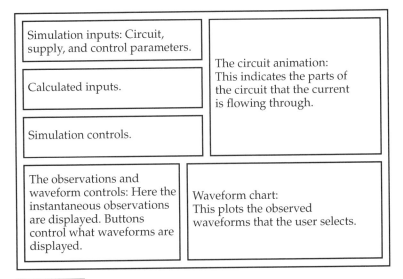

Figure 6-11
The front panel and brief user guide of `Single Phase AC Chopper.vi`.

E: Load-side voltage source
V_t: Thyristor voltage drop. Note that an ideal thyristor has $V_t = 0$
Amplitude of the supply
Frequency of the supply
Alpha 1: Triggering angle of the forward (or positive) thyristor, defined reference to the zero crossing point at 0 rad
Alpha 2: Triggering angle of the reverse (or negative) thyristor, defined reference to the zero crossing point at π rad

Waveform Chart

The waveform chart on the front panel plots all values observed during the simulation. Similar to the previous VI in Section 6-3, the waveform can only be altered by the per-unit value after the point in time where the change occurred and not before. In the VI the following waveforms can be observed on the chart.

V_s: Supply voltage
V_{t1}: Voltage across the thyristor Th1
V_{t2}: Voltage across the thyristor Th2
V_{ls}: Voltage across the supply inductance
I_1: Current through Th1
I_2: Current through Th2
I_o: Output (load) current
V_r: Voltage drop across the load resistance
V_l: Voltage drop across the load inductance
E: Voltage of the load-side source
V_o: Output voltage (across R, L, and E)

Calculated Values

There are a number of calculated values indicated on the front panel, including the rms and mean values for the output current and voltages and for the thyristors, and the turn-off angle. These values are provided at the end of each period. The commutation time can be calculated and displayed up to three times each period, and the beta values are provided when the appropriate current reaches zero.

Error Messages

As stated earlier, the control intervals for the triggering angles are limited under the inductive load operation. Therefore, it is very likely that the user may set triggering angles that are impossible to obtain under such conditions. When this occurs, an error message is displayed and the program is terminated.

6.4.2 Features of the VI

Although the principal structure of this VI is similar to the VI described in Section 6-3, there are a number of sub-VIs specifically designed to perform calculations in `Single Phase AC Chopper.vi`. Brief explanations of these sub-VIs are given in the following paragraphs.

Single Phase Chopper Calc Sub-VI

This sub-VI takes in most of the input values of the simulation, then determines what state the simulation is in (using the `Chopper Two State Determination.vi` explained next) and calculates the corresponding current output values. In the case when $L_s \neq 0$, the relevant differential equation solver VIs determine the current. When in state 0 only the forward thyristor is on, hence the current through the reverse thyristor is zero. Likewise when in state 1 only the reverse thyristor can be on and the forward thyristor current is zero. In the case of both $L_s = 0$ and $L = 0$, formula nodes are used to calculate the current.

Chopper Two State Determination Sub-VI

This sub-VI has three states. The third state is for the initialization of the simulation. The inputs to this sub-VI are the initial state, the supply voltage, and the initial alpha code, which outputs a state number, an alpha code, and an

Table 6-3 *Case statements for* `Chopper Two State Determination.vi`.

	Initial state = 0	
	Is v_s less than zero?	
	True	False
	Does initial alpha code = 0?	
T	F	
State = 1	state = 0	state = 0
error = (initial $i_1 \neq 0$)	error = False	error = False
alpha code = alpha 1	alpha code = initial alpha −1	alpha code = alpha 2
	Initial state = 1	
	Is v_s greater than or equal to zero?	
	True	False
	Does initial alpha code = 0?	
T	F	
state = 0	state = 1	state = 1
error = (initial $i_2 \neq 0$)	error = False	error = False
alpha code = alpha 2	alpha code = initial alpha −1	alpha code = alpha 1
	Initial state = 2	
	Is v_s greater than or equal to zero?	
	True	False
	Does initial alpha code = 0?	
T	F	
state = 0	state = 2	state = 2
error = (initial $i_2 \neq 0$)	error = False	error = False
alpha code = alpha 2	alpha code = initial alpha −1	alpha code = alpha 1

error raiser. The error raiser is true when the conditions for the thyristor to turn on are not correct when it is requested to be turned on. Table 6-3 shows the conditions and the outputs that the sub-VI is programmed on.

Voltage Across Thyristors Sub-VI

This VI calculates the voltage drop across the thyristors when they are both off, resulting in an open circuit. The VI involves the inputs of the current through the thyristors and the supply voltage. Table 6-4 shows the case statements for the implementation of this sub-VI.

Table 6-4 *Case statements for* `Voltage Across Thyristors.vi`.

Case when $v_s \geq 0$ Does $i_1 = i_2 = 0$?		Case when $v_s < 0$ Does $i_1 = i_2 = 0$?	
True	False	True	False
$V_{th1} = v_s$	$V_{th1} = 0$	$V_{th1} = 0$	$V_{th1} = 0$
$V_{th2} = 0$	$V_{th2} = 0$	$V_{th2} = v_s$	$V_{th2} = 0$

Beta Calculation Sub-VI

As illustrated in Fig. 6-10, β is the turn-off angle where the thyristor current becomes zero. In the VI, since the alpha values for the positive and reverse thyristors can be set to different values, the values of betas can be different. Beta 1 is the angle from when the supply voltage becomes positive to the time where the forward thyristor turns off. Beta 2 is the angle from when the supply voltage becomes negative to the time where the reverse thyristor turns off.

This is implemented by determining when the current through the corresponding resistor turns to zero in that state. There are 360 iterations for one period. The iteration number when the current through the resistor becomes zero is used to calculate the beta angle. For Beta 1 this is simply the iteration number. Beta 2 is equal to the iteration number less 180 when the thyristor turns off before the supply voltage becomes positive. When the thyristor turns off after the supply voltage becomes positive, Beta 2 equals the iteration number plus 180. Since the state does not change until the thyristor is turned on, it is not possible for the beta to be calculated early due to the thyristor not yet having been turned on with this implementation.

6.4.3 Self-Study Questions

As a learning exercise, to better understand the circuit, try to determine what has occurred to cause this error and how changing the input values so that the problem is rectified.

1. Set $\alpha_1 = \alpha_2 = 60°$, $f = 50$ Hz, Amplitude = 20 V, $R = 5\ \Omega$, $L = 0.01$ H, $R_s = 0\ \Omega$, $L_s = 0$ H, observe the turn-off angles β_1 and β_2, and verify the results analytically. *Note:* Use equation 6.21 and a trial-and-error method to determine the turn-off angles. Moreover, note that the simulation should run multiple periods to obtain correct values (where the calculated values do not change).

2. Use the same settings as in question 1, except $R = 1\,\Omega$, $\alpha_1 = \alpha_2 = 90°$, observe the values of β_1 and β_2, and comment. Determine the mean (average) and rms values of the current in the load and in a thyristor.
3. Use the same settings as in question 1, except $R = 0\,\Omega$, $L = 0.01$ H, $\alpha_1 = \alpha_2 = 90°$, observe the values of β_1 and β_2, and comment. *Note:* Observe that the picture frame has no zero state, and this state of operation is identical to an ac circuit having a pure inductance as a load, hence the current waveform lags the voltage 90°.
4. For the circuit given in Fig. 6-9, if the load is pure inductance ($R = 0\,\Omega$, $L = 5$ mH), plot voltage and current waveforms for $\alpha = 100°$ (symmetric control, $\alpha_{Th1} = \alpha_{Th2}$). The supply voltage is $v_s = 240\sin(\omega t)$, $f = 50$ Hz. What are your observations?

6.5 Cycloconverters

Cycloconverters are capable of directly, that is, without an intermediate dc link (as in the inverters), converting single- or multiple-phase ac power of a fixed frequency to single- or multiple-phase power of a variable frequency. Such converters may be used to link two ac power systems of different frequencies, to drive variable speed ac motors (as a direct frequency converter), or to convert the output of variable speed ac generators (such as generators driven by wind) to constant frequency.

Cycloconverters fabricate the output voltage of desired amplitude and frequency by sequentially applying appropriate segments of the input voltage waves to the output. Therefore, the output voltage waveforms are composed of segments of the input voltage waveforms. The transformation is accomplished by arrays of ac switches (triac or anti-parallel thyristors). To reduce distortion of the output voltage and input current waveforms, cycloconverters may be arranged in many different forms. A three-phase cycloconverter configuration used in this section is shown in Fig. 6-12.

The operating principle of the single-phase cycloconverter circuit is illustrated for a pure resistive load in Fig. 6-13. The output voltage is constructed by changing the triggering angle in successive pulses so as to obtain the desired waveform (highlighted). Note that the frequency of the output voltage is equal to 3/7 times the supply frequency.

The output frequency and the voltage can be made variable by changing the triggering angles of the thyristors. It should be noted here that there are

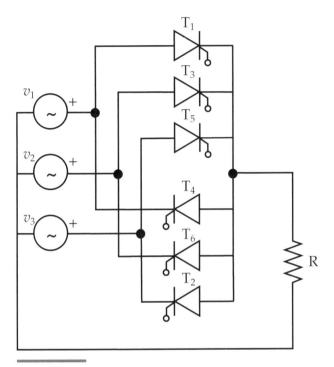

Figure 6-12
Single-phase output cycloconverter with a resistance load.

an unlimited number of output frequencies and voltages for a given three-phase supply at a fixed frequency.

In addition, examination of Figs. 6-12 and 6-13 reveals that if at any instant thyristors in both the positive and negative groups of opposite phases are conducting, then a short-circuit exists on the supply by means of the power switches. To avoid this in practice, a reactor can be inserted between the groups to limit the circulating current, or extra care should be taken during the generation of the triggering pulses.

6.5.1 Virtual Instrument Panel

Due to the unlimited number of variations of input frequency and voltage and output frequency and voltage that users may wish to observe, there are an infinite number of cases to program for a fully integrated cycloconverter simulation. Therefore, the VI provided here has limited features, programmed to

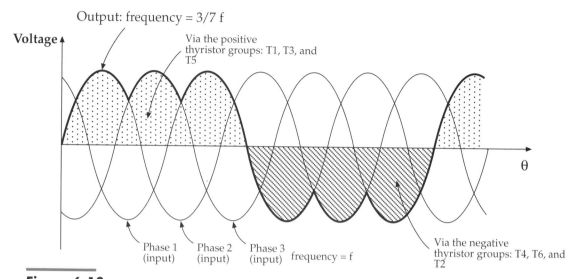

Figure 6-13
Waveforms illustrating the generation of single-phase output with variable frequency by the cycloconverter circuit given in Fig. 6-12.

generate an output voltage at 30 Hz for a three-phase input voltage at 50 Hz. The front panel of the VI Cycloconverter.vi is shown in Fig. 6-14.

6.5.2 Features of the VI

The VI begins with a pop-up message that remains visible until the user reads it and presses the Continue button. After the Continue button has been pressed, another pop-up screen asks the user if he or she wishes to use the default values for per-unit values or use per-unit values from a presaved source.

The VI is controlled by buttons similar to the previous VIs: Quit, Start, Pause, Reset, Alter Per Unit Value, and Simulation Speed. Note that this VI is programmed to display only one output period. Therefore, the Reset button can stop the simulation at the end of the current period that it is performing. The per-unit values remain the same and all calculated values remain showing until both the Reset button has been released and the Start button reactivated. If the Reset button is not pressed at the end of the output period, the simulation program stops.

The simulation allows the user to set a number of front panel parameters so that their implications can be observed. Here is the list of these parameters that the user sets as input values.

Chapter 6 • Introduction to Power Electronics Circuits

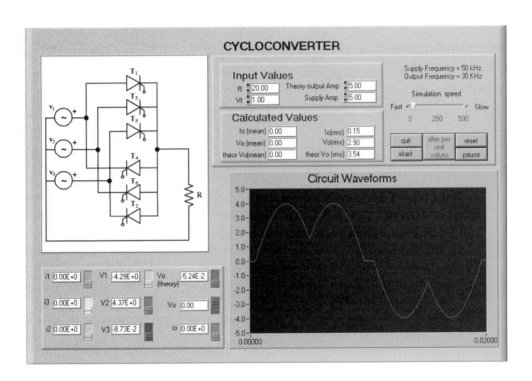

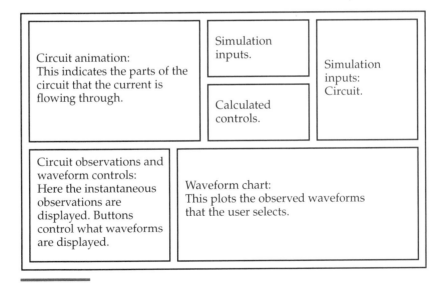

Figure 6-14
The front panel and brief user guide of `Cycloconverter.vi`.

R: Resistance of the load

V_t: The constant voltage drop that occurs across the thyristor when it is on. Note that an ideal thyristor has $V_t = 0$

Supply amplitude: The amplitude of the supply voltage

Output amplitude : The amplitude of the theoretical output voltage

Similar to the previous front panel charts, the chart in this VI can display all values observed during the simulation, which can be selected using waveform controls. If the waveform cannot be observed on the chart (i.e., too small or too large), the per-unit value can be altered by selecting the control as discussed earlier. Note that the waveform will only be altered by the per-unit value after the point in time where the change occurred and not before.

There are a number of calculated values that the user may wish to observe. These include the rms and mean values for the output current and voltages. The rms and mean values for the theoretical output wave are also provided so that the user can compare between theoretical and obtained values. The rms and mean values are provided at the end of each period.

6.6 PWM and Single-Phase Inverter (H-Bridge) Control Methods

Let's consider the simple switching circuit given in Fig. 6-15a, where the mechanical switch can be replaced with various types of switching transistors in power electronic circuits, which can be operated at extremely high frequencies. Essentially, we can identify three methods for controlling the switch.

- By varying t_{on} and keeping the switching period T constant, which is known as a Pulse Width Modulation, PWM (Fig. 6-15b)
- By varying T and keeping t_{on} constant (Frequency Modulation, FM)
- By a combination of the above methods

The choice of switching method in power electronics circuits very much depends on the application. One important consideration is the reduction of the switching losses to improve the efficiency of the power conversion.

In this section, we will consider the PWM method, which can be used for both voltage and current control purposes in practical systems. For example,

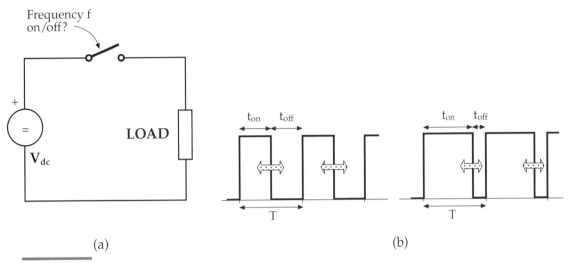

Figure 6-15
(a) Simple dc switching circuit, and (b) PWM switching method.

let's take an ideal transistor (which means that it can operate at infinitely high frequency and has no switching and conduction losses). If we apply a PWM control signal to this transistor, as shown in Fig. 6-15b, the mean voltage across the load can be given by

$$V_{\text{mean(load)}} = V_{\text{dc}} \frac{t_{\text{on}}}{t_{\text{on}} + t_{\text{off}}} = V_{\text{dc}} \frac{t_{\text{on}}}{T} = V_{\text{dc}} t_{\text{on}} f \quad (6.23)$$

where V_{dc} is the dc supply voltage, t_{on} is the on-time and t_{off} is the off-time of the transistor, f is the switching frequency, and T is the period of the switching. The ratio of t_{on}/T is known as the duty cycle δ of the switching signal and is usually given as a percentage.

Effectively, this switching method applies a square wave across the load. For some applications, this may be acceptable, such as in dc motor control. Most of the time, however, sinusoidal voltage waveforms with limited total harmonic distortion (in other words, elimination or reduction of some of the harmonics) are desired in applications, such as in ac motor control.

Although the limitation of the total harmonic distortion can be achieved by a simple filter circuit, the switching techniques are preferred in power electronic circuits due to the presence of additional power losses and increased size, specifically at medium and high power applications, where filters are used.

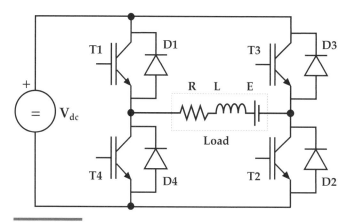

Figure 6-16
Single-phase inverter, H-Bridge circuit with a load (R, L, E).

The most popular converter topology used in dc motor, stepper motor, and ac motor applications is shown in Fig. 6-16 and is known as an H-bridge circuit. Note that in the figure there are reverse parallel diodes (also known as freewheeling or by-pass diodes) across each switching device to provide a current path, which are required in inductive load switching.

The PWM method is mainly used for shaping the output waveform in the H-bridge applications. The method is usually employed in dc motors and stepper motors to limit or regulate winding currents, and utilized in ac motor applications to approximate sinusoidal wave outputs by switching the power switches at a rate higher than the fundamental frequency. Note that the ultimate aims in ac or dc motor control are to control torque, speed, position, acceleration, and deceleration rates.

The virtual instrument designed here is a very comprehensive simulation tool, aimed to study PWM signal generation concepts and the operation of an H-bridge. The details of the simulation will be given in the following sections together with background knowledge.

6.6.1 Virtual Instrument Panel

The study presented here contains a very comprehensive simulation tool, which allows the user to program the PWM switching method and demonstrates the harmonic spectrum associated with each output. In addition, the impact of the programmed switching on the circuit variables can be studied by observing the waveforms of the voltage and current. There are two major VIs used in this section, which will be explained next.

Part 1

This part of the study (Part 1.vi) is intended to simulate a switching method applied to a general load (Fig. 6-16), which can represent various practical load cases, such as a winding of dc motors, permanent magnet stepper motors, and brushless dc motors. The user may alter the values of the parameters of the model circuit and the switching signal (square wave generator).

Note that although the simulation tool provided here can give accurate results, the user is still responsible for entering sensible values for parameters —for instance, a negative diode turn-on voltage would not make sense, although the simulation would still perform the calculation.

The following parameters can be adjusted using the controls on the front panel of the VI, which is illustrated in Fig. 6-17:

V_{dc}: Supply voltage
R: Load resistor
L: Load inductor
E: Load-side voltage source
V_t: Voltage drop across the transistor
V_{df}: On-state voltage drop across the freewheeling diode
$I_{l\,init}$: Initial current value of inductor
Duty cycle δ: The duty cycle of the switching signal, PWM
Frequency: The frequency of the switching signal

Note that when these values are changed, the circuit simulation will only use the new values on the next period of the switching waveform, which means that change is not necessarily instantaneous.

The simulation controls available on the front panel are similar to the controls mentioned previously:

Start: When pressed, the simulation commences.
Restart: When pressed, the simulation halts before the next period from the wave generator. The simulation can be restarted (from time $t = 0$) by pressing the Start button.
Quit: This aborts execution of the program.

262 LabVIEW for Electric Circuits, Machines, Drives, and Laboratories

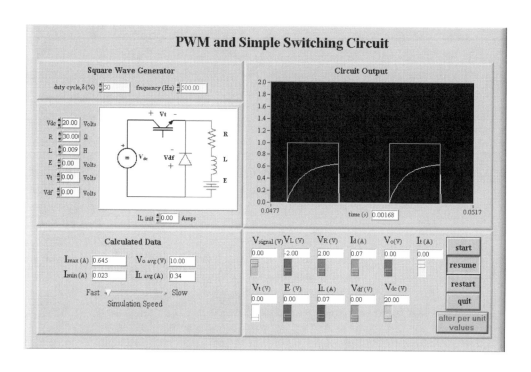

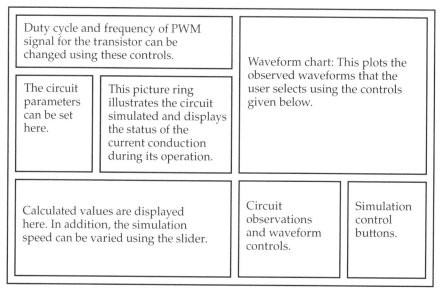

Figure 6-17
The front panel and brief user guide of `Part 1.vi`.

Pause/resume: This pauses execution of the simulation, which can be restarted simply by pressing the same button again. While the program is paused, although other controls may be changed, none will have any effect until the program is resumed.

Alter per-unit values: This allows the user to alter the per-unit values of each data item (primarily for comparison purposes only).

Simulation speed: This sliding control affects how quickly the simulation proceeds. By sliding the selector toward the slow end, the simulation will take longer to plot each sample point, hence observations can be made accurately.

The chart provided on the front panel can display various waveforms when associated buttons are pressed.

V_{signal}: Generated switching signal based on the frequency and the duty cycle settings

V_l: Voltage across the load inductor

V_r: Voltage across the load resistor

I_d: Freewheeling diode current

V_o: Output voltage, voltage across the load, $R + L + E$

I_t: Transistor current

V_{dc}: Supply voltage (dc)

V_t: Voltage drop across the transistor

E: Load-side voltage source

I_l: Load current

V_{df}: On-state voltage drop across the freewheeling diode

In addition, the front panel contains four indicators that display the calculated values, which are significant in the operation and sizing of the circuit.

I_{max}: The maximum value of the load current from the previous period

I_{min}: The minimum value of the load current from the previous period

$V_{o\ avg}$: The average value of the output voltage over the previous period

$I_{l\ avg}$: The average value of the load current over the previous period

Table 6-5 illustrates the structure of `Part 1.vi`, which can be utilized by potential developers. Note that this is not a LabVIEW code. It is a pseudocode giving the sequential formation of the program.

Table 6-5 *A pseudo-code structure of* `Part 1.vi` *to illustrate the sequential formation of the program.*

```
give user the option to read in per-unit values (puv)
do (while true)
   set stop and go buttons to off;
   wait until the start button is pressed, then start simulation;
   clear chart;
   plot the first points on the chart from given initial values;
   do (while restart button hasn't been pressed)
      case (restart button is pressed)
         do nothing, allow simulation to reset;
      case (restart button isn't pressed)
         read control values (e.g., R, L, etc.);
         for i=0 to 99
            check if puv need to be obtained;
            case (user requests change to puv)
               run value selector;
               give user the option to save these puvs on disk;
            get pulse value;
            set dx on chart;
            calculate coefficients a, b for diff equation solver;
            solve for the current at the end of the current time interval;
            case (current negative)
               set current to 0;
            case (current positive)
               leave current;
            case (pulse high)
               //transistor is on
               set trans on picture to on on the picture ring;
               calculate id, it, vo;
            case (pulse low)
               //diode is on
               set diode on picture to on on the picture ring;
               calculate id, it, vl;
            input selected values into chart (after per-unit value calcs);
            calculate Vr, Vl;
            case (pause pressed)
               wait until released;
            case (pause not pressed)
               do nothing;
            read graph selector switches;
            case (value selected)
               set plot color to that designated;
            case (value not selected)
               set plot color to background color, to disappear
            for i=0 to 10
               if value (i) is being plotted and puv (i)/=1
                  show puv indicator;
               else
                  do not show puv indicator;
            end for;
            update max, min, averages;
         end for;
      end case;
   while stop button hasn't been pressed;
while true;
```

Part 2

This program, `Launch.vi`, simulates the response of the circuit given to certain programmed control signals. `Part 2.vi` contains four key component sub-VIs: `Screen 1.vi`, `Screen 2.vi`, `Screen 3.vi`, and `Screen 4.vi`. Each sub-VI has a specific goal to achieve. After completing one sub-VI, the next is automatically executed.

`Screen 1.vi` allows the user to construct a desired load voltage waveform aiming to limit its harmonic contents. Based on this setting, `Screen 2.vi` constructs the transistor switching signals to achieve the desired output voltage (load voltage). Finally, `Screen 3.vi` uses the generated transistor signals to calculate the response of the circuit. In fact, `Screen 3.vi` is a modified version of `Part 1.vi`, which can allow the user to control various circuit parameters and view circuit associated waveforms. Finally, `Screen 4.vi` is executed simultaneously with `Screen 3.vi` to reexamine the original and the actual load voltage waveforms. Let's look at the details of each of these sub-VIs.

Screen 1.vi

The front panel of `Screen 1.vi` is given in Fig. 6-18. On the front panel, the user can program a desired load voltage waveform using two methods: by a programmed wave and by comparing a triangular wave to a sine wave. For each case, the number of samples must be set, which controls how many sample points the wave will be constructed from.

Note that the spectral analysis of this desired voltage waveform is also given on the front panel of `Screen 1.vi`. Once the user is satisfied about the harmonic contents of the programmed waveform above, pressing the Done button advances the program to `Screen 2.vi`.

In the Programmed Wave option, the period of the generated waveform and PWM features can be defined. In addition, there are three switch controls where the user can define various template features to the waveform generated. These switches and their states are Soft/Hard, Half/Full, and Non-inverted/Inverted.

To explain the Programmed Wave option, let's consider some sample settings and corresponding waveforms, which are given in Fig. 6-19. Note that the programmed output waveforms are displayed as a per-unit value. The actual magnitude of the output voltage depends on the dc supply voltage (also known as dc link voltage).

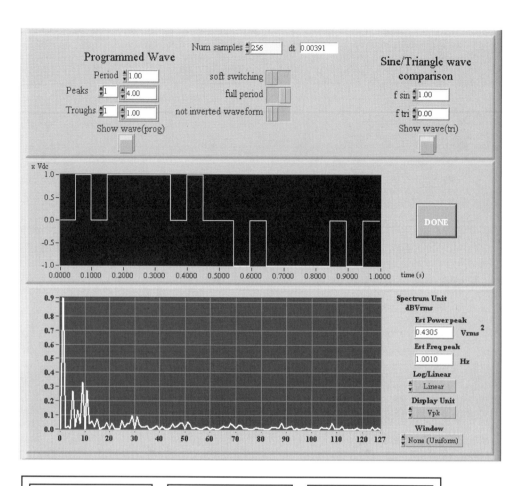

Figure 6-18
The front panel and brief user guide of Screen 1 of Part 2.vi.

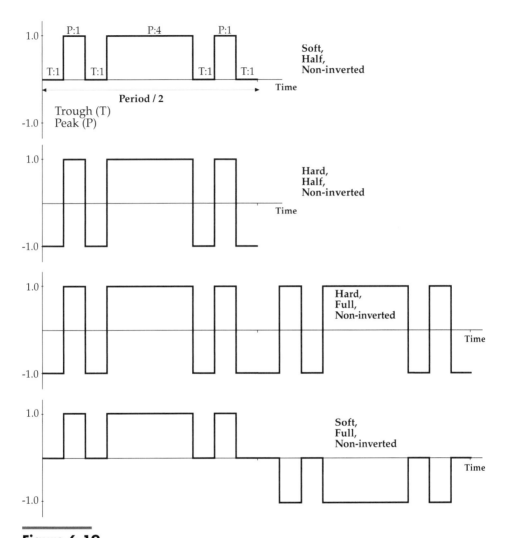

Figure 6-19
Sample programmed waveforms and their settings using the Programmed Wave option in `Screen 1.vi`, where Period : 1, Peaks : [1, 4, 1], Troughs : [1, 1, 1, 1].

The waveform is constructed by specifying the relative widths of peaks and troughs based on the total number of sample points and the total width of the waveform.

For the settings given in Fig. 6-19, the total width of the waveform is $1 + 4 + 1 + 1 + 1 + 1 + 1 = 10$. So the total number of the sample points is $10w$, where w is the number of sample points that each peak and trough is specified relative to.

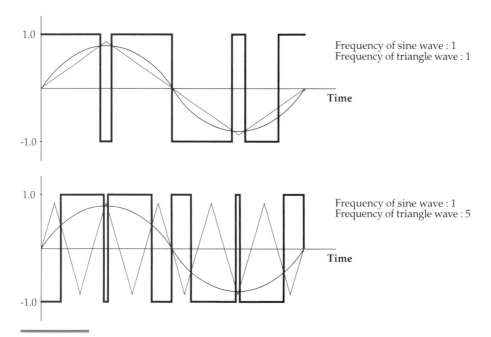

Figure 6-20
Sample programmed waveforms using the Sine/Triangle Wave Comparison option in `Screen 1.vi`.

$$w = \frac{\text{Number of samples}}{\text{Total width of the waveform}}$$

The period of the waveform is important in determining the fundamental frequency of the waveform.

In the Sine/Triangle Wave Comparison option, the output waveform is generated by comparing a sine wave of specified frequency to a triangle wave of specified frequency. Two example settings and corresponding waveforms are given in Fig. 6-20. As can be seen, when the triangle wave is greater than the sine wave, the load waveform is −1.0; when the sine wave is greater than the triangle wave, the load waveform is +1.0. The fundamental frequency of the programmed waveform is equal to the frequency of the sine wave. The frequency of the triangle wave should be an integer multiple of the sine wave.

`Screen 2.vi`

This sub-VI displays the switching states of the transistors. Fig. 6-21 illustrates two of the switching patterns that are given for the first and the fourth pro-

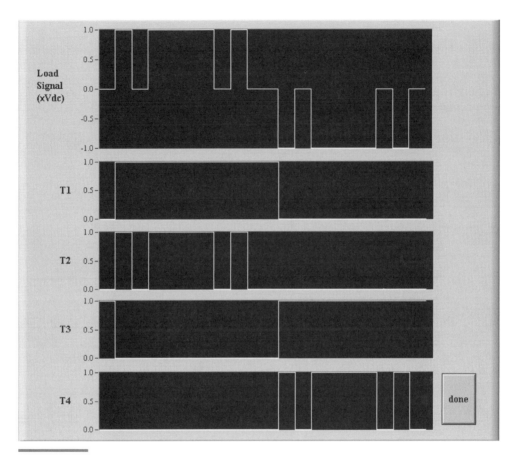

Figure 6-21
The front panel of Screen 2 of `Part 2.vi`.

grammed waveforms in Fig. 6-19. The front panel of `Screen 2.vi` is given in Fig. 6-22. The front panel has no user inputs. It simply takes the desired load voltage waveform and generates switching signals for transistors T1 to T4. After viewing the switching signals and pressing the Done button, the user can proceed to `Screen 3.vi`, where the dynamic response of the H-bridge circuit is tested.

`Screen 3.vi`

This sub-VI is similar to the VI described in `Part 1.vi`, which accepts the switching signals of the transistors as inputs and allows the user to study

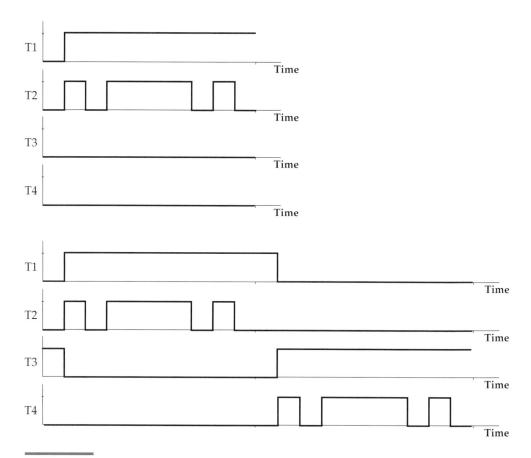

Figure 6-22
Transistors' switching states in Screen 2.vi, which are based on the circuit and the programmed waveforms (the first and the fourth) given in Fig. 6-19.

the dynamic behavior of the circuit. The front panel of this VI is given in Fig. 6-23.

The Circuit Output chart on the front panel displays the circuit data measurements selected on the Graph Selector. The suggested data to be displayed on the chart are the load current and the load voltage waveforms.

The second chart given on the panel displays the state of each diode D1 to D4 and each transistor T1 to T4. A red plot indicates that the particular component is conducting current. The comparison of these voltage and current waveforms and the status of the transistors and diodes can provide a powerful insight into the inductive load switching.

Chapter 6 • Introduction to Power Electronics Circuits

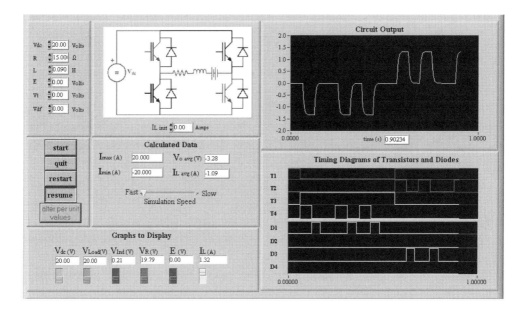

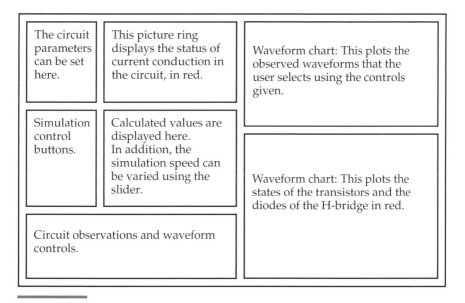

Figure 6-23
The front panel and brief user guide of Screen 3 of `Part 2.vi`.

`Screen 4.vi`

Note that the original switching signals of the transistors cannot force the load voltage waveform to be as specified. Since the current decay time in the circuit depends on the load parameters, the load voltage waveform may be different from that generated in `Screen 1.vi`.

At the end of one period of calculation in `Screen 3.vi`, `Screen 4.vi` displays the actual load voltage waveform and performs a harmonic spectrum analysis. Note that `Screen 3.vi` and `Screen 4.vi` run simultaneously. The user can select either panel, so a comparison can be made.

6.6.2 Some Implementation Details of the Sub-VIs

Because of the complexity of the VIs we are studying, it is important to provide some of the implementation details for LabVIEW users and potential developers.

A half wave is a wave whose peaks have the same sign. To generate the full wave, it is taken and appended to the original half wave that has been reflected in the line of the x-axis, and then in the line of the y-axis. This is the method of producing the full wave from the half wave by using negation and reversing arrays, and so on. To better understand the following explanations, I advise you to consider the diagrams of the VIs.

`Append If Necessary.vi` and `Append If Required.vi`

These VIs are intended to be used when constructing arrays of sample points. When an array is constructed by finding half of the array then deducing the second half of the array from the first, it may occur (where the desired number of sample points is odd) that the final array will have 1 less point in it. To solve this, the final point in the smaller array is appended to the end of the array to increase the array length by 1.

`Full Trans Wave.vi`

This sub-VI determines the half or full wave transistor signals required to achieve a specified load voltage output. The inputs and outputs to this sub-VI are listed next:

Inputs

   Load voltages: This is an array of load voltages representing the load voltage waveform.

   High load voltage: This is the value that is recognized as being the high value of load voltage.

   Switching type: This gives the option of selecting soft or hard switching. Soft switching occurs when, to modulate the amplitude of a waveform, one transistor remains on, while one is turned on and off. Hard switching is when both transistors are switched on and off.

   Low load voltage: This is the low value of load voltage.

   Number of samples: This is the number of samples required in the output voltage waveform.

   Half/full wave: This controls whether the wave is a half or full wave. False indicates a half wave.

Outputs

   T1, T2, T3, T4: These are arrays of signals to be used as inputs into transistors T1, T2, T3, T4 in order to achieve the load voltage waveform as in the input, Load Voltages.

This sub-VI does the bulk of the computation. If the wave is a half wave, then this can do all of the computation. If a full wave is to be calculated, then the first half wave is found and presented to the second half wave. The first half wave is found by using `Half Trans Wave.vi` with the load voltages as input. The second half wave is found by using the remaining sample points in Load Voltages as input into `Half Trans Wave.vi`.

`Half Trans Wave.vi`

This sub-VI determines the half wave transistor inputs required to achieve a specified load voltage output. Half wave means that all peaks have the same value. Note that the roles of T1 and T2 have been switched, as have the roles of T3 and T4. This should be used as a reference as to how the inputs are constructed.

Inputs
  Load voltages
  High load voltage

Switching type
Low load voltage

Outputs
T1, T2, T3, T4

The implementation of this sub-VI is broken into two cases depending on whether the high load voltage is positive or negative. It should be noted that when the peak value is positive, then transistors T1 and T2 are turned on. When the peak value is negative, transistors T3 and T4 are turned on. This is why there are two modes of operation.

Then for each of these cases the type of switching (hard or soft) must be considered. Sub-VIs `High Pos.vi` (which will be explained next) and `High Neg.vi` contain the logic, which decides on what value each transistor should have.

High Pos.vi

The purpose of this sub-VI is to encapsulate the logic needed to determine the signals to transistors T3 and T4 to achieve a specified load voltage value. This is only for the case where the load peak value is negative, and for sample points either on or after the first peak in the load voltage. Transistors T1 and T2 are off in these cases. The physical modes of operation are

Soft Switch
    Load voltage high: T3 and T4 on
    Load voltage low: T3 on, T4 off
Hard Switch
    Load voltage high: T3 and T4 on
    Load voltage low: T3 and T4 off

Triangle Sine Conv.vi

This sub-VI generates the signals into T1 to T3 given a load voltage known to be generated by comparing a sine wave to a triangular wave. The principal logic behind this comparison is

    Load voltage positive: T1 and T2 on; T3 and T4 off
    Load voltage negative: T1 and T2 off; T3 and T4 on

Triangle Sine Wave Extra.vi

This VI generates a PWM pulse by comparing a triangular waveform to a sinusoidal waveform. The waveform makes a transition from high to low (or vice versa) whenever the sine wave crosses the triangular wave. The sine waveform determines the fundamental frequency of the output (as long as the frequency of the triangular waveform is an integer multiple of the sine wave frequency).

The output wave is in the form of an array of sample points, which always contains points resulting from the comparison of the triangular wave with one sine wave. As a result of this, the theoretical fundamental period may be different from that given (based entirely on the sine frequency) if the frequency of the triangular wave is not an integral multiple of the sinusoid frequency.

The signal generator is implemented by first testing when the sine wave is greater than the triangular wave. This gives an array of Boolean values. Then this array of Boolean values is converted into a binary array where true is represented by 1 and false is represented by 0. So the constructed wave is high when the sine wave exceeds the triangle wave and low otherwise.

The binary array is shifted and scaled correspondingly to achieve the specified high and low voltage levels. There is a case for when the high value is zero and a case for when it is nonzero.

RL Solver.vi

This sub-VI solves the current of the load system (essentially the differential equation containing R, L, and E) to model a typical winding in various motor applications, given an initial condition of current and over a period of time.

The load voltage is calculated after taking into account the voltage drops across the transistor(s). Positive current is defined as ingoing current to the resistor at the node where the emitter of T1 is connected to the resistor. All other voltages are measured according to the passive sign convention. The VI uses a fourth order RK numerical method available in the G-Math Toolkit of Lab-VIEW. Finally, a circuit mode naming convention (P1–P9) is used to organize the picture ring and to display the circuit status, which are

- P1 T1 and T2 are on, in red
- P2 T1 and D3 are on, in red
- P3 D4 and T2 are on, in red
- P4 D3 and D4 are on, in red

P5 T3 and T4 are on, in red
P6 D1 and T3 are on, in red
P7 T4 and D2 are on, in red
P8 D1 and D2 are on, in red
P9 No conducting device, all black

6.7 References

Ertugrul, N. "Power Electronics Lecture Notes." Department of Electrical and Electronic Engineering. University of Adelaide, 1997.

Histand, M. B., and D. G. Alciatore. *Introduction to Mechatronics and Measurement Systems.* New York: McGraw-Hill, 1999.

Kenjo, T. *Stepping Motors and Their Microprocessor Controls.* New York: Oxford University Press, 1984.

Kugel, U., and M. Yorck. "Stepper Motor Model for Dynamic Simulation." [On-line, Sept. 2001] Available: http://users.bart.nl/~myorck/PAPER/stepper/stepper.html.

Lander, C. W. *Power Electronics,* 3rd ed. New York: McGraw-Hill, 1993.

Maas, J. *Industrial Electronics.* Englewood Cliffs, NJ: Prentice Hall, 1995.

Rossiter, D., G. Kandasamy, and E. Robinson. "Final Year Project Reports on Power Electronics Circuit Simulations." Department of Electrical and Electronic Engineering. University of Adelaide, 1999.

SCR Manual, 5th ed. General Electric, 1972.

Slutsky, E. B., and D. W. Messaros. *Introduction to Electrical Engineering Laboratories: Circuits, Electronics, and Digital Logic.* Englewood Cliffs, NJ: Prentice Hall, 1992.

Vu, L. "Final Year Project Report: Development of LabVIEW-Based Teaching Modules." Department of Electrical and Electronic Engineering. University of Adelaide, 2001.

Simulation of Electrical Machines and Systems

7

In the industrial world, the bulk of the energy produced is consumed by a vast number of electric motors that are rarely powered directly from the main supply and rarely operated at steady state. In many practical systems, an electric machine is often a member of a group of machines that are interlinked at the control level. The electric machines in such systems start and stop frequently, change their operating modes to motoring or generating, and, in most cases, are controlled by power electronics drive systems, which in reality never operate at steady state. Therefore, understanding the operation and the behavior of electric machines and drives in a dynamic environment requires special attention.

Most of the textbooks published in this area provide static models of electric machines and drives. However, since static models do not take time variations into account, they provide very limited insights under real, dynamic operating conditions. Although there are a few alternative software tools (other than LabVIEW) and textbooks available in the market, these are either expensive or require additional time and effort to make the software modules fully functional.

This chapter introduces simulation of various rotating electrical machines, drives, and systems with LabVIEW by providing working simulation models with highly interactive front panels. However, it should be noted that using a simulation without understanding the basic theory of electric machines and drives results in serious problems since it will be only a simple toy, not an essential tool. Therefore, each section in this chapter is devoted to a specific motor or system, and basic theoretical concepts are adequately reinforced with the associated LabVIEW simulation and with complete codes. However, remember that further study may be needed to comprehend some of the details.

The chapter provides a comprehensive framework for modeling and simulation, integrating the electric machines and drives by providing the fundamental concepts of modular and hierarchical model composition. The user interface of the simulations can display all the parameters that are utilized in each simulation, and the programming codes provide easy access points for further development. We believe that, in addition to students of electrical, electronic, and computer systems engineering, potential developers can benefit from the codes and implementation methods.

Most of the simulation models presented here are verified using practical measured data in the laboratory, and thus are believed to predict the behavior of the system correctly.

7.1 Rotating Field Simulation in AC Machines

The principle of rotating magnetic fields is the key to the operation of most three-phase ac motors, such as conventional synchronous, permanent magnet synchronous, and induction motors. The rotating magnetic field is usually developed in the stator of an ac motor in order to produce mechanical rotation of the rotor. The rotor field is made to follow this rotating stator field by being attracted and repelled, hence the free rotating rotor follows the rotating field in the stator.

However, a rotating magnetic field in three-phase ac motors is probably most difficult to understand since it cannot be visualized easily unless particular tools are used. Furthermore, unlike the gradually varying sinusoidal voltage, an additional difficulty arises when an ac motor is driven by an inverter that provides only programmed step voltages (high or low) to the motor windings.

Fig. 7-1 illustrates the stators of two three-phase ac motors, where three-phase windings (a, b, and c) are placed 120° mechanical apart. As illustrated, the phase windings are placed in slots in the housing of the ac motors. The figure also illustrates the three phasors of the mmf at a given time instant. When an ac voltage is applied to the stator windings, current flows through the wind-

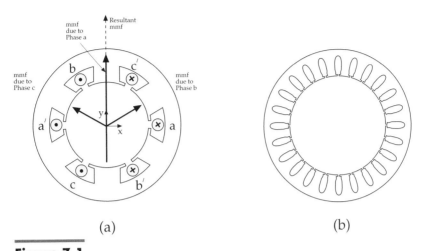

Figure 7-1
The stators of three-phase ac electrical machines: (a) a 6-slot machine showing the established mmf in three phases at a given time instant assuming that all three phases are conducting currents, and (b) 24-slot ac machine used in the simulation.

ings, which develops the magnetic field depending on the direction of current flow through that winding.

To establish a rotating magnetic field in the stator of a motor, the number of pole pairs must be the same as (or a multiple of) the number of phases in the applied voltage. The poles must then be displaced from each other by an angle equal to the phase angle between the individual phases of the applied voltage.

If the voltages applied to phases a, b, and c are 120° out of phase, the currents that flow in the phases are displaced from each other by 120°. Since the magnetic fields generated in the coils are in phase with their respective currents, the magnetic fields are also 120° out of phase with each other.

Referring to Fig. 7-1, if the three components of magnetic fields are broken into their **x** and **y** components (which are the unit vectors in the horizontal and vertical directions, respectively), the total magnetic flux density in the stator can be calculated by summing the three vectors. If this is done, the net magnetic flux density in the stator is found as

$$B_{net}(t) = (1.5B_m \sin \omega t)\mathbf{x} - (1.5B_m \cos \omega t)\mathbf{y} \tag{7.1}$$

where B_m is the maximum value of the magnetic flux density and ω is the angular velocity. Note that the magnitude of the field is constant, $1.5B_m$, and that the angle changes continuously in a direction at angular velocity of ω, which rotates one revolution for each cycle of ac voltage.

In general, if the windings of a three-phase ac motor are wound $2\pi/3$ electrical degrees apart in space and if the windings are excited by balanced three-phase currents, a sinusoidally distributed rotating field of constant magnitude is produced.

7.1.1 Virtual Instrument Panel

There are two VIs designed to visualize the rotating magnetic field in this section. In the first VI, Rotating ac Field.vi, the rotating field concept in three-phase ac motors is demonstrated using the vectors. In the second VI, the rotating field concept is applied to a three-phase induction motor, and the influence of the stator voltage shape is demonstrated.

Rotating Magnetic Field in Three-Phase AC Motors

To analyze the rotating magnetic field in a three-phase motor stator, refer to the graph shown in Fig. 7-2, the Rotating ac Field.vi front panel. The

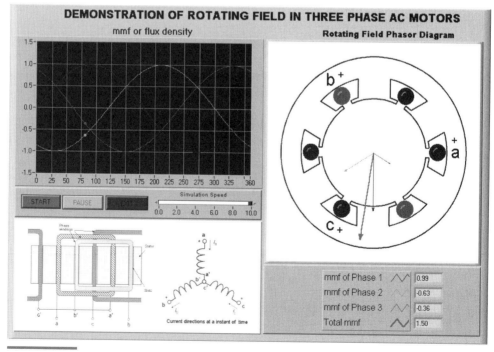

Figure 7-2
The front panel of Rotating ac Field.vi.

drawing in the bottom-left corner of the front panel illustrates the slots and the winding positions on the stator of a simple three-phase ac motor. As shown in the same location, the three-phase windings are tied together to form a star-connected stator. This drawing illustrates the direction of currents at a given time instant. Note that the sum of the currents in two of the phases is equal to the current in the third phase, hence their directions are opposite when the sign notation is considered.

The cross section of the simple motor is provided as a background image on the right-hand side where a pair of LED indicators is assigned to each phase winding. The LEDs are colored to distinguish the phases, and the intensity of their color will vary indicating the level of magnetic flux density.

The controls of the simulation are also provided on the front panel: Start/Stop, Resume/Pause, Exit, and Simulation Speed.

The waveforms in the graph are of the three input phases, displaced 120° because of the way they are generated in a three-phase ac generator. The waveforms are labeled to match their associated phase. Once the simulation is started, these waveforms representing the magnetic flux densities are plotted instantly. The moving dots on the waveforms indicate the instantaneous value at the corresponding electrical angle and mechanical position.

When the dot points start moving on the graph, the flux density vectors are displayed. The direction of these vectors can be determined using the right-hand rule at any given instant, and their lengths are proportional to the magnitudes of the moving dots.

All three-phase flux density vectors and the resultant vector are displayed on the diagram. Note that from one point to the next, the polarities are rotating from one pole to the next in a counterclockwise (CCW) manner. One complete cycle of input voltage (or current or flux, since they are all 120° apart in a three-phase balanced system) produces a 360° rotation of the pole polarities.

For example, if the field is evaluated at 60° electrical position from the starting point, Phase 3 has no current field, Phase 1 has current flow in a positive direction, Phase 2 has current flow in a negative direction, and two magnetic flux vectors of the current carrying windings will have equal magnitude with phase difference. The vector sum of these two vectors can easily be visualized: Its magnitude is 1.5 times the magnitude of each vector.

Rotating Magnetic Field in Induction Motors

The three-phase induction motors also operate on the principle of a rotating magnetic field. As already stated, the rotating magnetic field has a constant magnitude rotating at synchronous speed defined by the supply frequency.

In induction motors, there are two windings, one in the stator creating a magnetic field of constant magnitude utilizing the ac supply, and the second one on the rotor. You can easily see that if the winding on the rotor rotates at synchronous speed (i.e., the speed of the rotating stator-produced field), then no relative motion of field and rotor occurs, hence no emf will be induced in the rotor winding.

The conclusion drawn from this explanation is that the torque becomes zero at synchronous speed. This can also be seen in torque-versus-slip characteristics of the induction motor as studied in Chapter 5, where torque is zero when slip s is 1.0 ($n = n_s$).

At any other speed, however, an emf will be induced in the rotor windings and will differ as a function of speed. In induction motors, the induced voltage E_r in the rotor and its frequency f_r are given as

$$E_r = sE_o \tag{7.2}$$
$$f_r = sf_s \tag{7.3}$$

where E_o is the per-phase voltage induced in the rotor at standstill, and f_s is the supply frequency.

The front panel of `Rotating Field Induction.vi` is shown in Fig. 7-3. It allows users to visualize various aspects related to the rotating field and rotor induced waveforms in induction motors. As given in the figure, the sample three-phase induction motor has a 24-slot stator and a rotor with the same number of slots, where the rotor windings or squirrel cage conductors are placed. Similar to the previous simulation, LEDs are used to represent the coils in each slot.

There are two excitation options in this VI: Three-Phase Sinewave and Three-Phase Inverter Driven. In addition, the slip of the induction motor, the supply frequency, and the supply voltage can be varied using their controls. Furthermore, on the front panel of the VI, induced three-phase voltage on the rotor circuit can be displayed, the rotor LEDs can be hidden, or the inverter animation can be deactivated using the switch controls available.

7.2 Dynamic Simulation of Three-Phase Induction (Asynchronous) Motor

The dynamic behavior of electric motors is a difficult topic to study without a computer-aided tool. The VI `Induction motor simulation.vi` in this section investigates the dynamic operation of three-phase asynchronous

Chapter 7 • Simulation of Electrical Machines and Systems

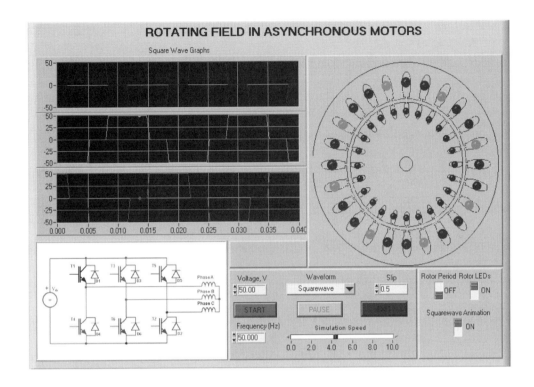

The plots in these graphs display the three-phase voltage waveforms applied to the stator windings.

This animation window illustrates the level of voltage applied to the stator windings, which use LED indicators. The color intensity is proportional to the level of voltage applied

This picture ring indicates the status of the inverter switches.

These are available simulation controls.

Figure 7-3
The front panel and brief user guide of Rotating Field Induction.vi.

motors. In the simulation, the asynchronous motor is powered from a three-phase sinusoidal voltage source and the motor is started from standstill.

After performing this simulation, students should be able to

- understand the dynamic behavior of the three-phase induction motor under no-load or load conditions.
- observe the starting performance of the motor.

In the VI, the motor is modeled in the d-q reference frame by five nonlinear differential equations, and the simulation uses the stationary reference frame of the induction motor as explained in reference 4. In the simulation, the fluxes are selected as the state space variables. The nonlinear differential equations are given next.

$$\frac{d\psi_{qs}}{dt} = \omega_b \left[v_{qs} - \frac{R_s X'_{rr}}{D}\psi_{qs} - \frac{\omega}{\omega_b}\psi_{ds} + \frac{R_s X_m}{D}\psi_{qr} \right] \qquad (7.4)$$

$$\frac{d\psi_{ds}}{dt} = \omega_b \left[v_{ds} + \frac{\omega}{\omega_b}\psi_{qs} - \frac{R_s X'_{rr}}{D}\psi_{ds} + \frac{R_s X_m}{D}\psi_{dr} \right] \qquad (7.5)$$

$$\frac{d\psi_{qr}}{dt} = \omega_b \left[v_{qr} + \frac{R'_r X_m}{D}\psi_{qs} - \frac{R'_r X_{ss}}{D}\psi_{qr} - \frac{\omega - \omega_r}{\omega_b}\psi_{dr} \right] \qquad (7.6)$$

$$\frac{d\psi_{dr}}{dt} = \omega_b \left[v_{dr} + \frac{R'_r X_m}{D}\psi_{ds} + \frac{\omega - \omega_r}{\omega_b}\psi_{qr} - \frac{R'_r X_{ss}}{D}\psi_{dr} \right] \qquad (7.7)$$

$$\frac{d\omega_r}{dt} = \left(T_e - T_L \right)\left(\frac{p}{2}\right)\left(\frac{1}{J}\right) \qquad (7.8)$$

where $D = X_{ss}X'_{rr} - X_m^2$, p is the number of poles of the motor, J is the moment of inertia, ω_b is the base speed in rad/s, subscripts d and q represent the d and q axes, and subscripts s and r indicate the stator and rotor quantities. The electromagnetic torque developed by the machine is given by

$$T_e = \left(\frac{3}{2}\right)\left(\frac{p}{2}\right)\frac{X_m}{D\omega_b}(\psi_{qs}\psi_{dr} - \psi_{qr}\psi_{ds}) \qquad (7.9)$$

These differential equations are solved by the Runge-Kutta numerical method that is implemented using the tools available in LabVIEW. The Runge-Kutta method does not need a special starting arrangement, the step width can be changed easily, and storage requirement is minimal. The Runge-Kutta formula used in the numerical solution involves weighted average values taken at different points in the interval $t_n \leq t \leq t_{n+1}$, and is given by

$$y_{n+1} = y_n + \frac{1}{6}(k_1 + 2k_2 + 2k_3 + k_4) \qquad (7.10)$$

where the coefficients are

$$k_1 = \Delta t f'(x_n, y_n)$$
$$k_2 = \Delta t f'(x_n + \Delta t/2, y_n + k_1/2)$$
$$k_3 = \Delta t f'(x_n + \Delta t/2, y_n + k_2/2)$$
$$k_4 = \Delta t f'(x_n + \Delta t, y_n + k_3)$$

Here n is the time step, $\Delta t = t_{n+1} - t_n$, $f'(x_n, y_n) = dy_n/dx_n$.

The simulation also provides an inverse transformation to determine the abc reference frame that corresponds to the real parameters of the motor for easy comparison. Conversion to the abc reference frame is achieved by using the following transformations.

$$f_{abcs} = (K^{-1})f_{qdos} \tag{7.11}$$

where $f_{abcs} = [f_{as} f_{bs} f_{cs}]$ and $f_{qdos} = [f_{qs} f_{ds} f_{os}]$. Here, f can be either the voltage, the current, or the flux linkage of the machine.

$$K^{-1} = \begin{bmatrix} \cos\theta & \sin\theta & 1 \\ \cos(\theta - 2\pi/3) & \sin(\theta - 2\pi/3) & 1 \\ \cos(\theta + 2\pi/3) & \sin(\theta + 2\pi/3) & 1 \end{bmatrix} \tag{7.12}$$

In equation 7.12, $\theta = \int \omega \, dt$ is used for the stator transformation, and $\theta = \int (\omega - \omega_r) \, dt$ is used for the rotor transformations.

7.2.1 Virtual Instrument Panel

The front panel of the VI is shown in Fig. 7-4. Before starting the simulation, the user should enter the motor and the load parameters, and set the time step. After the simulation is instigated, the graphs display the estimated values of the electromagnetic torque and rotor speed. However, make sure that the motor parameters you entered are practical. As a guide, the sample settings in Table 7-1, which were obtained using a real machine in our laboratory, can be used. The tests performed in Chapter 5 can provide all the equivalent circuit parameters and the total moment of inertia of the rotating system.

7.2.2 Self-Study Questions

Open and run the custom-written VI named `Induction Motor.vi` in Section 7.2 of the Chapter 7 folder, and investigate the following questions.

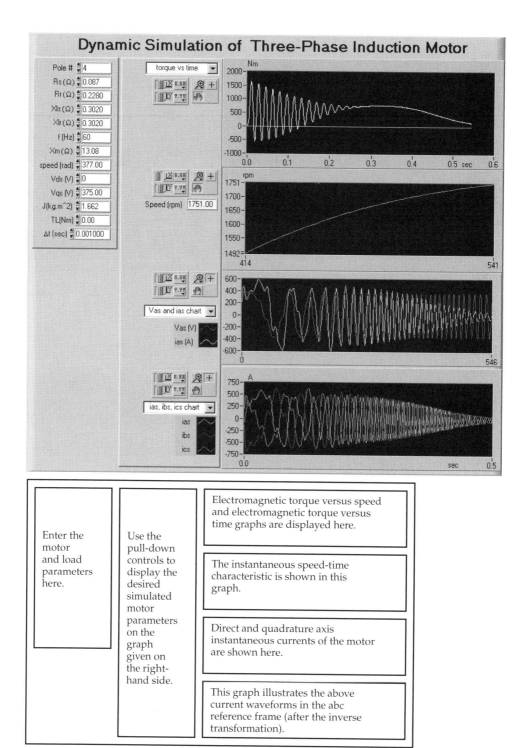

Figure 7-4
The front panel and brief user guide of Induction motor simulation.vi.

Table 7-1 *Sample settings for* `Induction motor simulation.vi`.

Number of poles: 4	$X_m = 43 \, \Omega$
$R_s = 1.05 \, \Omega$	Speed, $\omega_m = 157$ rad/s
$R_r = 0.16 \, \Omega$	$V_{ds} = 0$ V
$X_{ls} = 2.4 \, \Omega$	$V_{qs} = 225.2$ V
$X_{lr} = 0.37 \, \Omega$	$J = 0.39$ kg m^2
$f = 50$ Hz	$T_L = 30$ Nm
	$\Delta t = 0.001$ s

1. Input the equivalent circuit parameters given in Table 7-1, run the simulation, and observe the changes in the electromagnetic torque, the speed, and the line currents.
2. Vary the total moment of inertia and the load torque (only one at a time) and observe the changes in the speed-time characteristic.
3. Vary the equivalent rotor resistance of the motor and observe the changes in the torque-speed characteristic.
4. The electromagnetic torque can be plotted in two different forms: torque versus speed and torque versus slip. Run the simulation until the steady-state speed is reached and comment on the characteristics in questions 2 and 3. Have you noticed any changes in the value of the slip when the load (or the equivalent rotor resistance or the moment of inertia) is increased?
5. Print one set of the torque-speed graph and identify the starting torque and the breakdown torque on the graph.

7.3 Dynamic Simulation of Brushless Permanent Magnet AC Motor Drives

Recent advances in permanent magnet materials, power devices, and microelectronic technology have greatly contributed to new energy-efficient and high-performance electrical drives, such as Brushless Permanent Magnet (PM) Synchronous motor drives. These motors usually have three phases, with higher efficiency, higher power factors, higher output power per mass and volume, and better dynamic performance than their counterparts.

A very flexible computer simulation is provided here. This tool can be used to study the motor behavior and analyze the complete drive system (including the inverter and the controller) without implementing the hardware. In addition, the drive simulations can be forced to operate under extreme conditions without damaging the motor drive. In the simulation, the dynamic as well as the steady-state operation of the three-phase permanent magnet ac motor drive can be observed in a variety of operating conditions, such as with and without current control, with sinusoidal or rectangular current excitation, and with sinusoidal or trapezoidal back emf waveforms.

After performing this experiment, students should be able to

- understand the operation of the Brushless Permanent Magnet motor drives under transient as well as steady-state operating conditions.
- observe the behavior of the complete motor drive in a closed loop control system.

Motor and Drive Model

Depending on the stator winding arrangement and the shape and location of the permanent magnets on the rotor, PM ac motors can be broadly classified into two groups.

The first group possess trapezoidal back emf per phase and are called Brushless Trapezoidal Permanent Magnet (BTPM) motors (or simply brushless dc motors). The second group possess sinusoidal back emf and are called Brushless Sinusoidal Permanent Magnet (BSPM) motors (or brushless permanent magnet synchronous motors). In order to produce a constant ripple-free electromagnetic torque, the motor types require rectangular or sinusoidal winding currents, respectively. Any deviation from these ideal current excitations produces torque pulsations. Fig. 7-5 illustrates the back emf and desired excitation current per phase for PM ac motors.

Fig. 7-6 presents the block diagram of the complete drive. The principal differences between the two types of motors in terms of hardware are the accuracy of the rotor position sensor and their respective control requirements. In order to obtain a general dynamic model for the BSTM and BTPM motors, the three-phase abc modeling approach is used in the simulation model.

The three-phase star-connected PM ac motors can be modeled by a network (Fig. 7-7) consisting of a winding resistance, an equivalent winding inductance, and a back emf source per phase, all connected in series. In this study, it is assumed that the stator resistances of all the windings are equal and the

Chapter 7 • Simulation of Electrical Machines and Systems

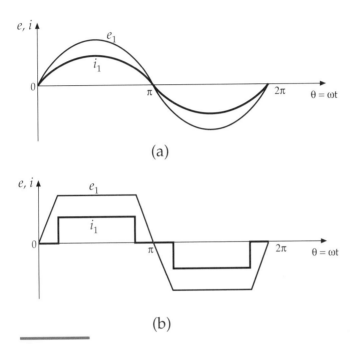

Figure 7-5
The ideal current waveforms reference to the back emfs of the PM ac motors (per phase): (a) for BSPM, and (b) for BTPM.

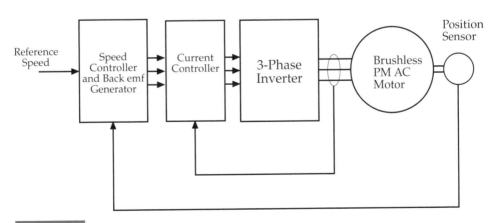

Figure 7-6
The block diagram of Brushless PM Motor Simulation.vi.

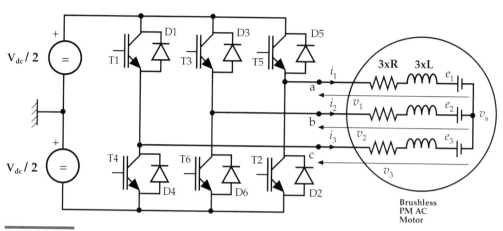

Figure 7-7
Power supply, inverter, and motor models used in the VI.

self- and the mutual inductances are constant. Therefore, the voltage equations in matrix form for a three-phase PM ac motor are expressed as

$$\begin{bmatrix} v_1 \\ v_2 \\ v_3 \end{bmatrix} = \begin{bmatrix} R & 0 & 0 \\ 0 & R & 0 \\ 0 & 0 & R \end{bmatrix} \begin{bmatrix} i_1 \\ i_2 \\ i_3 \end{bmatrix} + \begin{bmatrix} L & 0 & 0 \\ 0 & L & 0 \\ 0 & 0 & L \end{bmatrix} \frac{d}{dt} \begin{bmatrix} i_1 \\ i_2 \\ i_3 \end{bmatrix} + \begin{bmatrix} e_1 \\ e_2 \\ e_3 \end{bmatrix} \quad (7.13)$$

where v_1, v_2, and v_3 are the phase voltages; R is the winding resistance; i_1, i_2, and i_3 are the line currents; L is the equivalent winding inductance; and e_1, e_2, and e_3 are the back emfs of the phases.

In the PM ac motors, because of the large air-gap between stator and rotor, saturation is negligible. Therefore, the flux linkages become a linear function of the phase currents, and hence the electromagnetic torque is given by

$$T_e = \frac{1}{\omega_r}(e_1 i_1 + e_2 i_2 + e_3 i_3) \quad (7.14)$$

Here ω_r is the angular speed of the rotor.

However, in order to study the transient behavior of the PM ac motors, the mechanical state equation must also be known, which is given as

$$T_e - T_l = J \frac{d\omega_r}{dt} \quad (7.15)$$

Here T_l is the load torque, and J is the inertia of the motor and the connected load.

As mentioned earlier, the two groups of motors (BTPM and BSPM) possess different back emf waveforms. If the stator windings of the three-phase motor are symmetrically displaced, the ideal back emf equations of the BSPM motor can be given by

$$\begin{bmatrix} e_1 \\ e_2 \\ e_3 \end{bmatrix} = \begin{bmatrix} E_m \sin(\theta_e) \\ E_m \sin(\theta_e - 2\pi/3) \\ E_m \sin(\theta_e - 4\pi/3) \end{bmatrix} \quad (7.16)$$

For BTPM motors, however, the back emf waveform of Phase 1 is given in piecewise linear form as

$$e_1 = \begin{cases} \dfrac{E_m}{\pi/6}\theta_e & 0 < \theta_e \leq \dfrac{\pi}{6} \\ E_m & \dfrac{\pi}{6} < \theta_e \leq \dfrac{5\pi}{6} \\ -\dfrac{E_m}{\pi/6}(\theta_e - \pi) & \dfrac{5\pi}{6} < \theta_e \leq \dfrac{7\pi}{6} \\ -E_m & \dfrac{7\pi}{6} < \theta_e \leq \dfrac{11\pi}{6} \\ \dfrac{E_m}{\pi/6}(\theta_e - 2\pi) & \dfrac{11\pi}{6} < \theta_e \leq 2\pi \end{cases} \quad (7.17)$$

Note that equation 7.17 can be used to define the other two phases of the motor simply by shifting 120° electrical.

In addition, remember that the practical back emf waveforms of the motors always deviate from these ideal assumptions. In these cases, the back emf waveforms may be modeled by using a look-up table or by using multiple harmonic components of the real waveform.

In equation 7.16 and equation 7.17, E_m is the maximum value of the back emfs, which can be given by

$$E_m = k_e \omega_r \quad (7.18)$$

where k_e is the back emf constant, and θ_e is the electrical rotor position, given by

$$\theta_e = p\theta_r = p \int \omega_r \, dt \quad (7.19)$$

Here θ_r is the mechanical rotor position, and p is the number of pole pairs of the motor.

Generally, three-phase PM ac motors are powered from three-phase inverters. Therefore, to obtain a complete drive simulation, the inverter should also be modeled. The inverter that is controlled by the switching signal provides the desired terminal voltage to each phase winding of the motor. In practical motor drives, the motor is normally star-connected, and the star point is normally left floating (which means that the star point is not linked to anywhere in the power circuit; see Fig. 7-7). Therefore, due to the inverter switching, the star point voltage varies, and the effective winding voltage depends on the star-point voltage, which should also be determined in the simulation. Due to the symmetry in the motor windings, only one phase voltage (Phase 1) is analyzed below; it can be repeated for the other two phases.

If v_a is the terminal voltage of the motor and v_s is the floating star-point voltage of the motor, both relative to the midpoint of the dc link voltage of the inverter, the phase voltage v_1 can be given as

$$v_1 = v_a - v_s \tag{7.20}$$

The terminal voltage v_a is determined by the switching states of the phase, which can be $\pm V_{dc}/2$. For the star-connected PM ac motor, it is always true that the summation of the line currents equals zero. Therefore, when all three phases conduct current, the floating star-point voltage of the motor can easily be derived from the three voltage equations given here.

$$v_s = \left(\frac{v_a + v_b + v_c}{3}\right) - \left(\frac{e_1 + e_2 + e_3}{3}\right) \tag{7.21}$$

Similarly, if only two of the phases (say, Phase 1 and 2) are conducting currents, the floating star-point voltage of the motor can be derived as

$$v_s = \left(\frac{v_a + v_b}{2}\right) - \left(\frac{e_1 + e_2}{2}\right) \tag{7.22}$$

Table 7-2 summarizes the estimated star point and phase voltage values of the inverter-driven motors, which are used in the VI to model the motor's drive.

7.3.1 Virtual Instrument Panel

The main objective of the VI presented here is to create a general simulation tool for the PM ac motors that can be utilized to study the dynamic as well as the steady-state performance under various excitation and loading conditions. In addition, the simulation tool can be used to study further control and parameter estimation concepts in the motor drives.

Table 7-2 *Summary of the star point and the phase voltages.*

Trapezoidal Back emfs	Sinusoidal Back emfs
$e_1 + e_2 + e_3 \neq 0$ (except zero crossing instants)	$e_1 + e_2 + e_3 = 0$
$v_a, v_b, v_c = \pm V_{dc}/2$ $v_s = \mathbf{K}[(v_a + v_b + v_c) - (e_1 + e_2 + e_3)]$	
If $i_1 = 0$ $\mathbf{K} = 1/2$ then $v_1 = e_1$, $v_2 = v_b - v_s$, $v_3 = v_c - v_s$	
If $i_2 = 0$ $\mathbf{K} = 1/2$ then $v_1 = v_a - v_s$, $v_2 = e_2$, $v_3 = v_c - v_s$	
If $i_3 = 0$ $\mathbf{K} = 1/2$ then $v_1 = v_a - v_s$, $v_2 = v_b - v_s$, $v_3 = e_3$	
If $i_1 \neq 0$, $i_2 \neq 0$, $i_3 \neq 0$ $\mathbf{K} = 1/3$ then $v_1 = v_a - v_s$, $v_2 = v_b - v_s$, $v_3 = v_c - v_s$	

The principal front panels of the PM ac motor drive and their explanation diagrams are given in Fig. 7-8 and Fig. 7-9, respectively. When the programming structure of the VI is examined, you will see that the graphical programming simplifies the complex drive structure significantly and provides easy debugging and a very user-friendly interface.

A number of sub-VIs are implemented in this study. The Motor sub-VI is based on the motor's functions explained in the motor model above and consists of four computation subsections. The first subsection calculates the electrical rotor position based on equation 7.19. The second section produces the back emf waveforms using equation 7.16 and equation 7.17. Then the three-phase voltages are estimated in the third section, which utilizes Table 7-2. Finally, section four solves the differential equations (equation 7.13) and computes the electromagnetic torque of the motor (equation 7.14).

The Motor sub-VI is customized, and six inputs and four outputs are defined, which are used to link to the other sub-VIs in the program. One of the inputs is the array of the inputs for the switching signals, which are used to link the control signals of the power devices in the inverter. The parameters of the motor and the calculation interval are the other two inputs in the sub-VI.

In order to set the initial values at the top-level VI, the final step values of the currents and the rotor position are also provided as inputs to the Motor sub-VI. An input signal named Mode is defined to select the type of motor: BTPM or BSPM. The outputs of the Motor sub-VI include the line currents, the phase voltages, the torque, and the rotor position.

Similar to the practical motor drive system, the Control sub-VI is implemented as the current controller of the motor drive. From the control point of view, both the BSPM and BTPM motors can accommodate the identical current

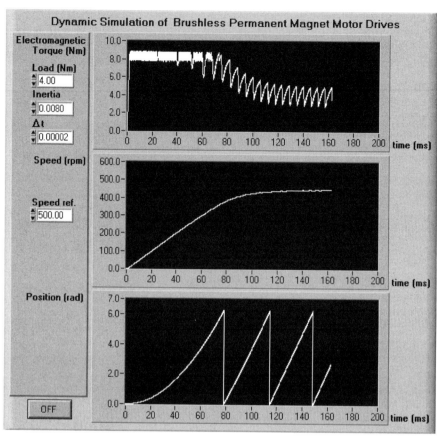
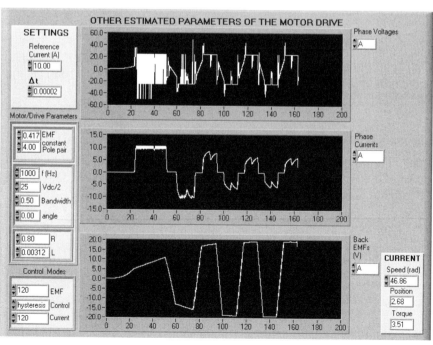

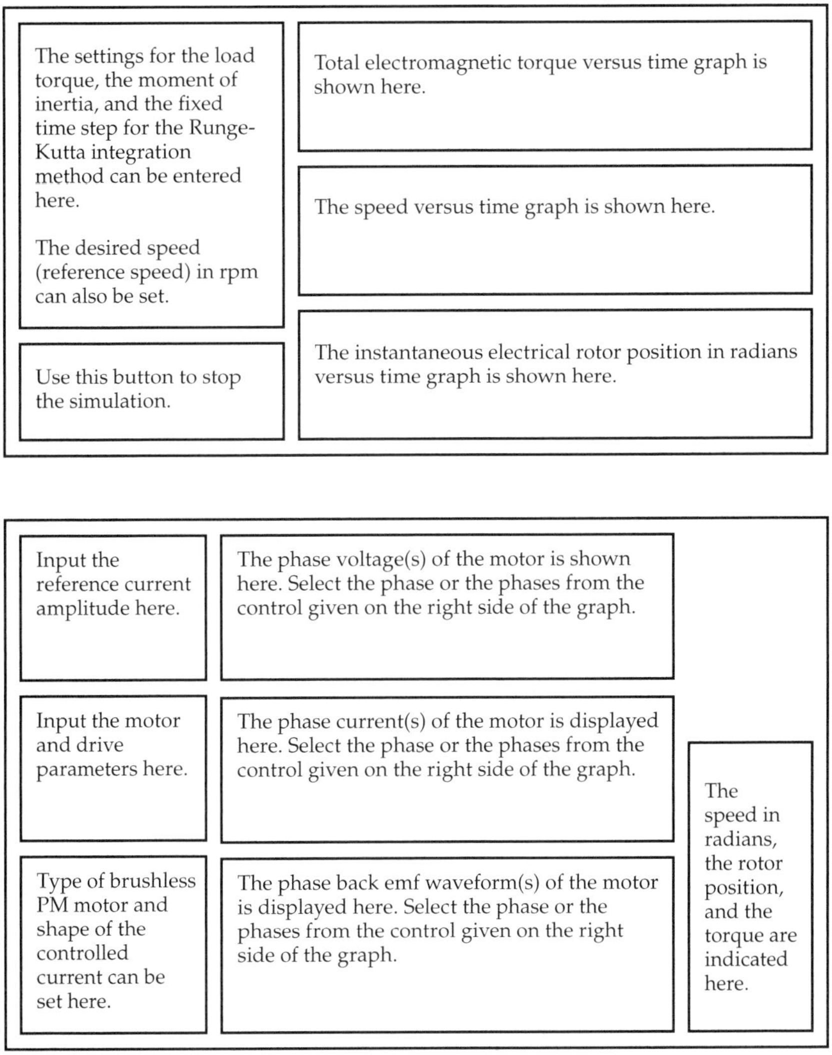

Chapter 7 • Simulation of Electrical Machines and Systems

The settings for the load torque, the moment of inertia, and the fixed time step for the Runge-Kutta integration method can be entered here.	Total electromagnetic torque versus time graph is shown here.
The desired speed (reference speed) in rpm can also be set.	The speed versus time graph is shown here.
Use this button to stop the simulation.	The instantaneous electrical rotor position in radians versus time graph is shown here.

Input the reference current amplitude here.	The phase voltage(s) of the motor is shown here. Select the phase or the phases from the control given on the right side of the graph.	
Input the motor and drive parameters here.	The phase current(s) of the motor is displayed here. Select the phase or the phases from the control given on the right side of the graph.	The speed in radians, the rotor position, and the torque are indicated here.
Type of brushless PM motor and shape of the controlled current can be set here.	The phase back emf waveform(s) of the motor is displayed here. Select the phase or the phases from the control given on the right side of the graph.	

Figure 7-9
The brief user guides for the front panels given in Fig. 7-8.

Figure 7-8
The front panels of `Brushless PM Motor Simulation.vi`.

controllers. The primary difference is in their reference current shape: sinusoidal or rectangular.

The Control sub-VI generates the three-phase current reference signals using the rotor position and the amplitude value of the current command. The reference current waveforms can be either as ideal sine waves or as piecewise rectangular waveforms, which can be expressed for Phase 1 as shown:

$$i_1 = I_m \sin(\theta_e) \tag{7.23}$$

$$i_1 = \begin{cases} 0 & 0 < \theta_e \leq \dfrac{\pi}{6} \\ I_m & \dfrac{\pi}{6} < \theta_e \leq \dfrac{5\pi}{6} \\ 0 & \dfrac{5\pi}{6} < \theta_e \leq \dfrac{7\pi}{6} \\ I_m & \dfrac{7\pi}{6} < \theta_e \leq \dfrac{11\pi}{6} \\ 0 & \dfrac{11\pi}{6} < \theta_e \leq 2\pi \end{cases} \tag{7.24}$$

where I_m is the amplitude of the stator current command.

Although there are various current control schemes used in practice that force the actual current to follow the reference signal, only two of the commonly used schemes are implemented in the Control sub-VI: Hysteresis and PWM current control, which generates the switching signals required by the Motor sub-VI.

There are four waveform graphs on the front panel of Brushless PM Motor Simulation.vi, which display the line currents, the phase voltages, the electromagnetic torque, and the rotor position. Moreover, the user has the option to display either single-phase or all three-phase currents and voltages.

In order to simulate the transient performance of the motor drive while maintaining simplicity, the Drive sub-VI is implemented, which has five inputs and three outputs. The inputs are the current command, the angular speed, the rotor position, the currents, and the calculation interval. The outputs include the torque, the rotor position, and the currents. Two additional sub-VIs are also included in the main VI to solve the mechanical equation of motion and to simulate a PI (proportional integral) speed regulator.

The motor and drive simulation presented here is verified comprehensively using real measured data and operating under steady-state as well as transient operating modes. The parameters of the real test motor (BTPM) are given in Table 7-3.

Table 7-3 *The test motor nameplate data and measured parameters.*

Torque constant	0.31 Nm/A
Back emf constant, k_e	0.417 V/rad/s
Moment of inertia, J	0.0008 kg m²
Number of poles	8
Winding resistance R	0.8 Ω
Equivalent winding inductance, L	3.12 mH

7.3.2 Self-Study Questions

Open and run the custom-written VI `Brushless PM Motor Simulation.vi` in Section 7.3 of the Chapter 7 folder, and investigate the following questions. The simulation can be started using the default values of the motor parameters. Alternative motor parameters can be obtained from any other textbook or from the motor manufacturers' catalogs. Note that the motor and controller settings are given in the second front panel.

1. Run the simulation and observe the changes in the total electromagnetic torque, the speed, and the rotor position, and identify the acceleration time of the motor.
2. Vary the total moment of inertia or the load torque and observe the changes in the speed-time characteristic.
3. For the operating conditions in question 2, observe the phase voltages, the line currents, and the back emf waveforms of the motor.
4. Study the transient behavior of the motor drive for the different motor types (sinusoidal and trapezoidal brushless PM motors) and for the different current controller settings (hysteresis and PWM current control).

7.4 Dynamic Simulation of Direct Current Motors

DC motors are the oldest and still most widely used motor types in servo drives covering a wide range of applications. In industrial drives containing dc motors, starting, braking, speed, and load changing commonly occur. To predict and understand the operation of dc motor drives (Motor + Power Electronics controller) under these transient conditions, you should either build a real motor drive or solve the dynamic equations of the motor numerically using the simulation model of the complete drive.

Building a real motor drive is an expensive solution. In addition, the real system cannot easily imitate the potential faults or extreme operating conditions that might occur in the real world. One solution to this problem is to build a computer simulation model of the drive system. If such a simulation tool is carefully designed, the information obtained from it can be used to determine the behavior of the motor under transient as well as at steady-state operating conditions. Furthermore, the simulation results may be utilized to select protection devices and to determine the approximate settings of the feedback controllers.

After performing this experiment, students should be able to:

- understand the operation of a dc motor under transient and steady-state operating conditions.
- understand the starting difficulties of dc motors.
- observe the behavior of the dc motor drive when it is operating in the single quadrant control system using a simple step-down dc-dc converter.

DC Motor Drive Model

The principal speed control method in dc motors is that if the armature voltage is varied, the speed will change proportionally. With a step-down dc-dc converter, the input dc voltage is converted to another voltage that is lower than the input, while the output current gets larger. As explained in Section 6.6, an elementary dc converter with PWM switching technique can provide a variable voltage output for dc motors.

In the dynamic simulation, the following equations are used: the voltage equation of the armature circuit of the motor, the dynamic equation of the motor-load (mechanical) system, and the electromagnetic torque equation of the motor, respectively.

$$v_a = R_a i_a + L_a \frac{di_a}{dt} + e_a \tag{7.25}$$

$$T_a = J \frac{d\omega_m}{dt} + B\omega_m + T_L \tag{7.26}$$

$$T_e = k_e i_a \tag{7.27}$$

where v_a is the supply voltage, i_a is the winding current, R_a is the resistance of the armature circuit, L_a is the inductance of the armature circuit, J is the moment of inertia of the system (including the motor and the load), the back emf is $e_a = k_e \cdot \omega_m$ (here k_e is known as the back emf constant or torque constant), ω_m is the angular speed of the system, T_e is the electromagnetic torque developed by the motor, T_L is the load torque, and B is the damping coefficient, which is usually ignored.

Note that the voltage and the load equations are valid for any type of dc motor. In this study, a Permanent Magnet excited dc motor is considered, and the state variables i_a and ω_m in the linear differential equations are solved simultaneously using the Runge-Kutta method available in the G-Math Toolkit of LabVIEW.

The PWM signal generated in this VI controls the switching states of the transistor, hence the average value of the output voltage can be varied. Although high-performance applications require a closed-loop control, an open-loop control is adequate in many applications, as will be demonstrated here.

The controller implemented here is a single-quadrant controller employing a single switching transistor. Therefore, it can deliver to the motor only positive (or only negative) voltage and current, hence control speed is only in one direction with the load torque as an opposing torque.

7.4.1 Virtual Instrument Panel

In the VI `Brush DC Motor Simulation.vi` (in Section 7.4 of the Chapter 7 folder), a Permanent Magnet excited (which is analogous to a conventional motor operating under a constant field current) dc motor drive simulation is provided. The front panel of the VI is given in Fig. 7-10.

The dc step-down converter is implemented to provide a PWM voltage waveform to the motor winding, which operates in open loop. The duty cycles of the PWM signal can be varied by the controls provided on the front panel.

It should be noted here that since the drive system is an open-loop system, there is no speed or current feedback in the drive. However, a closed-loop system can be achieved easily as shown in the previous section.

The user can vary the frequency as well as the duty cycle of the control signal from the front panel, to observe continuous as well as discontinuous current conduction modes, which are highly dependent on the motor parameters, operating speed, and switching conditions.

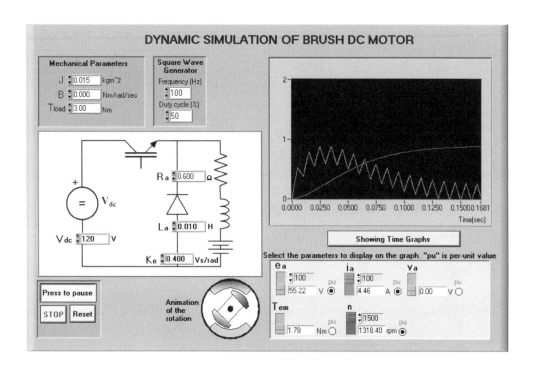

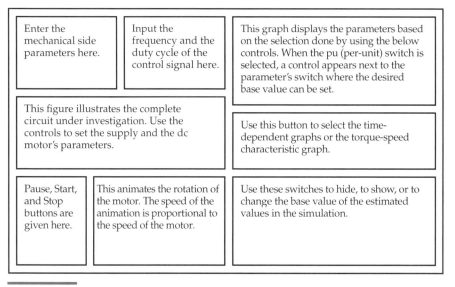

Figure 7-10
Front panel and brief user guide of DC motor simulation.vi.

7.4.2 Self-Study Questions

Open and run the custom-written VI named `DC Motor Simulation.vi`, and investigate the following questions.

1. Investigate the motor's starting behavior from standstill with and without a mechanical load attached. Observe that the motor starts only after its developed torque exceeds the load torque, and the acceleration rate depends on the inertia of the system.
2. When the motor is started, the initial values of the current and the back emf are zero. Due to the inductance of the circuit, the current rises and falls exponentially within the control interval of the converter's switch (the transistor). Vary the duty cycle and the frequency of the control signal and obtain the continuous and discontinuous current conduction modes in the motor drive.
3. Investigate how the motor current, the torque, and the speed vary under the operating conditions in question 2.
4. Observe CCW (counterclockwise) rotation by reversing the supply voltage.
5. The motor and system parameters can be varied at any instant while the simulation is running. Vary only one parameter at a time and observe the estimated parameters.
6. To study the retardation, simply change the value of the supply voltage to zero with a 0% duty cycle while the motor is operating at a steady-state speed.
7. The starting behavior of the dc motor can be observed by keeping the transistor on all the time (by setting the duty cycle to 100%) at start. Repeat this test by reducing the duty cycle and observe the maximum current value obtained and the time taken to reach the steady-state speed.
8. When the switching frequency of the inductance of the motor winding is high, the motor current is continuous and reaches steady state after a certain number of cycles. Observe this condition displaying the estimated current waveform of the motor.

7.5 Simulation of Stepper Motors

Stepper motors are simple, robust, and reliable and are well suited for open- or closed-loop control applications. They can be used as actuators of pointing mechanisms for antennas, mirrors, or telescopes, and in machine tools, robotics, typewriters, printers, watches, process control systems, tape and disk drive systems, and programmable controllers. They are usually used in connection with gears; hence a gear model as well as electrical and mechanical stiffness, friction, and resistance should be considered in the design of a complete stepper motor drive system.

Although there are a number of stepper motor and drive types available for use in these applications, a commonly used motor and drive type may be sufficient to explain and demonstrate the principal operations of stepper motors. There are basically four types of stepper motors:

- Variable Reluctance (VR)
- Permanent Magnet (PM)
- Hybrid
- Linear

Other types of stepper motors also exist, such as solenoid-ratchet, phase-pulsed synchronous, electro-hydraulic, and harmonic drive. However, in this study we will consider PM stepper motors only and will provide a basic computer simulation tool. But first, to provide an introduction, let's briefly describe the listed stepper motors and then extend the discussion to explain and develop the mathematical model of a PM stepper motor and its open-loop drive.

VR Stepper Motors

The stator of a variable-reluctance stepper motor has a magnetic core constructed with a stack of steel laminations. The rotor is made of unmagnetized soft steel with teeth and slots. Since there is no permanent-magnet rotor, there is no residual torque in such motors to hold the rotor at one position when turned off.

When the stator coils are energized in VR stepper motors, the rotor teeth will align with the energized stator poles. This type of motor operates on the principle of minimizing the reluctance along the path of the applied magnetic field. By alternating the windings that are energized in the stator, the stator field changes, and the rotor is moved to a new position.

The VR stepper motor mentioned here is a single-stack motor. That is, all the phases are arranged in a single stack, or plane. The disadvantage of this design for a stepper motor is that the steps are generally quite large (above

15°). To reduce the step size, multistack (magnetically isolated sections excited by a separate winding or phase) stepper motors are used.

PM Stepper Motors

Permanent-magnet stepper motors have a permanent-magnet rotor, armature magnetized perpendicular to the rotation axis. These types of motors can maintain the holding torque indefinitely when the rotor is stopped. The force or the torque that exists without the excitation of the windings of the motor is known as residual, or detent, or cogging torque (which is generally about one-tenth of the holding torque). It is noticeable by turning its rotor manually.

Hybrid Stepper Motors

The hybrid stepper motor consists of two pieces of soft iron, as well as an axially magnetized, round PM rotor. The term *hybrid* is derived from the fact that the motor is operated under the combined principles of the permanent-magnet and variable-reluctance stepper motors. The stator core structure of a hybrid motor is essentially the same as its VR counterpart. The main difference is that in the VR motor, only one of the two coils of one phase is wound on one pole, while a typical hybrid motor will have coils of two different phases wound on the same pole. The two coils at a pole are wound in a configuration known as a bifilar connection.

Linear Stepper Motors

The linear stepper motor has been made flat instead of round so its motion will be along a straight line instead of rotary. Applications for this kind of motor are widespread, ranging from a coil winding system to transporting a silicon semiconductor wafer through a laser inspection station.

Step Angle

As the name suggests, each revolution of the stepper motor's shaft is made up of a series of discrete individual steps. A step in a stepper motor is defined as the angular rotation produced by the output shaft each time the motor receives a step pulse that causes the shaft to rotate a certain number of degrees. The sequence and rate of the step pulses control the position and speed of the rotor.

The number of poles on the rotor and stator determine the step angle that will occur each time the polarity of the winding is reversed. The greater the number of poles, the smaller the step angle that can be achieved. The number of steps per revolution of the rotor can be given by

$$s = n_r p_s \qquad (7.28)$$

where n_r is the number of rotor teeth, and p_s is the number of phases of the stepper motor. Then the step size $\Delta \theta$ is determined by

$$\Delta \theta = 360°/s \qquad (7.29)$$

Note that for the motor configuration explained in Section 7.5.1, $n_r = 2$ and $p_s = 2$. Hence the full step size is 90°, and the half step size is 45°.

There are three possibilities to increase the number of steps in a stepper motor:

- Increasing the number of phases (which increases complexity of the drive circuit and its cost, and is limited by manufacturing capabilities)
- Increasing the number of poles of the rotor (which also is limited by manufacturing capabilities)
- Using various control techniques such as half-stepping and micro-stepping.

7.5.1 Mathematical Model

One of the distinct features of the stepper motors is that the pole faces on the rotor and the stator are offset so that there will be only a limited number of rotor teeth aligning themselves with an energized stator pole.

Due to the atypical structure of these motors, most of the time the explanations involve visual aids, such as rotor state diagrams. For simplicity, we will consider a two-phase PM stepper motor in this study and provide explanations for the computer simulation study. The two-phase PM stepper motor and its inert rotor states are shown in Fig. 7-11. Note that the rotor states displayed in the figure include half-steps but not micro-steps, which will be explained later.

In Fig. 7-11, the PM rotor is shown at the center. The outer circle represents the stator, with embedded Phase a and Phase b windings. Two opposite wind-

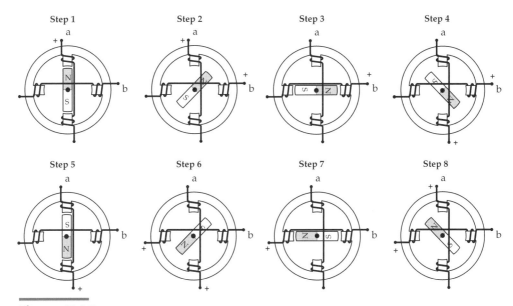

Figure 7-11
The rotor position states in the two-phase stepper motor (excluding micro-steps).

ings shown in the figure are connected in series. As illustrated, the rotor states depend on the status of the excited windings, which create the magnetic field.

The magnetic field, as a result of the excitation, can be found by using the right-hand rule. Therefore, the PM rotor moves and aligns itself with this field. If, for example, Phase a is active (Phase b is switched off), the series windings produce an electrical south pole and north pole in the stator, which brings the rotor into a stable position. However, if multiple phases are excited, then the rotor position can be determined by the vector summation of the magnetic fields of the excited motor phases, which is illustrated by the rotor positions between the main stator poles.

Note that this motor configuration, known as a bipolar motor, is designed with separate coils that need to be driven in either direction (the polarity needs to be reversed during operation) for proper stepping to occur.

For a two-phase PM stepper motor, the following voltage equations can be given for Phases a and b.

$$v_a = e_a + Ri_a(t) + L\frac{di_a(t)}{dt} \tag{7.30}$$

$$v_b = e_b + Ri_b(t) + L\frac{di_b(t)}{dt} \tag{7.31}$$

where v_a and v_b are the voltages applied to the windings, i_a and i_b are the winding currents, R is the resistance and L the inductance of the windings, and e_a and e_b are the back emf of the windings induced by the rotation of the rotor. They are given as

$$e_a = k_e \omega \sin \theta \quad (7.32)$$
$$e_b = -k_e \omega \cos \theta \quad (7.33)$$

where k_e is the peak value of flux linkage (or the back emf constant), ω is the rotational velocity of the motor, and θ is the angular rotor position.

The electromagnetic torque generated by each phase is then given by

$$T_a = (-k_e \sin \theta) i_a(t) \quad (7.34)$$
$$T_b = (k_e \cos \theta) i_b(t) \quad (7.35)$$

Hence the total torque $T_a + T_b$ can be included in the equation of motion of the motor, where the state variable is the rotational velocity ω of the motor.

$$T_a + T_b = J \frac{d\omega}{dt} + B\omega + T_{\text{load}} \quad (7.36)$$

The rotor position θ can be found from

$$\frac{d\theta}{dt} = \omega \quad (7.37)$$

The preceding differential equations can be used to observe the dynamic behavior of the PM or Hybrid stepper motor. In this study, this is achieved by solving these linear differential equations numerically using LabVIEW.

7.5.2 Control of Stepper Motors

It should be emphasized here that the power circuits required to achieve the rotor positions shown in Fig. 7-12 are important in supplying the correct excitation sequence. In addition, the number of switching devices and their voltage and current ratings may change significantly based on the switching circuit topology used.

As can be seen in Fig. 7-11, the current in the winding is bi-directional—which can only be achieved if an H-bridge configuration is utilized—and should accommodate eight switching devices in total (Fig. 7-12). If the phases and the stator coils are energized in the sequences required, it is possible to control the steps and the direction of the motor.

Three possible operating modes of the two-phase stepper motor, corresponding switching states for each step, and timing diagram and correspond-

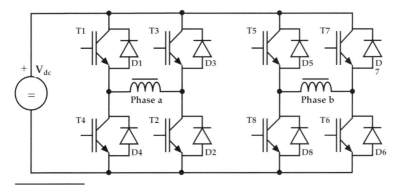

Figure 7-12
H-bridge drive circuit for the two-phase bipolar PM stepper motor.

ing logic level signals for each step are summarized in Table 7-4. Note that there are current conduction states in the drive circuit when the freewheeling diodes conduct currents. However, they are not given in the table.

- Wave-drive mode (one phase ON at a given time)
- Full-step mode (two phase ON at a given time, highest torque)
- Half-step mode (combination of both, results in half-step size)

As also shown in the table, the sequence of pulses causes the motor shaft to rotate clockwise (CW) or counterclockwise (CCW) in 45° steps. As stated earlier, the direction of the currents in each phase is bi-directional (positive and negative). Hence the H-bridge drive circuit can provide bi-directional current and half stepping when the correct pulse sequence is provided to its switching devices.

Stepping motors come in a wide range of angular resolution. The coarsest motors typically turn 90° per step, while high-resolution permanent-magnet motors are commonly able to handle 1.8° or even 0.72° per step. With an appropriate controller, most permanent magnet and hybrid stepper motors can be run in half-steps, and some controllers can handle smaller fractional steps or micro-steps.

Note that Table 7-4 does not indicate the micro-stepping operation of the stepper motor. In the micro-stepping mode of operation, the current varies as a sine wave, and that is always out of phase with the other phase. Hence the rotor in this operating mode can reach hundreds of intermediate steps. In Fig. 7-11, if the vector summations of the magnetic fields of the phases are considered, the micro-step positions of the rotor can easily be visualized.

Table 7-4 Switching sequences to obtain various operating modes in a two-phase PM stepper motor.

Wave Drive (one phase ON at a given time)

Voltage in the Phases	Step #: Direction	\	Status of the Supply								Step #: θ_m:	Control Signals							
			1	2	3	4	5	6	7	8		1 0°	2 45°	3 90°	4 135°	5 180°	6 225°	7 270°	8 315°
a,+	CW ↓		1	0	0	0	0	0	0	0	T1	ON	–	–	–	–	–	–	–
b,+			0	0	1	0	0	0	0	0	T2	ON	–	–	–	–	–	–	–
a,–			0	0	0	0	1	0	0	0	T3	–	–	ON	–	–	–	–	–
b,–	CCW ↑		0	0	0	0	0	0	1	0	T4	–	–	ON	–	–	–	–	–
											T5	–	–	–	–	ON	–	–	–
											T6	–	–	–	–	ON	–	–	–
											T7	–	–	–	–	–	–	ON	–
											T8	–	–	–	–	–	–	ON	–
												CW →							←CCW

Full Step (two phases ON at a given time, highest torque)

Voltage in the Phases	Step #: Direction		Status of the Supply								Step #: θ_m:	Control Signals							
			1	2	3	4	5	6	7	8		1 0°	2 45°	3 90°	4 135°	5 180°	6 225°	7 270°	8 315°
a,+	CW ↓		0	1	0	0	0	0	0	1	T1	–	ON	–	–	–	–	–	ON
b,+			0	1	1	0	0	0	0	0	T2	–	ON	–	–	–	–	–	ON
a,–			0	0	0	1	1	0	0	0	T3	–	ON	–	ON	–	–	–	–
b,–	CCW ↑		0	0	0	0	1	1	0	1	T4	–	ON	–	ON	–	–	–	–
											T5	–	–	–	ON	–	ON	–	–
											T6	–	–	–	ON	–	ON	–	–
											T7	–	–	–	–	–	ON	–	ON
											T8	–	–	–	–	–	ON	–	ON
												CW →							←CCW

Half Step (combination of both)

Voltage in the Phases	Step #: Direction		Status of the Supply								Step #: θ_m:	Control Signals							
			1	2	3	4	5	6	7	8		1 0°	2 45°	3 90°	4 135°	5 180°	6 225°	7 270°	8 315°
a,+	CW ↓		1	1	0	0	0	0	0	1	T1	ON	ON	–	–	–	–	–	ON
b,+			0	1	1	1	0	0	0	0	T2	ON	ON	–	–	–	–	–	ON
a,–			0	0	0	1	1	1	0	0	T3	–	–	ON	ON	–	–	–	–
b,–	CCW ↑		0	0	0	0	0	1	1	1	T4	–	–	ON	ON	–	–	–	–
											T5	–	–	–	ON	ON	ON	–	–
											T6	–	–	–	ON	ON	ON	–	–
											T7	–	–	–	–	–	ON	ON	ON
											T8	–	–	–	–	–	ON	ON	ON
												CW →							←CCW

7.5.3 Virtual Instrument Panel

The foremost aim of the VI illustrated in Fig. 7-13 is to provide some insights, first into the operation principles, and second for the performance prediction of the stepper motors. The custom-written VI is named `Stepper Motor Simulation.vi` and is located in Section 7.5 of the Chapter 7 folder.

In the simulation, the rotation of the stepper motor is performed as a function of a set of motor parameters, drive parameters, and simulation settings, all of which are specified by the user. The motor parameters include the winding parameters; the drive and simulation settings include the polar moment of inertia, the damping coefficient, the load torque, the supply voltage, the voltage drop across the switching devices of the H-bridge, the desired number of steps, the integration time step, and the pulse length.

A number of simulation control buttons are also provided on the front panel of the VI: Start/Stop, Pause/Resume, Exit, CW/CCW, and a sliding control for Simulation Speed.

The rotor's position and speed, the voltages, currents, and torques of the windings, and the total torque can be displayed in the chart provided. Note that these parameters require scaling to be viewed on the same graph, which can be done using the controls provided.

On the front panel, two picture rings are utilized to animate the rotation of the motor. The diagrams of each of the drive circuitry states and motor schematics for every 5° turn of the rotor are included in the picture rings, which are continuously updated to reflect the current status of the motor drive. It should be emphasized here that the rotor states at every 5° rotation may not be ideal, but this method is preferred to reduce the complexity of the programming.

However, for continuous rotor position images, a picture control VI can be designed where the rotor position can be plotted from the instantaneous position data. In such implementation, the stationary part of the stepper motor, the stator, can be a background image for the plotting area.

As the stepper motor simulation executes, the graphs of the rotor angle, the velocity, the winding voltages, current, and torques can be displayed. These can be selected using the switches provided at the bottom of the graph. Note that since these estimated values may be very diverse, a control is provided at the bottom of each graph display button that can be used to scale up or down the parameters.

Let's now briefly examine one step in the clockwise rotation of the stepper motor to illustrate the operation of the driving circuitry and the subsequent rotor rotation.

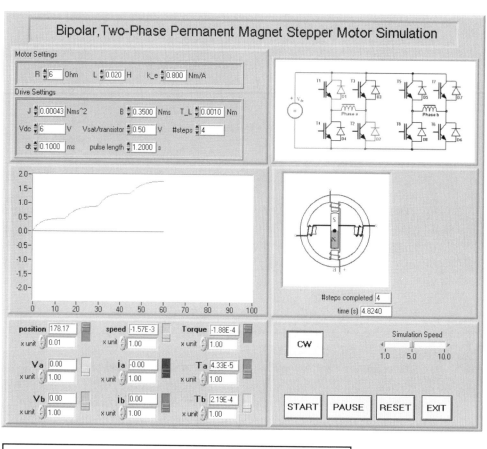

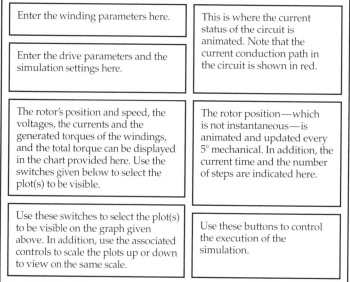

Figure 7-13
The front panel and brief user guide of the stepper motor simulation.

Referring to the initial state (unexcited) of the drive circuit, if Phase a and Phase b are connected to the dc supply, the phase windings will create a resultant magnetic field that rotates the rotor and aligns it with this field. When the permanent magnet rotor is aligned, an equilibrium condition exists, even with the windings still energized. The rotor will realign itself with each new stator field. For the next step, only one phase winding should be excited (to rotate the rotor in a defined direction), so that the resulting magnetic field rotates the rotor by another 45°.

Programming Details

Due to the complexity of the dynamic simulation of the stepper motor drive, the angular position of the rotor is estimated using separate sub-VIs that are interlinked to complete an open-loop motor drive. Therefore, note that the validity of the estimated values very much depends on the correct parameter settings of the stepper motor and the drive system. The critical parameters in the simulation can be listed as the moment of inertia, the damping coefficient as well as the time step used in the integration, and the pulse width (which is assumed to be sufficient for the rotor movement to take place).

Note that to start the simulation, you have to press the Start button. Simulation time is incremented and the calculations for a new rotor position are performed only if the Start button has the value True (and the Pause button has not been pressed). A particular simulation stops automatically when the number of steps, as specified by the user, have been simulated through time. When the Start button is pressed again, rotation continues from the last rotor position. To set the rotor position to the initial position (the position where the north pole of the rotor is vertically upward), the Reset button has to be pressed.

Due to the presence of the Exit button, there is an outermost while loop in the main VI that executes conditional on the value of this button. This loop contains a set of shift registers that store all the values that need to be passed through from the previous time instant to the current one.

In the VI, the rotor position is calculated at each discrete time instant using the mathematical model provided earlier, which takes place in a separate sub-VI and is also used to animate the rotor position. The Next Phase Signal sub-VI returns the correct sequence of Phase a and Phase b voltages given input parameters such as time, direction of rotation (CW/CCW), and the length of time for each single pulse.

The Next Circuit sub-VI animates the driving circuit of the motor. Its inputs are the current value of the Phase a and Phase b voltages and the current diagram it is displaying (an unsigned integer value). The current conduction path of the circuit is indicated in red in the displayed frames. For example, if Phase a had a positive voltage, the animation panel displays a circuit in which transistors T1 and T2 are highlighted in red. The display is conditional on the sign of the Phase a and Phase b voltages. In addition, pairs of freewheeling diodes may be highlighted in red during the current decay period of the outgoing phase, which is due to the inductance of the windings.

To solve the differential equations of the stepper motor drive, the trapezoidal discrete time integration rule is used in the VI, which is implemented in a sub-VI named Single Time Step Calc. At each instant of the simulation time, the calculated values of the state variables are evaluated to generate the discrete values of the induced voltages, the winding currents, the electromagnetic torques, the speed, and the rotor position.

7.6 Steering and Control of Four-Wheel Direct-Drive Electric Vehicles

Since the 1990s, electric vehicles (EVs) have begun to improve considerably and have started to compete with the internal combustion engine–powered vehicle. This is mainly due to advances in the design and manufacture of power electronics devices, improvements in energy storage technology, emerging alternative energy sources (such as fuel cells), and the discovery of new and more efficient electric motor configurations (such as axial field and permanent magnet ac motors). Many believe that we are in the transition stage of this development toward future all-electric vehicles.

Table 7-5 illustrates the power requirements of standard and all-electric vehicles as well as transition stage vehicles. There are various components or systems in an EV: energy storage unit, energy control system (power electron-

Table 7-5 *Standard to electric vehicle migration.*

Configuration	Electric (kW)	Mechanical (kW)	Generator (kW)
• Standard car	1	50–100	–
• Standard with integrated starter alternator (ISG)	4–6	50–100	–
• Parallel hybrid	10–30	20–60	–
• Series hybrid	–	50–100	20–60
• All-electric vehicle	50–100	–	–

ics and control hardware), electric propulsion systems (engines, motors), devices to provide ancillary functions (air conditioning, power steering, etc.), and mechanical drive configurations. By incorporating these systems in an EV, control strategies can be extended and optimized, mainly to reduce power losses and to increase efficiency.

In the following subsections, the current status of EV research is discussed briefly and a detailed LabVIEW-based analysis tool is presented for an advanced four-wheel drive and four-wheel steering electric vehicle with in-hub direct drive motors.

The need to develop control strategies for steering and speed in such a system was identified from the literature and resulted in the development of an educational tool to demonstrate the control functions required in such a complex system.

Drive Train Options in Road Vehicles

In a conventional motor vehicle, the power generated by the internal combustion engine (ICE) is transferred to the wheels via a gearbox, which matches the speed-torque characteristics of the ICE to the vehicle characteristics.

From the gearbox output, either a tail-shaft, differential, and half-shaft combination for rear-wheel drive vehicles, or a differential and half-shaft combination for front-wheel drive vehicles is used to transfer the mechanical torque to the road wheels and hence to the ground. The action of the differential is to allow the two driven wheels to rotate at different speeds while the vehicle is traveling in a curved path and still transmit the torque to each wheel, thereby generating forward (or reverse) motion.

An electric motor can be fitted to a vehicle in several ways:

- Simple replacement of the ICE with one electric motor (Fig. 7-14a)
- Hybrid (series or parallel) connection in ICE and electric motor configuration (Fig. 7-14b, c)
- Direct connection of electric motors to the drive wheels (Fig. 7-14d, e)

Direct Drive

As illustrated in the direct connection (Fig. 7-14d, e), the differential is removed completely and the functionality of the differential is generated through the direct control of each electric motor. In fact, when the motors are mounted inside the road wheel hubs (Fig. 7-14e), drive train weight is

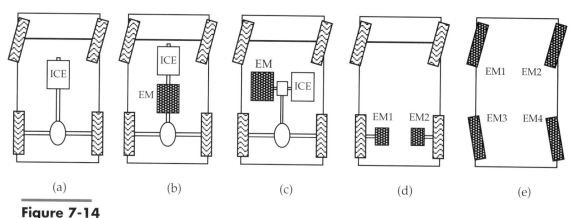

Figure 7-14
Drive options in electric vehicles: (a) simple conversion, (b) series hybrid, (c) parallel hybrid, (d) direct drive, and (e) direct drive in-hub. (ICE: internal combustion engine; EM: electric motor)

reduced to effectively nil. One disadvantage of this system is that the sprung mass of the wheel is increased, requiring further modification to the vehicle suspension. The principal advantage of this type of drive is that such a wheel can effectively turn greater than $\pm 90°$, increasing vehicle maneuverability for parking, and so on, as there is no mechanical drive coupling limiting the rotational steering action.

However, one result of the in-hub motor drive configuration is the need to develop a control strategy to replace the mechanical torque sharing previously provided by the differential. Note that the same control strategy can also be extended to incorporate control of individual in-hub motor steering angles, allowing the vehicle to turn and maneuver much more easily than existing conventional vehicles.

Engineers believe that among the various EV configurations, those illustrated in Fig. 7-14d, e will provide the most compatible and safest vehicle and control configurations for future systems, and may include various advanced features such as drive by wire, active steering, adaptability to smart roads, and trip computers for navigation and communications.

7.6.1 Criteria Used to Develop a LabVIEW Simulation

Development of a suitable control strategy for the in-hub direct motor drive and development of the LabVIEW-based algorithms and tools are provided here.

Although there are several motor types that may be used in direct drive configurations previously mentioned, the PM ac motor is selected for the simulation due to its higher efficiency, superior starting and response characteristics, and its regenerative operation, which allows energy to flow back into the storage system.

Furthermore, an additional drive option is also provided in the simulation using two- or four-wheel steering along with four direct-drive electric motors, one in each wheel.

As stated previously, the benefits gained by mounting the drive motors directly in the road wheels are partially offset by the need for additional complexity in the control system, particularly evident when the vehicle is traveling in a curved path.

As each wheel is traveling along a path with a different radius, they have different linear and rotational velocities as well as different loads. Determining the desired set points for these three parameters that should be supplied to each motor controller is the principal task for this simulation. In addition, the aim is to provide a clear insight into the operation of such a complex system using the capabilities of LabVIEW programming.

The parameters listed in Table 7-6 are identified as contributing to a simulation solution, and the following conventions were used throughout the simulation:

- Steering Angle: clockwise is positive
- Vehicle Velocity: forward is positive (head winds are negative)

The relationships between the first three parameters in the table, for two-wheel steering, are shown in Fig. 7-15.

The effective vehicle turning radius, R in the figure, can be calculated using the steering angle θ_s and the vehicle wheel base length B.

$$R = B \cot(\theta_s) \tag{7.38}$$

Table 7-6 *Vehicle parameters used in the simulation.*

Parameter	Symbol	Unit
• Steering Wheel Angle	θ_s	Degrees
• Vehicle Wheel Base Length	B	Meters
• Vehicle Wheel Base Width	A	Meters
• Two- or Four-Wheel Steering	True ≥ 4	Boolean value
• Road Wheel Diameter	D_w	Meters
• Accelerator Pedal Position	V_s	Meters/second

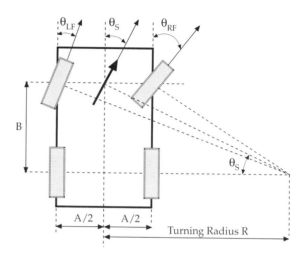

Figure 7-15
Two-wheel steering geometry.

Furthermore, the left (θ_{LF}) and right (θ_{RF}) front-wheel steering angles can be calculated by increasing or decreasing the turning radius by $A/2$, respectively, and recalculating the angle for each position as given in the first column of Table 7-6.

For four-wheel steering, the geometry changes so that instead of the vehicle turning point being constrained to move along a line common to the rear wheel axles, the turning point can now be located at any point in the x-y plane. Examples of this are illustrated in Fig. 7-16, where the horizontal lines in each

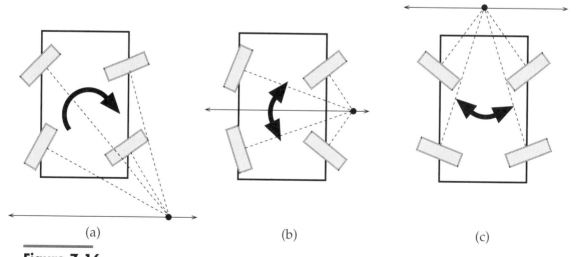

Figure 7-16
Various turning point positions for four-wheel steering geometry showing the resultant road-wheel angles.

example show one possible range of turning point positions. While this considerably increases the maneuverability of the vehicle as can be seen by the arrows indicating the direction of movement in Fig. 7-16, it would also probably be expected to significantly increase the risk of accidents.

Due to the potential risks associated with the turning point positions, it was decided that for four-wheel steering, the user will not have the ability to vary the position of the turning point line. Hence, the turning point in the simulation is permanently located midway between the front and rear wheel axles (Fig. 7-16b). From this position, the location of the turning point can move along a line perpendicular to the vehicle front/rear axis as the steering angle varies. This allows the vehicle direction to vary from travel in a straight line through to effectively spinning around its center point as the steering angle changes from 0° to 90°.

Various other measures can also be taken to reduce the risk of accidents due to the turning point positions, such as

- The maximum speed in the four-wheel steering mode can be limited to approximately 5 km/h.
- The four wheels can remain parallel as they turn for steering, facilitating parking or maneuvering in difficult situations.

Using the specifications developed above and by extending the formulas for two-wheel steering (as given in Table 7-7, first column), the road-wheel angles for four-wheel steering can easily be obtained. This is done by using the value of $B/2$ in place of B to allow for the reduced effective vehicle length (the distance from the line describing the range of turning point positions to the road wheel axis), with the rear-wheel steering angles on each side being equal but opposite in sign to the respective front-wheel steering angle.

Using the values determined above in conjunction with the road wheel diameter and the accelerator pedal position parameters, the algorithm can be extended to generate linear and angular velocity values at each wheel position.

The accelerator pedal position value was limited to the range of -5 to 33 m/s, resulting in a maximum forward vehicle velocity (V_A) of approxi-

Table 7-7 *Road-wheel steering angles for two- and four-wheel steering.*

Road-Wheel Angle (degrees)	Two-Wheel Steering	Four-Wheel Steering
Left Front, θ_{LF}	$\tan^{-1} B/(R + (A/2))$	$\tan^{-1}(B/2)/((R/2) + (A/2))$
Right Front, θ_{RF}	$\tan^{-1} B/(R - (A/2))$	$\tan^{-1}(B/2)/((R/2) - (A/2))$
Left Rear, θ_{LR}	0	$-1 \cdot \theta_{LF}$
Right Rear, θ_{RR}	0	$-1 \cdot \theta_{RF}$

mately 120 km/h (118.8 km/h exactly), and maximum reverse velocity of approximately 20 km/h (18 km/h exactly). This velocity is considered to be applied at a point centrally placed between the two front wheels, effectively at the steering position that is selected to allow it to be used as the reference position for later calculation of the linear velocities at the individual road wheel positions.

This velocity was further limited by multiplying it by the cosine of the steering angle (giving a velocity of $V_A \cos \theta_s$), effectively limiting the vehicle forward velocity as the steering angle increased. This reduction was incorporated as a safety measure, effectively reducing the maximum amount of kinetic energy available to tip the vehicle as the angle of turn increases (and the turning radius decreases).

The maximum velocity can be further limited by some other function that can incorporate the vehicle mass, wheel base width, and turning radius. However, this is not implemented in the simulation study. Using this approach, the forward velocity would always be limited to a value below that amount where the centripetal acceleration caused by the vehicle turning is sufficient to tip the vehicle, or even reduce the frictional load on the wheels on the inside of the turn.

The linear velocity for each wheel position is calculated using the ratio of the wheel position turning radius to the steering wheel position turning radius. This ratio is applied to the scaled forward velocity developed earlier generating individual wheel velocities at each position. The linear velocities are then converted to angular velocities, based on the diameter of the road wheels. Table 7-8 summarizes all the calculations mentioned earlier.

The remaining parameter to be calculated is the load torque at each wheel motor position. It is assumed that the total load generated by the vehicle is shared equally among the four wheel motors, hence by calculating the total load and dividing by four, the individual load at each axle is determined. Then the individual load is converted to a torque value by incorporating the road wheel diameter. Based on these assumptions, total vehicle road load can be given by

$$F_W = f_{RO} + f_{AE} + f_{CR} \qquad (7.39)$$

where f_{RO} is the rolling resistance, f_{AE} is the aerodynamic resistance, and f_{CR} is the climbing resistance.

The rolling resistance is proportional to the effort required to overcome tire distortion as the tires contact the road surface, and it is equal to

$$f_{RO} = fmg \qquad (7.40)$$

Table 7-8 *Road wheel linear and angular velocities for two- and four-wheel steering.*

Road Wheel Linear Velocity	Two-Wheel Steering (m/s and rpm)	Four-Wheel Steering (m/s and rpm)
Left Front, V_{LF}	$V_A \cos\theta_S \dfrac{(R+(A/2))/\cos\theta_{LF}}{B/\sin\theta_S}$	$V_A \cos\theta_S \dfrac{((R/2)+(A/2))/\cos\theta_{LF}}{(B/2)/\sin\theta_S}$
	$(60\,V_{LF})/(\pi D_W)$, in rpm	
Right Front, V_{RF}	$V_A \cos\theta_S \dfrac{(R-(A/2))/\cos\theta_{RF}}{B/\sin\theta_S}$	$V_A \cos\theta_S \dfrac{((R/2)-(A/2))/\cos\theta_{RF}}{(B/2)/\sin\theta_S}$
	$(60\,V_{RF})/(\pi D_W)$, in rpm	
Left Rear, V_{LR}	$V_A \cos\theta_S \dfrac{(R+(A/2))}{B/\sin\theta_S}$	V_{LF}
	$(60\,V_{LR})/(\pi D_W)$, in rpm	
Right Rear, V_{RR}	$V_A \cos\theta_S \dfrac{(R-(A/2))}{B/\sin\theta_S}$	V_{RF}
	$(60\,V_{RR})/(\pi D_W)$, in rpm	

where f is the rolling resistance coefficient, which is dependent on the vehicle velocity and the turning angle, m is the vehicle mass in kg, and g is the acceleration due to gravity.

The aerodynamic resistance results from the viscous frictional load of air on the front surface of the vehicle and is equal to

$$f_{AE} = 0.5\zeta C_W A(\nu + \nu_O)^2 \tag{7.41}$$

where ζ is the air density, C_W is the aerodynamic drag coefficient, A is front surface area of the vehicle, ν is the vehicle forward velocity, and ν_O is the head wind velocity.

The climbing resistance f_{CR} is due to the energy that has to be expended to move a vehicle up a slope (gain in potential energy), or the energy that is released when the vehicle moves down a slope. It is considered to be positive for increasing slopes and negative for decreasing slopes.

$$f_{CR} = mg\sin\alpha \tag{7.42}$$

where α is the grade angle in rads.

Some additional assumptions are also made to reduce the complexity of the LabVIEW simulation:

- The rolling resistance coefficient f is assumed to be constant (independent of the vehicle velocity and the steering angle).
- The ground is assumed to be level, hence $f_{CR} = 0$.

In addition to the described loads, the vehicle may accelerate or decelerate if the total motor propulsive force F_M is greater or less than the road load F_W, respectively. The resultant acceleration is given by

$$a = \frac{F_M - F_W}{k_M m} \tag{7.43}$$

where a is the acceleration, and k_M is a factor that increases the effective mass of the vehicle to include the effect of the rotational inertia of the rotating masses in the vehicle structure. However, for simplicity, the effects of acceleration are also ignored in the simulation (meaning the vehicle travels at a constant velocity).

Considering all the previous assumptions, the resultant total road load is given by

$$F_W = (fmg) + (0.5\zeta C_W A (v + v_O)^2) \tag{7.44}$$

Hence the axle load at each motor position is equal to

$$F_A = \frac{F_W}{4} \tag{7.45}$$

The individual motor torque T_A at each of the four motor positions can be calculated by multiplying by the road wheel radius $D_W/2$.

$$F_A = \frac{F_W D_W}{8} \tag{7.46}$$

Further Analysis

Analysis and discussion of the vehicle geometry show that a further control function should be implemented to prevent unbalanced load application between the four drive wheels.

When the steering angle is not equal to zero, the front left or front right road wheel becomes the position with the highest speed set-point (designated as the fourth motor for this discussion). The front motors are selected here since they are used for both two- and four-wheel steering modes, while the rear motors only have such limitation in four-wheel steering mode. Since the four motors are identical, it is reasonable to expect that the other three motors will achieve their set speeds before the fourth as they all start from a similar starting state. If this occurs, then the model that the control system uses to calculate the parameters will not match real-world conditions, resulting in unbalanced loads and possibly unstable vehicle operation (e.g., skidding or dragging).

In this case, the actual speed achieved by the fourth motor could be used in conjunction with its set-point speed to develop a ratio (actual/desired) that can be used to scale the set-points of the other three motors.

Therefore, whatever speed the fourth motor does actually achieve, the set-points for the remaining three motors would be scaled to follow that speed in the correct relationship according to the vehicle geometry. Once the fourth motor eventually achieves its set-point value, the scaling ratio becomes one, and the remaining three motors are able to run at the correct speed.

The selection of which of the two front motors to use is determined based on the sign of the steering angle. For positive steering angles, the front left wheel becomes the limiting factor, while for negative angles, the front right wheel provides the limiting function.

Motor Simulation

The previously designed PM ac motor simulation is also utilized in this study to implement the in-hub motor drives. However, parameters of the system are modified to reflect a real motor, which are obtained from a manufacturer's datasheet.

One major modification of the motor drive done in this study is the new user interface, which accepts only the three-system set-point parameters and generates outputs on the front panel showing the actual generated torque and the actual achieved speed only.

7.6.2 Virtual Instrument Panel

The Picture Control Toolkit of LabVIEW has been utilized in the VI development since it provides the ability to draw pictures on the computer screen and to modify those picture elements as a result of the program functions. The custom-written VI is named `Four Wheel Steer and Drive.vi` and is located in Section 7.6 of the Chapter 7 folder. These pictures are used to display the following parameters in the VI.

- An outline of the vehicle body, showing the steering angle and the resultant road-wheel angles for both two- and four-wheel steering

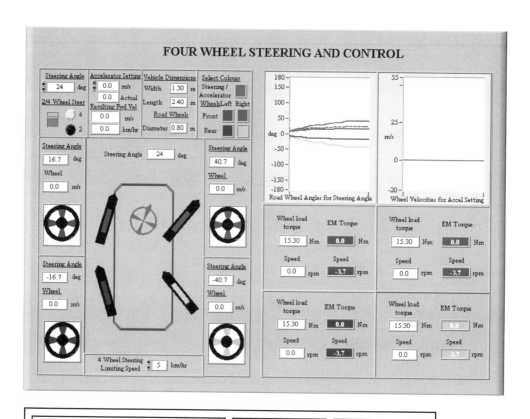

Figure 7-17
Front panel and brief user guide of the VI.

- A set of rotating wheel displays, with the rotational speed of the display directly proportional to the rpm set-point for that wheel position

Fig. 7-17 shows the complete front panel of the VI. Other typical features that were implemented in the VI are

- Provision for limiting the maximum speed achievable in four-wheel steering mode to a specified value (the control is shown at the bottom center of the display).
- Allowing the user to select the color associated with each of the four wheel positions and the steering wheel. This identifying color appears in the rotating wheel display, the wheel angle display, and in two graphs used to show the relationships between steering angle and wheel angle, and vehicle velocity and wheel axle linear velocity.

As stated earlier, the front panel of the VI accepts the three-system set-point parameters of the motor and generates outputs on the front panel showing the actual generated torque of the motor and also the actual achieved speed, which are also shown in the color selected by the user for the individual wheel positions.

However, it should be noted here that the speed of the simulation may be rather slow (depending on the computer platform), which is mainly due to the integration of four motor simulations and the rotating wheel displays.

7.6.3 Self-Study Questions

Open and run the custom-written VI `Four Wheel Steer and Drive.vi` in Section 7.6 of the Chapter 7 folder. Use the front panel displays with two- and four-wheel steering selected, respectively; enter the following combinations of accelerator setting and steering angle; and observe the operation of the VI.

- For two-wheel steering with 45° steering angle and 6 m/s accelerator setting
- For four-wheel steering with −25° steering angle and 2 m/s accelerator setting

Some of the distinct features of the simulation are listed next. I suggest that you study these features to get a better view of the simulation:

- Operation of the simulation by reduction of the computational load
- Investigation of a set-point limiting function to prevent unstable conditions from developing due to delay in achieving the set speed
- Investigation of a suitable speed limiting function to reduce the risk of rolling or sliding during turning
- Development of a comprehensive overseer function to identify when the actual motor speeds do not match the predicted performance, due either to delay or to slipping

7.7 Fault-Tolerant Motor Drive for Critical Applications

Some of the recent research activities in the area of electric motor drives for critical applications (such as aerospace and nuclear power plants) are focused on looking at various motor and drive topologies. Researchers feel that the concept of a fault-tolerant motor drive has now reached a level where it is feasible to be used in practice by the help of recent technological advances and developments in the area of power electronics and motor control.

In this section, a fault-tolerant motor drive system suitable for critical applications is simulated and motor/drive systems are studied, and a VI is presented to allow the user to examine the system thoroughly.

7.7.1 Fault-Tolerant Motor Drive System

Although a fault-tolerant system with a single motor drive may overcome most of the problems in critical applications, the nonexistent redundancy in the case of a complete motor failure is of primary concern. In addition, the redundancy of the control and feedback systems has to be considered for reliable operation.

It is suggested that the use of direct-drive multiple motors on the same shaft improves redundancy (Fig. 7-18). An axial field version of the motor may also provide a better utilization of the power density and the space, and high saliency interior PM motor designs may generate low currents under winding short-circuit conditions. As demonstrated in the figure, due to the slim design of the segments, if a higher power rating is needed, another segment(s) can be added to the same shaft without any difficulty.

In the active redundancy system in Fig. 7-18, either motor drive unit may sustain the function. Although the complexity of the circuit and the cost in-

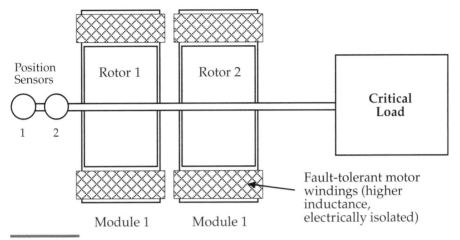

Figure 7-18
Multiple segments/modular motor drive with redundancy.

crease due to the multiple and independent controllers, even if one of the motor drive segments is partially or completely out of order, the remaining motor drive can continue to operate and may provide sufficient power for safe shutdown.

Definition of a Fault-Tolerant System

A high performance and a complete fault-tolerant system can be obtained if all of the components in the motor drive system are made fault-tolerant individually. Switched Reluctance (SR) motors are inherently fault-tolerant. However, they have significantly less torque density and efficiency than their PM ac counterparts, hence are not preferred in this study.

An alternative fault-tolerant motor configuration for PM ac motors can be obtained by separating the three-phase windings and driving each motor phase from a separate single-phase inverter (Fig. 7-19). It is evident that this new configuration doubles the number of power devices. However, the device voltage ratings are reduced since the devices withstand the phase voltage rather than the line voltage. As a result, the switching losses of the inverter will be reduced, in turn reducing the heat sink requirements, which also means less weight.

Although there is a marginal advantage in increasing the number of phases of the motor under open-circuit fault, there is no overall benefit since the com-

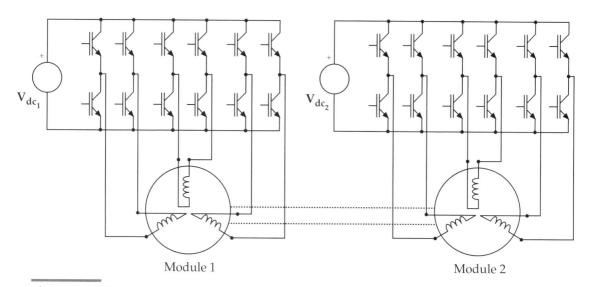

Figure 7-19
An alternative motor/inverter configuration against the device and winding failure of a two-segment motor.

plexity of the drive circuit increases, which reduces reliability. Therefore, a three-phase PM ac motor configuration is selected.

It should be emphasized here that, in the event of the failure of one motor phase, the reduction of the developed average torque can be compensated for by overrating each phase of the motor by a fault-tolerant factor, which is a function of the number of phases.

It was found that the motor system (modular structure), which is suitable as a direct drive in critical applications, should have the specifications summarized in Table 7-9.

Potential Faults in the Motor Drive

As mentioned previously, under any fault, the motor drive system should be able to continue to operate reliably and safely. One way to achieve this is to make the overall system fault-tolerant. Fault scenarios that may occur in a high-performance drive are identified and listed in Table 7-10. To address the faults in the drive, various methods and technologies can be implemented. Let's examine these faults briefly.

The motor drive is prone to open-circuit faults due to the failure of power switches or winding failures. These faults can be observed by current trans-

Table 7-9 *Specifications of the fault-tolerant motor.*

- Higher redundancy, by using identical motor segments on the same shaft
- Higher efficiency design over the operating speed range to better utilize the power supply
- Electrically isolated phases to prevent phase-to-phase short-circuit and reduce inverter faults
- Magnetically uncoupled windings to avoid reduction of performance in the case of a failure of the other phases
- Physically isolated phases to prevent propagation of the fault into the neighboring phases and to increase the thermal isolation
- Higher winding inductance to limit the winding short-circuit current
- Minimum weight and power loss design utilizing cobalt-iron based laminations
- Effective cooling

Table 7-10 *Fault scenarios in the drive.*

- Winding open-circuit
- Winding short-circuit (partial turn-to-turn or complete)
- Inverter switch open-circuit (analogous to winding terminal open-circuit)
- Inverter switch short-circuit (analogous to winding terminal short-circuit)
- Power supply failure
- Position sensor (which can also be used for speed sensing) failure
- Combination of the above faults
- Controller failure

ducers located in each phase, while the remaining switches in that phase(s) are turned off.

One of the most critical faults in the motor drive is a partial (turn-to-turn) winding short-circuit or a complete winding short-circuit due to the failure of two switches or a terminal short-circuit. The fault current in this case can be limited by the increased self-inductance of the windings in the design stage. It is desired that the short-circuit fault current be limited to the rated steady-state current. This fault can be detected and controlled by the controller if the fault is a complete short-circuit and if it occurs before the current sensor of the phase. However, it is impossible to eliminate in the case of a turn-to-turn short-circuit, in which case the solution is to shut down the motor segment completely and run the remaining segment at maximum (for a short period) or at rated (continuously) condition.

Inverter switch open-circuit faults are analogous to winding terminal becoming open-circuit. Hence similar measures previously explained can be undertaken to eliminate such faults.

Similarly, inverter switch/diode short-circuits are analogous to winding terminal short-circuits but can easily be eliminated by turning off the other switch.

Power supply failure is probably the most critical fault. The solution is to use multiple and electrically isolated power sources for every critical section of the drive instead of a single supply.

Failure of position sensors (which can also be used for speed sensing) in PM ac motors can be avoided by using multiple position sensors, together with indirect position detection methods operating simultaneously.

A diagram of the main components of the drive system is shown in Fig. 7-20. The design of a complete modular motor drive system is a relatively complex task as there is a wide array of design factors that need to be considered. Therefore, it is important to design the motor and drive together in an integrated manner. This means that for optimum performance, the motor, power electronic circuits, control circuits and algorithms, and measuring circuits should be considered together, and not in isolation.

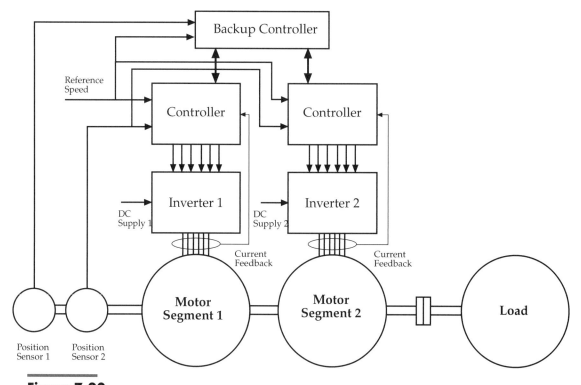

Figure 7-20
The complete block diagram of the fault-tolerant motor drive system with redundancy.

Finally, power processing and control units in the practical system should be designed carefully to take many challenging integration and control issues into account, such as

- Development of a supervisory system as a backup, and for condition monitoring
- Implementation of electromagnetic torque sharing and torque control

Note that in this study, a computer simulation model of the system has been developed based on the system diagram in Fig. 7-20, which provides a useful tool for studying the drive in detail and allows the system's reaction to synthetic faults to be observed.

System Equations of the Two-Segment Motor

As studied earlier, PM ac motor segments can be assumed to be balanced and have identical resistance R and equivalent winding inductance L. Hence the terminal voltage v for each phase is given by

$$v_n(t) = Ri_n(t) + L\frac{di_n(t)}{dt} + e_n(t), \quad n = 1, 2, 3, 4, 5, 6 \quad (7.47)$$

where e is the back emf of the phase windings. The solution of differential equation 7.47 is obtained in LabVIEW by using a numerical integration technique, the trapezoidal approximation, which obeys the rule given below. Note that the other integration methods (such as Simpson's or the Runge-Kutta method) may provide more accurate results. However, they are mathematically intensive and hence not used here.

$$\int_{t_1}^{t_2} i(t)\, dt = \Delta t \left(\frac{1}{2}i_1 + \frac{1}{2}i_2\right) \quad (7.48)$$

where $i(t)$ is the time-dependent function of the current, Δt is the time step in seconds, and i_1 and i_2 are the values of the current at two consecutive time steps.

The electromagnetic torque, T_e, produced by one phase of one motor is calculated by

$$T_e = \frac{e_n i_n}{\omega}, \quad n = 1, 2, 3, 4, 5, 6 \quad (7.49)$$

The total electromagnetic torque produced by the two motor segments (1 and 2) operating on a single shaft can be calculated by adding together the output torque of each motor (a and b).

$$T_{e(\text{total})} = \frac{(e_{a1}i_{a1} + e_{a2}i_{a2} + e_{a3}i_{a3}) + (e_{b4}i_{b4} + e_{b5}i_{b5} + e_{b6}i_{b6})}{\omega} \quad (7.50)$$

In each motor segment, the back emf waveforms must be out of phase with the adjacent phase by 120°, which is calculated for sine waveforms using the following relationship.

$$\begin{bmatrix} e_1 \\ e_2 \\ e_3 \end{bmatrix} = \begin{bmatrix} E_m \sin(\theta_e) \\ E_m \sin\left(\theta_e - \dfrac{2\pi}{3}\right) \\ E_m \sin\left(\theta_e - \dfrac{4\pi}{3}\right) \end{bmatrix} \quad (7.51)$$

As can be seen in equation 7.51, the torque ripple-free operation of the two-segment motor drive can be achieved by providing two segments having not only identical electrical parameters but also back emfs in phase. Furthermore, as discussed in the section on PM ac motors, the phase currents must be regulated in reference to the corresponding back emfs.

To calculate the rotor's angular velocity ω, the mechanical equation of motion must be solved:

$$T_{e(\text{total})} = J\frac{d\omega}{dt} + B\omega + T_L \quad (7.52)$$

Here, J is the polar moment of inertia of the whole system, B is the damping coefficient, and T_L is the load torque.

7.7.2 Virtual Instrument Panel

The custom-written VI is named `Fault Tolerant Drive.vi` and is located in Section 7.7 of the Chapter 7 folder. The front panel of the VI given in Fig. 7-21 provides the user with a choice between Hysteresis or PWM current control. Hysteresis current control uses a bandwidth centered on the desired current waveform. The actual current is constantly observed, and, at the instant the actual current becomes equal to the hysteresis bandwidth limit, the status of the terminal voltage is changed to control the current using the inverter switches. The PWM method, however, employs a fixed sampling frequency and thus constant time step to sample the actual current produced by the motor. The inverter is only switched at the instant when the current is sampled, according to whether the actual current is above or below the reference current value.

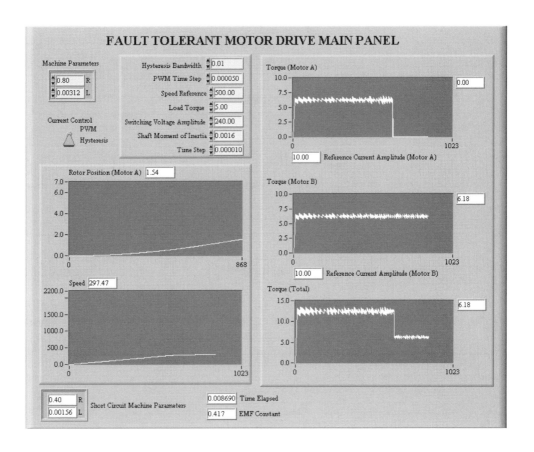

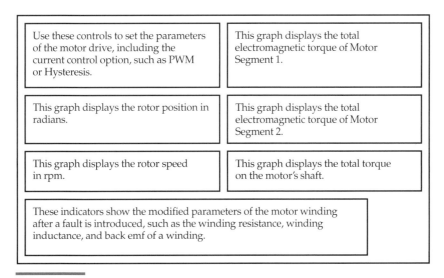

Figure 7-21
Front panel and brief user guide of the VI.

In the VI, the proportional integral (PI) controller for each machine is responsible for current regulation. A number of user controls are provided in the user panel, which accommodates the complete motor and system parameters, current control types, faulted winding parameters, and saving data options.

Since the dual motor system is required to be highly fault-tolerant, the VI should provide a means of testing the fault tolerance of the system. The Faults sub-VI allows any of the six motor phases to be short-circuited or open-circuited and allows the user to vary the two motors' back emf constants and specify the machine parameters for phases that become short-circuited. It also facilitates the complete shutdown of one of the two motors. The switches included in the sub-VI can be adjusted by the user as the system is running, which allows the user to watch the result of the synthetic faults.

Rather than concatenating all the waveform graphs into a single global user interface and creating a cluttered front panel, two sub-VIs have been created to allow the user to view the emf waveforms and actual currents in each of the six motor phases.

7.8 References

Ehsani, M., K. M. Rahman, and H. A. Toliyat. "Propulsion System Design of Electric and Hybrid Vehicles." IEEE Transactions on Industrial Electronics, 19–27, 1997.

Ertugrul, N. "Position Estimation and Performance Prediction for Permanent-Magnet Motor Drives." Ph.D. thesis, University of Newcastle upon Tyne, 1993.

Ertugrul, N. "Power Electronics in Mechantronics Lecture Notes." Department of Electrical and Electronic Engineering. University of Adelaide, 2001.

Ertugrul, N. "Power Electronics Lecture Notes." Department of Electrical and Electronic Engineering. University of Adelaide, 1997.

Ertugrul, N., and E. Chong. "Modeling and Simulation of an Axial Field Brushless Permanent Magnet Motor Drive." European Power Electronics Conference, Trondheim, Norway, 1997.

Ertugrul, N., W. L. Soong, S. Valtenbergs, and H. C. P. Ng. "Investigation of a Fault-Tolerant and High-Performance Motor Drive for Critical Applications." IEEE Region 10 International Conference on Electrical and Electronic Technology (TENCON 2001). Singapore, 19–22 August 2001.

Gray, C. B. *Electrical Machines and Drive Systems.* New York: Longman Scientific and Technical (Wiley), 1989.

Handbook of Small Motor, Gearmoter, and Control, 5th ed. Bodine Electric, 1993.

Jacek, F., and W. Mitchell. *Permanent Magnet Motor Technology.* New York: Marcel Dekker, 1997.

Kenjo, T. *Electric Motors and Their Controls: An Introduction.* New York: Oxford Science Publications, Oxford University Press, 1991.

Kenjo, T. *Stepping Motors and Their Microprocessor Controls.* New York: Oxford University Press, 1984.

Krause, P. C. *Analysis of Electrical Machinery.* New York: McGraw-Hill, 1987.

Maas, J. *Industrial Electronics.* Englewood Cliffs, NJ: Prentice Hall, 1995.

Telford, J. "Simulation of Steering and Speed Control for Four Wheel Drive Electric Vehicles, Final Year Student Project." Department of Electrical and Electronic Engineering. University of Adelaide, 1999.

Vu, L. "Final Year Project Report: Development of LabVIEW-Based Teaching Modules." Department of Electrical and Electronic Engineering. University of Adelaide, 2001.

Ying, L., and N. Ertugrul. "The Dynamic Simulation of the Three-Phase Brushless Permanent Magnet AC Motor Drives with LabVIEW." Australasian Universities Power Engineering Conference (AUPEC '99), 11–16. Darwin, 26–29 September 1999.

Real-Time Control of Electrical Machines

8

This chapter introduces real-time control of selected electrical machines using LabVIEW programming and custom-built hardware. The real-time control of electrical machines requires a significant amount of processing power. However, this chapter aims to provide fully functional LabVIEW user interfaces and primarily targets limited power data acquisition hardware available from National Instruments.

Custom-built hardware significantly reduces the need for processing power of the data acquisition (DAQ) card. The details of the hardware are all included in the Appendix for potential developers.

Five distinct real-time motor control concepts are covered in the chapter: dc motor control, stepper motor control, brushless trapezoidal PM motor control, starting of wound-rotor asynchronous motor, and Switched Reluctance motor control.

In the first, second, and fourth control modules, in addition to the software support, custom-built hardware solutions, which are required for the LabVIEW VI interfacing, are provided. However, all the VIs in this chapter assume that the user has access to a lower-end DAQ card available from National Instruments.

8.1 DC Motor Control

DC motors combine many positive characteristics that are expected from an ideal motor drive, such as slightly sloped speed-torque characteristic allowing high-speed stiffness and good transient behavior, and operation over a wide speed range without complex control techniques. Therefore, such motors are favored in servo motor applications. Although there are various types of dc motor configurations, in this study we will discuss the motor type with a PM excitation only.

Permanent Magnet Brush DC Motors

In these motors, the field poles are supplied by permanent magnets. The semicircular PMs are assembled on the stator frame to provide two or more field poles. The equivalent circuit of the motor is given in Fig. 8-1a.

The dc PM motors have several advantages, such as smaller size, less expensive, higher efficiency (no field coil power loss), high stall (starting) torque,

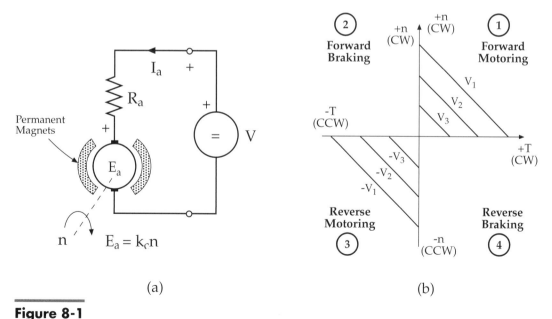

Figure 8-1
(a) Equivalent circuit of PM brush dc motor, and (b) four quadrants of motor operation and its speed-torque characteristics at different voltage inputs, $V_1 > V_2 > V_3$.

and linear speed-torque characteristics (Fig. 8-1b). The principal disadvantages of the PM brush dc motors are absence of field control and susceptibility to demagnetization (if they are overloaded).

The motor's equations can easily be obtained from the equivalent circuit given in the figure.

$$V = R_a I_a + E_a \tag{8.1}$$
$$E_a = k_c n \tag{8.2}$$
$$T = k_c I_a \tag{8.3}$$

where k_c is the back emf or torque constant in the PM brush dc motor

$$T = \frac{k_c(V - k_c n)}{R_a} \tag{8.4}$$

Equation 8.4 indicates that the speed versus torque curve in a PM brush dc motor is a straight line. At standstill, $n = 0$, hence $T = T_{start} = k_c V / R_a$, and the maximum speed occurs when $T = 0$.

8.1.1 Control of PM Brush DC Motors

Due to the absence of the field winding, PM brush dc motors can only be controlled using the armature voltage and current. In fact, current control is often employed in the dc motor control for both torque control and protection purposes.

From the control point of view, two methods can be given: open-loop control, and closed-loop speed control, which provides a great flexibility in operational characteristics. The control system shown in Fig. 8-2 illustrates a cascade closed-loop speed control system, which consists of an inner current loop and an outer speed loop.

Feedback Devices

For higher quality speed control, a tachometer (tacho-generator) is attached to the motor shaft. The tachometer is a small ac or dc generator that outputs an ac or dc voltage proportional to the speed of the motor. Since tachometer feedback can change instantaneously with speed change, it allows faster correction and tighter regulation from a controller. Usually the permanent-magnet dc generators are preferred for their compactness and reliability.

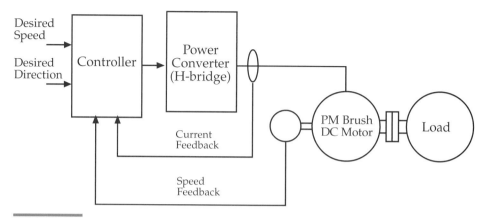

Figure 8-2
A block diagram of a closed-loop control system for PM brush dc motors.

The ideal tacho-generator would have a linear output voltage versus speed characteristic. However, the characteristic may deviate from ideal specifically at high speeds and may have a high level of noise due to the commutators' arrangement. Therefore, the tacho-generators should be selected carefully so that the output voltage noise is not significant.

There are various types of instantaneous current measurement methods in motor control applications: using low-value resistance or Hall-effect devices and measuring via intelligent switching devices in the power control section. The selection of the method depends on various factors, such as isolation requirement, bandwidth of the actual current, maximum level of the current, and so forth.

Power Circuit, H-Bridge

As mentioned previously, the H-bridge (Fig. 8-3) is perhaps the most versatile and popular circuit topology that is also used in dc motor applications. Today, by using MOSFETs or IGBTs, it is now easy to design power circuit switching above the audible frequency and with much reduced switching losses. As studied previously, the switching of the H-bridge may be done in various ways: varying switching frequency, varying duty cycle (PWM), or varying both.

Furthermore, the switches in the H-bridge (single-phase switching bridge) can be controlled in a choice of sequences to provide the desired voltage to the motor. Two of the common modes of switching sequences are called Bipolar

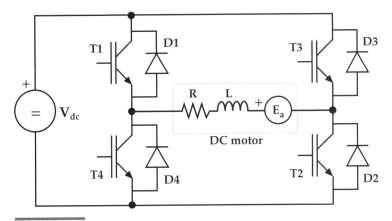

Figure 8-3
H-bridge configuration for PM brush dc motors.

and Unipolar driving methods. An H-bridge meets the bipolar current requirements and can facilitate four-quadrant control for dc motors (Fig. 8-1b).

The Choice of Feedback Control

As illustrated in Fig. 8-2, the control loop of the motor contains two transfer (feedback) elements: current and speed. For stable operation within the desired limits, these feedback signals have to be processed before inputting into the controller where the control decisions are made.

In fact, the processing of the feedback signals means assigning a transfer element with time-dependent behavior to the measured signal, which may have certain characteristics. The most important characteristics are known as Proportional (P), Proportional Integral (PI), and Proportional Integral and Differential (PID).

Note that the controllers with differentiating elements (such as D, PD, PID) are rarely used in the dc motor drive applications, since the D element amplifies any ripple in the measured signal and thus adversely affects the controlled quantity, such as speed.

The PI controller is the most important controller type used in the motor control applications. Fig. 8-4 illustrates five different controller characteristics and summarizes the system performance of two of the controllers, P and I, when an error occurs in the measured data. However, remember that every application has to be analyzed carefully and the design should be considered at the system level, not in isolation.

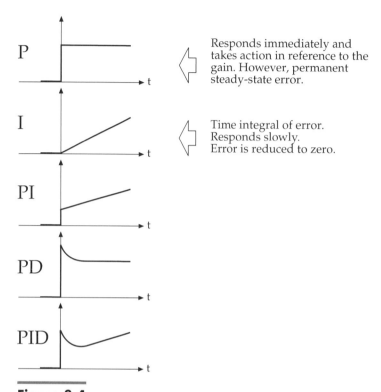

Figure 8-4
Controller characteristics of P, I, PI, PD, and PID, and control system performance of P and I controllers.

If we consider the overall system behavior of a dc servo motor drive, we can see that the most crucial characteristic of the motor drive is not the steady-state output power, but the maximum rotor acceleration/deceleration, which very much depends on the response of the feedback devices and associated control circuits. Table 8-1 provides an overview of tacho-generator based speed sensing and the processing method utilized in the dc motor control.

Operating Quadrants and Speed (Velocity) Profiles

All dc motors can operate in the first quadrant, and if the terminal voltage is reversible, they can operate in the first and third quadrants. Servo applications, however, may require a greater degree of control. For example, the motor may be required to reverse while rotating, say, in the positive direction,

Table 8-1 *Overview of the speed sensing and processing method.*

Characteristics	Speed Sensing DC Tacho-Generator	Signal Processing P	PI	PID
Control accuracy	Medium		Medium	Medium
Extended control range	Yes	Yes		
Control reaction	Fast	Fast		Fast
Low-speed operation	Good	Good		Good
Servo drive application	Yes	Yes	Yes	Yes

hence generating a negative torque at the positive speed. To achieve this the motor controller should be able to operate in the second or fourth quadrant where the load torque is in the direction of rotation. Fig. 8-5 illustrates the possible control modes of a dc motor with linear speed-torque characteristic for a given armature voltage.

In order to better utilize the machine cycle time, to control loads accurately, and to achieve smooth high-speed motion control without overloading the motor, the controller must direct the motor to achieve optimum results (in terms of response time and efficiency of the system). In practice, this is accomplished using shaped speed profiles to limit the energy dissipation (Fig. 8-6).

Speed profiles are given as a function of velocity versus time, where the area under the curve is position (since position is the integral of velocity). In practice, the trajectory generators are used to generate a speed profile that controls the acceleration, velocity, and deceleration so that the axis comes to a stop exactly at the programmed target position.

Note that a parabolic profile is used in more precise control applications, which minimizes the vibrations caused in a mechanical system by a moving mass. Moreover, if a triangular profile is used, individual settings for acceleration and deceleration may cause some system disturbances at the corners that translate to small vibrations, which extend the settling time.

8.1.2 Hardware Implementation Details

This section describes a complete control system for the closed-loop control of a PM brush dc motor including a custom-built VI. The principal components of the system are illustrated in Fig. 8-7.

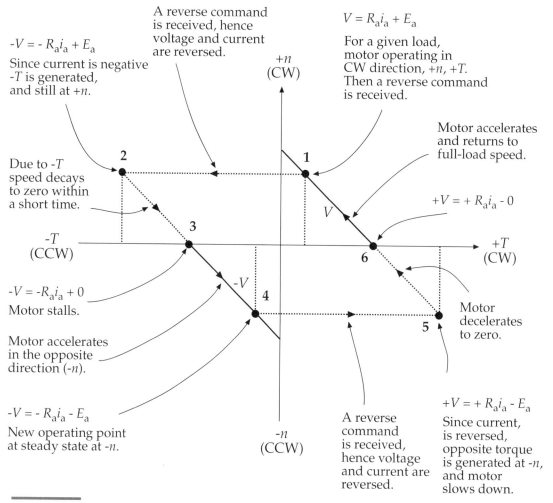

Figure 8-5
Four-quadrant operation of a dc motor and possible operating points.

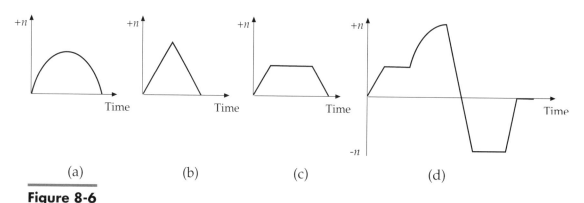

Figure 8-6
Equivalent common speed profiles in the motor drives: (a) parabolic, (b) triangular, (c) trapezoidal, and (d) combination.

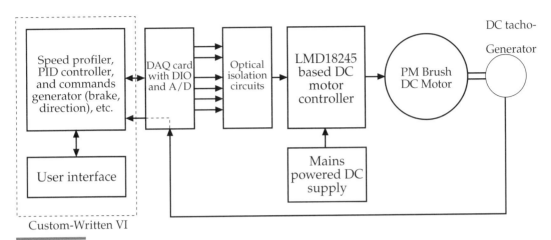

Figure 8-7
The closed-loop PM brush dc motor control block diagram implemented in this study.

Building a complete H-bridge with current feedback is an expensive and laborious task that should also include the driver circuits and the current sensing circuits. In addition, generating the high-speed switching signals for the current control may require significant processing power in a closed-loop system.

Therefore, in this study a monolithic solution using the LMD18245 H-bridge power amplifier from National Semiconductor is to implement a cost-effective PM brush dc servo motor controller. As schematically shown in Fig. 8-7, the purpose of the custom-built VI is to measure the analog speed, to implement a PID speed controller, and to generate a series of digital output signals to control the LMD18245.

The integrated circuit (IC), LMD18245, incorporates all the circuit blocks required to drive a PM brush dc motor. In addition, the IC can accept digital signals to control the direction of the rotation and braking action.

The motor's current control in the IC is achieved via a fixed off-time chopper technique. The typical features of LMD18245 are listed in Table 8-2, and the functional block diagram is given in Fig. 8-8. The IC incorporates all the circuit blocks needed to implement a fixed off-time control circuit. The blocks include an H-bridge with clamp diodes, an amplifier for sensing the load current, a comparator, a monostable, and a DAC for digital current control.

The H-bridge consists of four DMOS switches with freewheeling diodes. The motor's armature terminals are connected to Output 1 and Output 2. As stated earlier, external digital signals control direction, braking, and current levels.

Table 8-2 *Features of LMD18245.*

DMOS H-bridge power stage
Rated 55 V and 3 A continuous (6 A peak)
Low RDS (ON) of typically 0.3 Ω per power switch
Internal clamp diodes
Low-loss current sensing method
Digital or analog control of motor current
TTL and CMOS compatible inputs
Thermal shutdown (outputs off) at $T_J = 155°C$
Overcurrent protection
No shoot-through currents
15-lead TO-220 molded power package

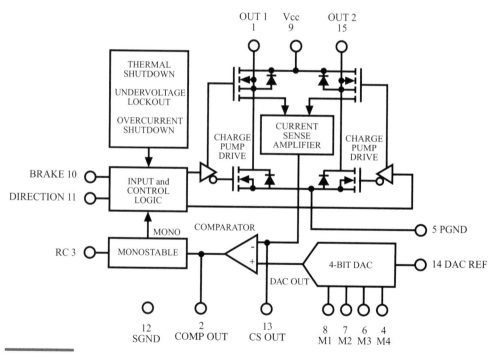

Figure 8-8
The functional block diagram of LMD18245.

The four-bit digital-to-analog converter (DAC) in the IC provides a digital path for controlling the motor's current via the DAQ card and the VI. The DAC sets the threshold voltage for chopping at $V_{DAC(ref)} \cdot D/16$, where D is the decimal equivalent (0–15) of the binary number applied at M4 through M1, the digital inputs of the DAC. M4 is the most significant bit (MSB). For applications that require higher resolution, an external DAC can drive the DAC REF input.

In the LMD18245, a parallel resistor-capacitor network is connected between RC (pin #3) and ground, which sets the off-time that is about 1.1 RC seconds.

The speed feedback signal with a dc tacho-generator yields a direct voltage that is proportional to speed. However, as mentioned earlier, a constant voltage with low-ripple (high precision) tacho-generator signal is difficult to achieve.

Depending on the quality of a dc tacho-generator used as a speed feedback, a filter may be required. For example, if the dc tacho-generator has few commutator segments, the output signal contains highly fluctuating voltage output specifically at low motor speeds, which is not suitable as a feedback in a closed-loop system.

Therefore, a simple filter is implemented in this study, which is based on the real tests on a tacho-generator. The principal aim of the filter is to minimize the fast varying voltage peaks, which works by taking two consecutive data points of the measured speed and subtracting them. If the resultant value is greater than a specified threshold value of 0.2 (which was determined by a real test run) the smallest data is inserted into the output array. Although this method was found to be effective in the test setup, some limitations were observed since the small values of the data from the input array take preference over the larger ones, even if the larger values are closer to being correct. For example, if comparing 0.3 and 0.01, 0.01 would be inserted into the output as it is lower.

A top-view photo of the circuit board for the dc motor control is given in Fig. 8-9. The hardware details including the circuit diagrams and printed circuit board layout are provided in the Appendix. Note that the motor's rating should not exceed the rating of the driver.

The custom-built circuit board has its own in-built mains power supply. In addition, two BNC connectors are provided to observe the motor's voltage and current in real time. The current measurement is implemented using a low value resistance that is connected in series with the armature winding.

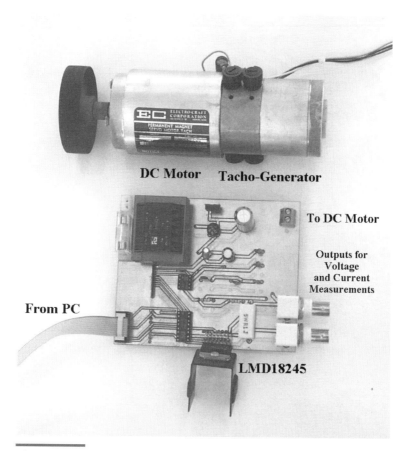

Figure 8-9
Photo of the custom-built dc motor controller including a Permanent Magnet Servo Motor and Tacho-Generator assembly (Electrocraft Corp.).

8.1.3 Details of the Virtual Instruments and Front Panels

The custom-written VIs named `Speed Profiling Panel.vi` and `DC Motor Control.vi` are located in Section 8.1 of the Chapter 8 folder of the accompanying CD-ROM. The front panels and their associated description diagrams of the VIs are given in Fig. 8-10 and Fig. 8-11. As provided in the main VI, `DC Motor Control.vi`, there is one analog input and there are six digital outputs.

A closed-loop current control system providing braking and acceleration independent of the supply voltage and the internal motor resistance is easy to

Chapter 8 • Real-Time Control of Electrical Machines

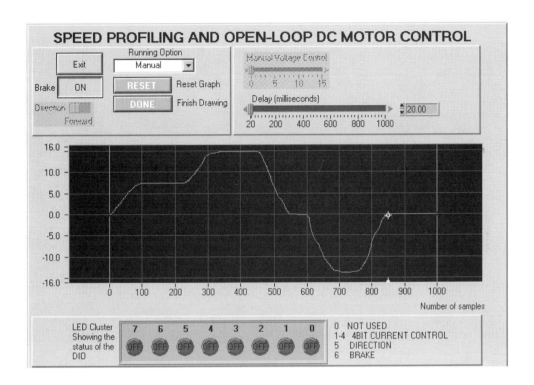

Figure 8-10
The opening front panel and brief user guide of the PM brush dc motor control VI (LabVIEW 5.1).

LabVIEW for Electric Circuits, Machines, Drives, and Laboratories

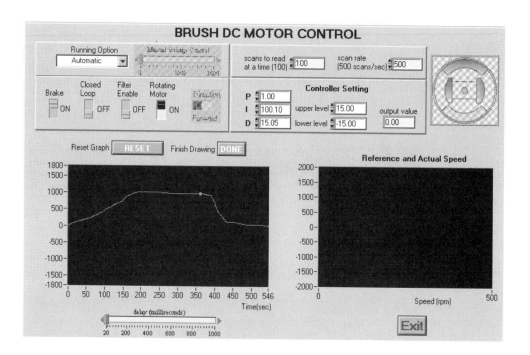

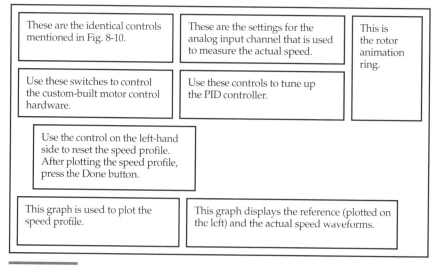

Figure 8-11
The main front panel and brief user guide of the PM brush dc motor control VI (LabVIEW 5.1).

superimpose on the motor control circuit provided. The IC controls the current to the dc motor, hence controlling the speed. While the IC has its own safety shutdowns for peak input voltages, it is best to limit the supply voltage. There are two running options in the VI, manual and automatic control, which can be switched to at any stage except when the brake is on.

Manual Control

This option gives the user direct real-time control of the speed, which is controlled by sliding the scroll bar labeled Manual Control. The scroll bar is initially set to zero—that is, the motor speed is zero—unless the closed-loop control is on. With this option the drawing graph is disabled along with the buttons associated with the graph. However, the direction control is enabled. The brake can be used at any time to pause the motor and the program. Note that when the brake is on, there will be no response from any buttons, or from the motor until the brake has been turned off.

Automatic Control

For this option the speed profile drawing graph and associated buttons are enabled, while the manual control and the direction control have been disabled. The idea behind this option is to allow the user to specify a speed profile that the motor control should perform over a period of time.

To achieve the desired speed profile, place the mouse pointer on the graph cursor, and, while the left button is pressed, move the cursor to create the desired pattern, including CW and CCW rotation regions.

When you are satisfied with the speed profile pattern, press the Done key. An array containing the pattern is stored and followed by the motor. The whereabouts of the speed in the execution is displayed on the speed pattern by using a dot pointer. The time axis of the speed pattern graph can be scaled depending on the value of delay, which can generate the same profile for any desired time period.

At any time, while the pattern is being executed, the user may terminate the graph using the Reset button. Then, the drawing graph returns and the motor ceases rotating. If the graph is allowed to finish executing, it automatically returns to the drawing graph. In each case the motor speed returns to zero.

Closed-Loop Speed Response

As stated earlier, the principal criteria governing the selection of P, PI, or PID can vary. Some are given as follows: stability of the control loop, reaction time, transient response, load behavior, and speed range.

To obtain an optimum response, the PID controller settings must be adjusted carefully, which can be done using the control parameters provided on the front panel. The parameters may be adjusted to change the response of the controller and hence the motor. The upper and lower level output controls have been specified to account for the maximum digital input to the controller chip (i.e., the IC takes a four-bit number for the speed control, hence the maximum value is $2^4 - 1 = 15$).

When the Closed-Loop switch is turned on, the closed-loop response runs the desired value through a simple PID speed controller. The PID controller compares the desired value of the speed with the measured value. It then sends to the motor control IC a current value (specified by the four-bit digital output) such that the motor follows the desired value that was specified by the speed profile graph.

A control switch has also been included in the front panel of the main VI to remove the speed feedback filter if the dc tacho-generator has low ripple. When the Closed-Loop switch is off, the desired value is sent straight to the motor control IC regardless of the actual value, effectively operating in the open-loop mode.

Scan Rates

To obtain real-time speed data from the analog input, a circular buffer was used. The current buffer size is 4000 points long. Every second 500 new points are placed directly from the input to the buffer. Each loop that the program takes removes and analyzes 100 points from the buffer. Once the buffer is full, it automatically starts adding the latest data to the front of the buffer. Therefore, the amount of data to be read in each iteration must be fairly high so data is not overwritten before it has been analyzed.

Rotating Motor Animation

The rotating motor animation aims to provide some insight for the user on how quickly the motor is rotating at any one time. The animation is connected directly to the execution loop, which rotates one step each loop iteration. However, note that the animation does not function in the manual control mode.

Programming Structure

When entering the program all the parameters are initialized to zero. Some of the controls and indicators in the front panel are also made invisible if they are not active.

The main loop is used to control how long the program executes. A switch called Exit is provided in the main loop so that when pressed it takes the program out of the loop. There are three sequences in the main loop.

Sequence 1

This is the stage where the desired speed acquisition (speed profile) is determined. Depending on what control is selected, the program determines what speed the user wants the motor to run at.

Sequence 2

The execution loop is located in this sequence. This is where the desired data is compared with the actual data and a control signal is generated to be utilized by the controller IC, which can modify the action of the motor.

The data is collected from the input and sent to the motor simultaneously. The input data is taken from the analog port and passed through the filter and then onto the `My Data Processing.vi`, which takes an average of the 100 data points. The output is a two-dimensional array containing the single average. This array is broken down into a single number and passed to the PID and to the `Loop For DC Control.vi`.

This sub-VI takes in the desired value specified by the user and passes it to the digital output port. As stated earlier, there are six bits of the digital channel being used: bits 1–4 for the speed control, bit 5 for the brake, and bit 6 for the direction. Note that bit 0 has not been used.

Finally, the desired value and the actual value are graphed on the speed graph for comparison purposes, where the red plot displays the actual speed and the blue illustrates the desired value.

The execution loop in Sequence 2 can exit when any of the following occur: the Exit button is pressed, the Reset button is pressed, there is an error in the data acquisition or data transmission, the manual control option is on, or all the data points from the drawn graph have been processed.

Sequence 3

In this sequence, the motor output is set to zero, such that the motor is not running while a new graph is being drawn.

8.2 Stepper Motor Control

In the real world, the step commands of stepper motors are usually generated in reference to a counter by an Application Specific Integrated Circuit (ASIC) or a microprocessor that is linked to a suitable drive circuit. Therefore, the switching devices in the drive circuit turn on and off to control the winding currents, which cause the rotor's movement.

In the VI presented here, a stepper motor control solution is given that utilizes complete custom-built hardware and a LabVIEW user interface. We'll assume that the fundamentals of the stepper motor and its control were well understood in the previous chapter. Therefore, the focus of this section will be mostly on the hardware details and on some implementation aspects.

It is always desirable to have a stepper motor drive that offers the best performance and the best price for an application. Unfortunately, this cannot be achieved easily since the motor drive design (including the processor, the controller, and the motor) depends on a number of factors. Therefore, it is difficult to recommend a general step-by-step design procedure, which is an iterative process and involves experience and practical tests. Table 8-3 provides a list of parameters that can be considered and utilized in the design of a stepper motor drive.

In this study, an open-loop two-phase bipolar stepper motor control system is considered. The principal components of the system are shown in Fig. 8-12. The details of the hardware including the circuit diagrams and printed circuit board (PCB) layout can be found in the Appendix.

One of the problems with commercially available stepping motor control chips is that many of them have relatively short market lifetimes. Therefore, we selected a commonly available stepper motor controller and designed the complete circuit around it. It should be emphasized that the custom-built circuit can be controlled either via a LabVIEW/DAQ card combination or via a parallel port of a PC.

Table 8-3 *Possible design criteria for stepper motor drives.*

Commonly Used Characteristics and Limits to System Performance	
Torque and output power	Winding resistance
Damping and resonance	Static step accuracy
Degrees per step	Micro-step accuracy
Positioning accuracy	Dynamic characteristics
Holding torque	Design time
Voltage rating	Cost

General Driver Aspects
Power supply design
Snubbing and protection circuits
Electrical losses (specifically Hysteresis losses in low-inductance motors)
Addressing electromagnetic interference problems

System Aspects
Analyzing the load
Friction or inertia loads
Friction torque/load power consumption
Damping

Motor Selection Aspects
Output power
Mechanical aspects
Cost
Customizing the motor as an option

Driver Design Aspects
Selecting driver type
Selecting driver mode (full/half, micro-stepping)

The principal unit in Fig. 8-12 is the block that contains two key integrated circuits, L297 and L298 (from SGS Thompson Microelectronics), which is the core in the design.

A typical circuit configuration utilizing L297 and L298, which can be used in two-phase bipolar stepper motors, is given in Fig. 8-13. In this circuit configuration, the L297 integrates all the control circuitry required to control two-phase bipolar and unipolar stepper motors, which is primarily intended for use with the L298N in stepping motor applications.

The L297 can take control signals from the system's controller, the VI, and provides all the necessary drive signals for the L298 dual H-bridge. In addi-

Stepper Motor Control Block Diagram

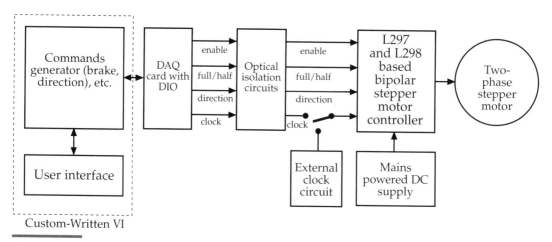

Figure 8-12
The open-loop stepper motor control block diagram.

tion, L297 includes two PWM chopper circuits to regulate the current in the motor windings. The features of the ICs, L297, and L298, are summarized in Table 8-4. The L298 contains dual H-bridge drivers each controlled by two logic inputs and an enable input. External connections are provided for current sense resistors. Although the L298 does not include protection diodes (freewheeling diodes), this circuit configuration allows the use of an external series resistor to be put in the current circulation path to vary the current decay time.

A top-view photo of the circuit board for the stepper motor control is given in Fig. 8-14. The hardware details including the circuit diagrams and PCB layout are provided in the Appendix. The stepper motor used in the circuit has the specifications TEC, SPH-39B-12, 12 V DC, 3.6 Deg/Step, 15 Ω. Note that if an alternative motor is used, the motor's rating should not exceed the rating of the dual H-bridge, L298.

Basically, this study and its design procedure is similar to the dc motor controller explained in the previous section. The custom-built circuit board has its own in-built mains power supply. In addition, four BNC connectors are provided as easy-to-access measurement points, such as to observe the phase voltages and the phase currents in real time. The current measurements are implemented using two low-value resistances in series with the windings.

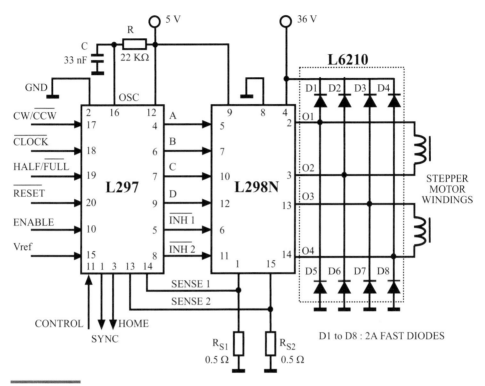

Figure 8-13
A typical configuration of an L297 stepper motor controller and L298 dual bridge driver for two-phase bipolar stepper motors.

Table 8-4 *Features of the L297 and L298.*

L297	L298
Normal/wave drive	Maximum operating supply voltage: 46 V
Half/full-step modes	
CW and CCW rotation options	Total dc current of 4 A
Motor current regulation	Low saturation voltage
Programmable load current	Overtemperature protection
Reset input and home output	High noise immunity (Logical 0 input voltage up to 1.5 V)
Enable input	

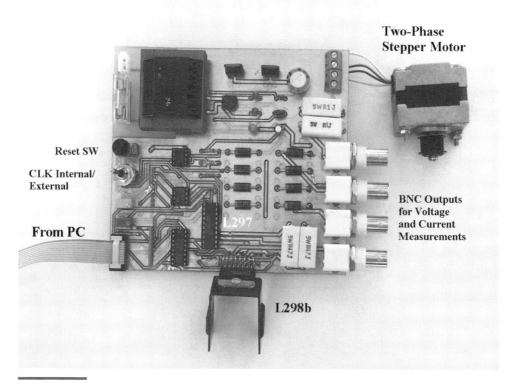

Figure 8-14
Photo of the custom-built stepper motor controller including a two-phase stepper motor assembly.

In the control hardware, there is an onboard clock circuit (via a timer on the circuit board) and a reset circuit, which effectively can make the circuit operate independently. Two clock options available on the circuit board, Internal and External, are selectable using a toggle switch. Note that in the latter option, the external clock should be provided via an optically isolated input (for example, from a PC's parallel port).

8.2.1 Virtual Instrument Panel

The custom-written VI named `Stepper Motor Control.vi` is located in Section 8.2 of the Chapter 8 folder. The VI provides a very simple user interface for the stepper motor drive. The front panel of the VI is given in Fig. 8-15.

Chapter 8 • Real-Time Control of Electrical Machines

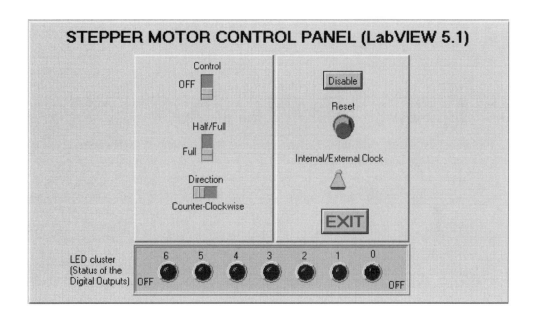

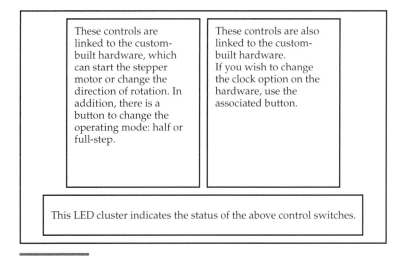

These controls are linked to the custom-built hardware, which can start the stepper motor or change the direction of rotation. In addition, there is a button to change the operating mode: half or full-step.

These controls are also linked to the custom-built hardware. If you wish to change the clock option on the hardware, use the associated button.

This LED cluster indicates the status of the above control switches.

Figure 8-15
The front panel and brief user guide of the VI (LabVIEW 5.1).

Note that the controls available on the front panel simply generate logic level signals via a DAQ card, which can also be generated externally.

8.3 Brushless Trapezoidal PM Motor Control

Three-phase brushless PM motors are widely used in industrial and domestic applications mainly due to their high efficiency and ideal operating characteristics as a servo drive. As stated in the previous chapter, brushless PM ac motors can be classified under two groups. The first group possess trapezoidal back emf per phase and are called Brushless Trapezoidal Permanent Magnet (BTPM) motors (or simply brushless dc motors). The second group possess sinusoidal back emf and are called Brushless Sinusoidal Permanent Magnet (BSPM) motors (or brushless permanent magnet synchronous motors). In this section, however, we will study the operation principles of BTPMs using two separate VIs: a computer simulation module and a real-time control module.

As in other three-phase ac motors, in brushless PM ac motors, the rotating field is created by the stator windings. The magnets in the rotor cause motor action by following the rotating magnetic field. These motors are effectively synchronous motors that should always rotate at synchronous speed. However, it is not possible to rotate a brushless PM motor without knowing where its rotor position is (unless at very low supply frequency and at no-load), which also indicates the shape of the back emfs.

The motors with trapezoidal back emf waveforms, also known as brushless dc motors, can be construed as an inside-out conventional brush dc motor with six commutators that are replaced by six inverter switches and rotor position sensors. In practice, however, three position sensors are sufficient to determine six current-commutation instants, as illustrated in Fig. 8-16.

Torque-ripple-free operation of the brushless PM motor is possible if the phase currents follow the profiles illustrated in the figure. In practice, however, neither the back emfs of the motors nor the real phase currents follow such ideal waveforms. Hence a degree of torque ripple is unavoidable. It may be argued that brushless PM ac motors have a higher degree of torque ripple than a conventional brush dc motor. However, it is possible to achieve a very low torque-ripple in brushless PM motors if high-resolution position sensors are utilized (such as resolvers), which in turn alter the phase current waveforms.

Chapter 8 • Real-Time Control of Electrical Machines

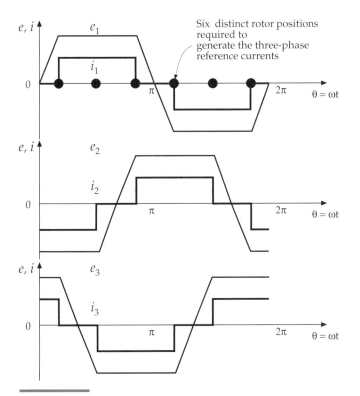

Figure 8-16
Ideal back emf waveforms and phase currents in BTPM motors and required rotor position data for correct commutation.

Fig. 8-17 illustrates the inverter topology and corresponding switching pattern for the BTPM motor. The rotor position sensing devices provide the signals to electronically switch the stator windings at the right time and in the correct sequence, to maintain smooth rotation.

Note that the controller in the figure contains logic circuitry, so that for every set of rotor position inputs, H1, H2, H3, it will produce a set of switching signal outputs for the transistors, T1, T2, . . . , T6. The logic circuit used to generate the output signals is given in Fig. 8-18. Note that the rotation of the motor can be changed within the controller, which in turn reverses the phase energizing sequence. Table 8-5 indicates the complete commutation sequence and the truth table of the switches of a BTPM motor drive in CW and CCW rotation.

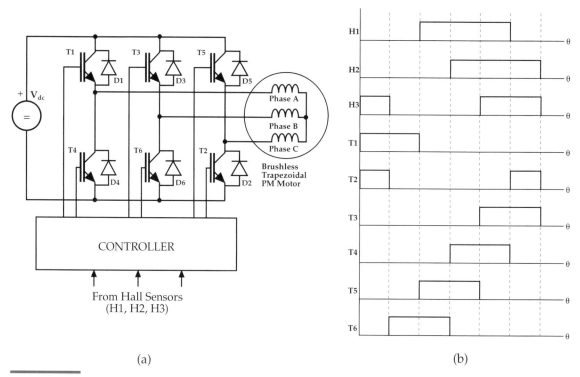

Figure 8-17
(a) Inverter topology, and (b) a switching pattern for CW rotation.

In keeping with the main aim of the LabVIEW programming in this book, the operation of a BTPM is best explained by means of animated diagrams illustrating the rotor positions, the states of the rotor position sensors (Hall-effect devices here), and the switches of the inverter.

Fig. 8-19 shows six distinct rotor positions for one revolution of a simple motor configuration, which can be obtained by the switching sequence given earlier in Fig. 8-17b.

The shaded half circle corresponds to one of the poles of the permanent magnet in the rotor. H1, H2, and H3 are the Hall-effect sensors, which change their states as a result of the rotor poles. The three-phase stator windings are wound perpendicular to the plane of the paper. For example, the labels a and a' represent the cross section of the Phase A winding. Similarly, b and b' and c and c' represent the cross sections of the phase B and the phase C windings, respectively.

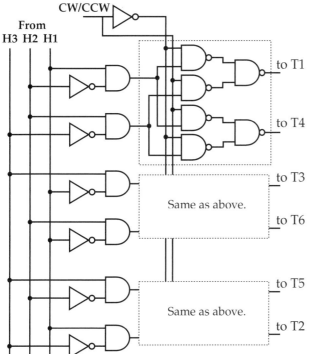

Figure 8-18
A simple logic circuit used to generate the switching signals.

Table 8-5 *Truth table for both CW and CCW commutation.*

H1	H2	H3	Switches ON		Phase Currents		
					a	b	c
			CW				
0	0	1	T1	T2	+	off	−
0	0	0	T1	T6	+	−	off
1	0	0	T5	T6	off	−	+
1	1	0	T5	T4	−	off	+
1	1	1	T3	T4	−	+	off
0	1	1	T3	T2	off	+	−
			CCW				
0	1	1	T5	T6	off	−	+
1	1	1	T1	T6	−	−	off
1	1	0	T1	T2	−	off	−
1	0	0	T3	T2	off	+	−
0	0	0	T3	T4	−	+	off
0	0	1	T5	T4	−	off	+

Note: High voltage: 1; Low voltage: 0

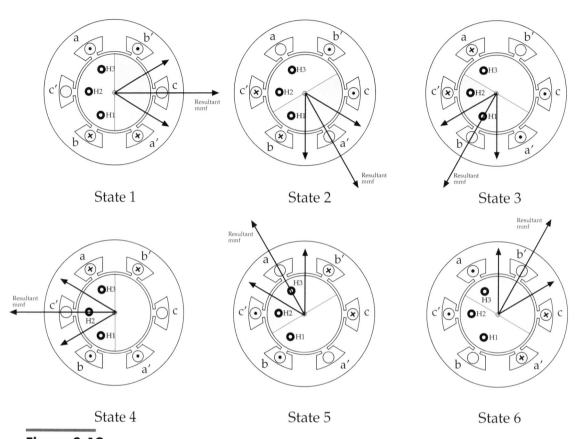

Figure 8-19
Six distinct rotor positions of a three-phase six-slot BTPM motor for CW rotation.

8.3.1 Virtual Instrument Panel

Computer Simulation

In this VI, the previously described motor configuration, drive, and control circuits are simulated simply from the viewpoint of the input (Hall-effect sensors) and the output (controller) signals.

The aim of this simulation VI is to show the rotation of the motor, alongside its position and driving signals and the switching states of the inverter circuitry. The front panel of the VI is given in Fig. 8-20. The VI is simple as the only motor parameters to have as controls are a single CW/CCW flag. The

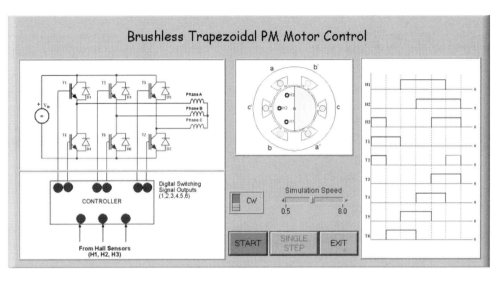

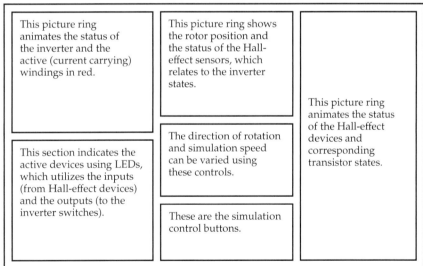

Figure 8-20
The front panel and brief user guide of the simulation VI.

custom-written VI is named `Brushless PM Simulation.vi` and is located in Section 8.3 of the Chapter 8 folder.

The display in the front panel is to show the states of the inverter, the motor schematic, the states of Hall-effect sensors, and the transistor signals versus rotor angle. To effect animation, these displays are periodically updated to reflect a new rotor position, or state. There are six diagrams for each of the CW and CCW rotations, and one zero state, which indicates that all of the inverter switches are inactive and hence the stator windings are unenergized.

The front panel has a number of user controlled buttons for starting/stopping the simulation, single stepping through the simulation, and exiting the VI. In addition, there is a sliding control to alter the speed of the simulation.

To explain the operation of the simulation, we now consider one step command in the clockwise rotation of the motor. First, let's assume the controller is turned on and the rotor is in the position State 1 of Fig. 8-19. The Hall-effect sensors produce a zero signal. Such an input makes the controller produce outputs to switch on just T6 and T1.

By following the active loop (shown in red) in the animated inverter diagram, you can see that current flows from the positive terminal of the dc voltage source, through T1, through the Phase A winding in the positive direction, through the Phase B winding in the negative direction, and through T6, to the negative terminal of the dc supply.

Hence, the rotor is forced to rotate clockwise and, subsequently, into the position shown in Fig. 8-19 (State 2). As illustrated in the same figure, for the excitation of the Phase A and B windings, using the right-hand rule and considering the resultant magnetic field, the final position of the rotor can be determined easily.

In this new position, H1 produces a high signal, with H2 and H3 low. To this input, the controller logic drives T5 and T6 to make Phase B active in the negative direction and Phase C active in the positive direction, which results in the magnetic field shown in Fig. 8-19 (State 3), thus continuing the clockwise rotation.

As stated earlier, in CCW rotation the inputs into the controller are just the inverted Hall sensor signals. The CCW rotation can be studied simply by selecting the CCW mode from the button available on the front panel, and then following the preceding explanations and Table 8-5.

Inspect the VI's programming diagram and you will see that the code is directly reliant on the mechanical action of the user buttons. The Single-Step button latches when released so that the button is highlighted from after it is pressed until when the VI reads it once, after which it reverts to its original value of false.

The foremost programming structure in this VI is the outermost loop, which exits when the Exit button is pressed and keeps track of the instantaneous state of the simulated motor. The pseudo-code for the outermost loop of the complete module is given in Table 8-6. Note that there are alternative ways for numbering the states.

Table 8-6 *A pseudo-code structure of the outer loop.*

```
initialize the state to zero // an inert motor
while (not exit) loop
   // STEP 1: calculate the next state
   if (CW rotation)
      decrement the state // effectively modulo 6
      display the correct control panel diagram // Hall sensor
      signals as the input
   else // CCW rotation
      increment the state // effectively modulo 6
      display the correct control panel diagram // inverted Hall
      sensor signals
   end if
   //STEP 2: propagate the appropriate state to the indicators
   if (start)
      set the simulation speed
      propagate the calculated next state
   else // stop
      if (single step)
         propagate the calculated next state
      else // do nothing
         propagate the existing state
      end if
   end if
   // STEP 3: update the indicators/display
   update the display of the graphs (of transistor signals versus
   rotor angle)
   update the circuit diagram showing the active transistors
   update the motor schematic
   update the LEDs associated with the H1, H2, H3, and T1, T2,...,
   T6 signals
end loop
```

Real-Time Control

It should be emphasized here that practical motor drives require current feedback in order to regulate the phase currents as closely as possible to the ideal reference currents that are generated by the rotor position data. This VI, however (which is analogous to the simulation VI), has a very limited power design based on slow speed digital I/O ports available in most of the data acquisition cards of National Instruments and does not use current feedback from the real motor (which should utilize analog inputs of the data acquisition card and ideally should operate simultaneously with the digital I/O ports). Furthermore, note that the minimum number of digital I/O required in this study is nine, three assigned as inputs from Hall-effect sensors and six assigned as outputs for the inverter switches.

The custom-written VI is named `Brushless real-time.vi` and is located in Section 8.3 of the Chapter 8 folder. The front panel of the VI is shown in Fig. 8-21. The VI simply accepts the Hall-effect sensor inputs, which should be linked to an individual line of two digital I/O ports. In this VI there is no longer any need to store the motor states. Each iteration of the outermost loop reads the current value of the Hall-effect sensor inputs and, dependent on these values, produces the required output transistor switching signals. Similar to the simulation VI, the module shows the corresponding animation of the rotor, the inverter, and the truth table.

8.4 Starting Wound-Rotor Asynchronous Motors

Slip-ring motors are widely used in pump drives, steel mill drives, tower cranes, and conveyors, all of which require a high starting torque. Starting slip-ring induction motors have been of interest to engineers for some time. In a standard squirrel cage asynchronous motor, for a constant supply voltage and constant frequency, the shape of the torque-speed curve does not change, and the starting current is high. In wound-rotor (slip-ring) induction motors, however, it is possible to change the torque-speed characteristic by inserting external resistances into the rotor circuit that can be accessed via slip-rings and brushes.

Although adding external resistances greatly diminishes the efficiency of the motor since the slip increases, it reduces the starting current significantly. It should be noted here that these aspects can be simulated in the Asynchronous Motor Experiment presented earlier in Chapter 5.

8.4.1 Principles of the Starting

As stated above, wound-rotor induction motors are ideally suited to accelerate high-inertia loads with minimum effort. The most common starting methods are to use a mechanically varying resistance bank (or a water tank with conducting blades).

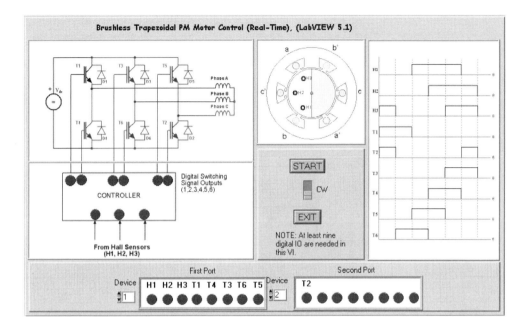

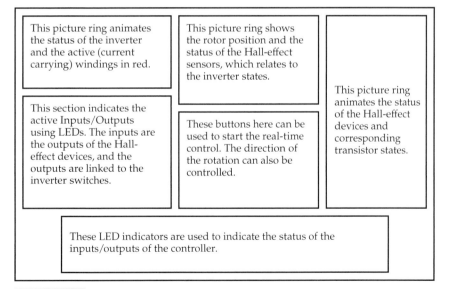

Figure 8-21
The front panel and brief user guide of the real-time control VI.

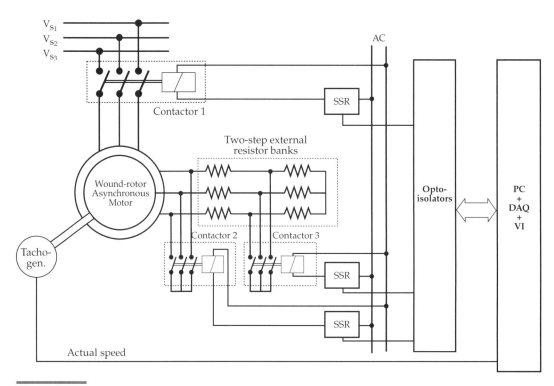

Figure 8-22
Control block diagram for a wound-rotor asynchronous motor with external resistances.

The stator of the asynchronous motor is connected to the main supply. Three-phase (usually star-connected) rotor windings are connected to three-phase star-connected external resistors via a set of slip rings and brushes. The principal wiring diagram of the motor with a two-step external resistance is shown in Fig. 8-22.

Fig. 8-23 illustrates the torque-speed characteristics of a wound-rotor induction motor for three different rotor resistances, including the no-external resistance case (which corresponds to a state of short-circuited rotor terminals). In addition, the figure shows a load characteristic and the corresponding speed values or slip values at the time of transitions to the new characteristics.

The slip at which maximum torque occurs is directly proportional to the rotor circuit resistance, which can be obtained from the per-phase equivalent circuit of the asynchronous motor given in Chapter 5.

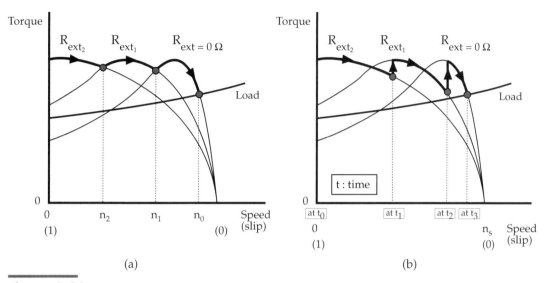

Figure 8-23
Torque-speed characteristics of wound-rotor asynchronous motor: (a) when the external resistor is varied in reference to the measured speed (to maintain the maximum torque profile); (b) when the external resistor is varied in reference to a preset time.

$$s_{Tmax} = \frac{(R'_2 + R'_{ext})}{\left[\left(\frac{X_m}{X_1 + X_m}\right)^2 R_1 + (X_1 + X'_2)^2\right]^{1/2}} \quad (8.5)$$

where R'_2 is the rotor resistance of the phase winding and R'_{ext} is the external resistance added to the rotor phase.

Normally, the value of the external resistance is chosen to make sure that the maximum torque occurs at start (when $s_{Tmax} = 1$). Note that the external resistance given in equation 8.5 is the referred value to the stator. The real value of the external resistor can be calculated if the turn ratio of the stator and the rotor windings are known. Then, the values of the resistances in each resistance bank (three power resistances used in each step) can be estimated simply by dividing the total value of the external resistance by the number of steps (which is 2 in Fig. 8-22 and Fig. 8-23).

Let's now reconsider Fig. 8-22 and Fig. 8-23 and explain the operating principles of the starting concept. Automatic starting of the three-phase wound-rotor asynchronous motor is initiated by energizing the main contactor first (contactor 1). The motor runs up to speed n_2 (or until the time t_1) with the full rotor resistance R_{ext2}. When the speed n_2 is reached (or the specified time is reached), the rotor contactor 3 closes. Hence the motor follows the new torque-speed characteristic that corresponds to the external rotor resistance R_{ext1}.

When the speed n_1 is reached (or when the set time t_2 has expired), contactor 2 is energized, and the motor runs with no external rotor resistance. This operating condition is analogous to the squirrel cage induction motor. Under this condition, the motor speed increases again up to the speed n_0, which is determined by the load characteristic. When there is no external resistance in the rotor, the full-load speed is reached, whereupon the rotor terminals are short-circuited. At this point, rated conditions, the only resistance is the rotor winding resistance. The stopping process occurs when the main contactor and all other rotor contactors are deenergized.

By correct timing of the switching of the external resistance banks, the torque characteristics can be made to follow the solid curve as shown in Fig. 8-23, which can make the maximum torque available over the acceleration range, up to the operating point where the motor torque is equal to the load torque. As seen in the figure, if the developed torque is kept at maximum up to the continuous operating speed, a fast acceleration time can be achieved.

As illustrated in Fig. 8-23, for two distinct values of external resistances, when the rotor resistance is increased, the pullout speed of the motor decreases. However, the maximum torque remains constant. In fact, for a certain value of external resistance, the maximum torque can occur at the starting conditions, which provides the ideal starting condition to accelerate heavy inertial loads. The external resistances are normally set to the highest values (which can provide the above starting condition).

The conventional rotor control circuits used in wound-rotor asynchronous motors involve either sliding power resistors or a number of resistor banks controlled by a timer circuit or a Programmable Logic Controller. However, as illustrated in Fig. 8-23a, a simple, reliable, and correct control scheme can be achieved using LabVIEW if the rotor speed is used as a feedback signal, which provides the best starting torque profile available. Furthermore, by limiting the current in the rotor circuit using external resistors, hence providing a higher power factor and torque at the instant of starting, the stator line current can be reduced considerably.

8.4.2 Hardware Details

The complete hardware implemented in this study has a block diagram given in Fig. 8-22. As shown, the hardware consists of switching circuits to operate three-phase contactors, isolation circuits to interface with a PC via a DAQ card, and various other devices such as solid-state relays and contactors.

In principle, a circuit board contains an onboard power supply to power low-voltage devices, which accepts multiple digital inputs from the DAQ card and outputs high-voltage control signals to activate the contactors that short-circuit the terminals on the rotor circuit.

In this study, the switching and interface (isolation) circuit is constructed on a common PCB that has its own built-in power supply. The circuit diagram and the PCB layout are given in the Appendix.

Let's briefly explain the subsections of the complete hardware. The motor control systems require a high degree of electrical isolation. The PC is isolated from the high-voltage side by using optical isolators. The optical isolators receive the signals from the digital output of the DAQ card. The solid-state relays (SSR) and the optical isolators require a floating power supply, which is implemented using a small mains transformer, a bridge rectifier circuit, and a voltage regulator.

The interface circuit provided in this study is capable of controlling up to five high-voltage circuits (five contactors). However, only three outputs are utilized (since there are two resistance banks and one main contactor in the motor circuit). The specification of the power contactors and the resistances depend on the motor under test.

The specifications of the solid-state relays and the opto-couplers are summarized in Table 8-7. Note that if the opto-couplers in the circuit are not chosen carefully, the TTL levels from the output of the DAQ board may degrade significantly (much less than 5 V). In this case additional buffering circuits

Table 8-7 *Specifications of the solid-state relays and opto-couplers.*

Solid-State Relay, RS-348-431	Opto-Coupler, Toshiba TLP521-1
Load Side	Transistor Output
Load current (rated): 2.5 A rms	$V_{CE} = 8$ V, $I_c = 20$ mA
Single cycle peak current overload: 30 A	Isolation voltage: 2500 V_{rms}
Voltage range: 28 V–280 V rms	Turn-on time: 2–3 ms
Off-state leakage (max): 6.5 mA rms	Diode Side
Typical min load current: 100 mA rms	Forward current: $I_F = 20$ mA
Control Side	Forward voltage: $V_F = 1.15$ V
Input voltage range: 0 V–24 V dc	*Note:* Hence the input resistance
Release voltage: 1.0 V	is selected as $R_{input} = 385$ Ω
Response time: 10 ms	

have to be built to restore the 5 V logic levels, powered from the 5 V power supply of the DAQ card or from the power supply of the PCB.

The AC contactors normally require high voltages to actuate, which should be compatible with the relay's load circuit voltage. For correct sizing of the solid-state relay current, a simple test can be carried out: The current required to actuate the contactor is measured by an ammeter, which can be used to determine the load current of the SSR. However, it is always advisable to have a margin in the current rating of the SSR, as the transient current during the actuation is substantially higher than the rated control current. Since the electromechanical contactors are very slow in comparison with the solid-state relays, the response time of the relays is not considered in the design.

The high-voltage side of the SSR is connected to the control winding of the contactor, in series with the ac mains voltage supply. The value of the per-phase resistance that should be added to the rotor to achieve maximum torque at startup was determined to be 0.68 Ω for the wound-rotor asynchronous motor used in our study. The power rating of each resistance is about 2 kW. Note that the details of the hardware are provided in the Appendix.

8.4.3 Virtual Instrument Panel

As stated earlier, a good strategy for determining the switching instants for the contactors supplying external resistance to the rotor should be based on the motor speed. As shown earlier in Fig. 8-23a, such a system can optimize the torque characteristics of the motor with the external resistance automatically. However, in this study only a user-defined time control is presented.

A LabVIEW program interface is written to control the contactors via opto-couplers and solid-state relays. The custom-written VI is named `Wound Rotor Induction Starting.vi` and is located in Section 8.4 of the Chapter 8 folder. The front panel of the VI is given in Fig. 8-24. The VI constructed in this study is a simple digital I/O program, writing to the three digital output ports of a DAQ card. Once the VI writes to the digital output ports after waiting for the correct amount of time, the opto-couplers receive the digital signal that supply control signals to the solid-state relays. The relays then close to complete the control circuit for the contactors. Following this action, the contactors actuate and short-circuit the relevant terminals, which, in turn, vary the resistance of the rotor circuit.

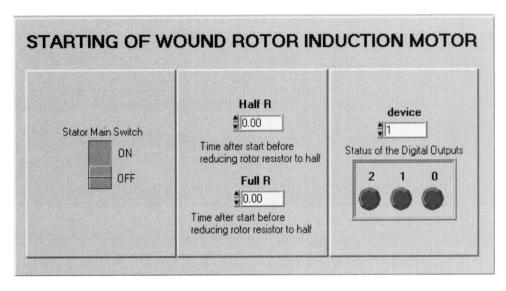

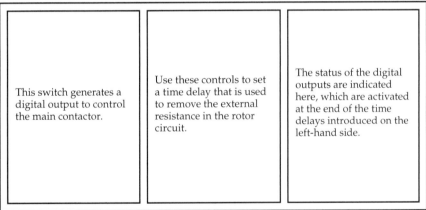

Figure 8-24
The front panel and brief user guide of the VI.

8.5 Switched Reluctance Motor Control

Because of its intrinsic simplicity and inexpensiveness, the Switched Reluctance (SR) motor has become the subject of great interest in the field of electric motor drives. The motor is well suited to many applications, mainly due to its inherent fault-tolerant feature, operating at very high speed (which is limited by the type of bearings used), and its high reliability and robustness.

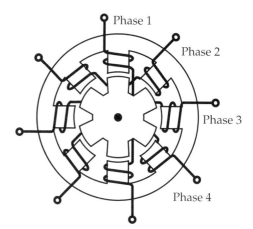

Figure 8-25
A four-phase 8/6 SR motor topology and its winding arrangements.

A four-phase 8/6 (8 stator and 6 rotor poles) SR motor is illustrated in Fig. 8-25. The SR motor is fundamentally simple in its construction. The stator and rotor resemble that of a variable reluctance stepper motor. Both the stator and the rotor have salient poles, and the number of rotor and stator poles is always unequal. Although many combinations of the number of stator and rotor poles are possible, 8/6 four-phase and 6/4 three-phase types have found widespread use.

There are no windings or permanent magnets on the rotor. The phase windings on the stator of the SR motor consist of concentrated windings wrapped around each stator pole. In the conventional arrangement, each stator pole winding is connected with that of the diametrically opposite pole to form a stator phase.

The position when the inter-pole axis, or the axis of the center of the inter-polar space in the rotor, is in line with the energized stator phase is known as the unaligned position, which corresponds to the position of minimum inductance. It can therefore be seen that the rotor experiences a torque attracting it to the minimum reluctance aligned position, when current is applied to the stator winding. Hence, the torque production in the SR motor can be viewed as resulting from the change of the magnetic reluctance seen by a motor phase with the change in rotor position.

8.5.1 Principles of Motor Control

In SR motors, when current flows in one of the stator phases and produces a magnetic field, the nearest rotor pole tends to position itself with the direction of the developed magnetic field. The new position, which is termed the *aligned position*, is reached when the rotor pole center axis is aligned with the stator pole center axis (assuming symmetrical poles). The aligned position also corresponds to the position of minimum reluctance, and hence the position of maximum inductance.

The operation of an SR motor is compatible with the type of control used with rectangular current excited BTPM motors studied earlier in this chapter. SR motors can operate from a voltage source or a current source. When the SR motor operates from a voltage source, which is normally the case in low-power SR motor drives, the controlled parameters are the voltage and current conduction interval in one phase. The voltage control can be achieved by using a variable dc supply implemented by either a controlled rectifier or a dc chopper.

The profiles of phase current depend on the drive operation. Two basic operations are, first, when the back EMF is smaller than the terminal supply voltage (low-speed operation) and second, the opposite case. Fig. 8-26b illustrates the variation of phase inductance as a function of the rotor position. In practice, the inductance-versus-position characteristics can be obtained by differentiating the flux linkage and the current curves, as given in Fig. 8-26a. In Fig. 8-26c, a typical low-speed, current chopping operation is shown together with the terminal voltage levels.

Despite their simple structure, SR motors have a highly coupled and nonlinear mathematical model due to the saliency of both stator and rotor. In SR motors, basically the torque is developed for the rotor to move into line (aligned position) with the stator poles and minimize the reluctance (or maximize the inductance) of the magnetic circuit.

The single-phase voltage equation of a SR motor can be given in terms of flux linkages by

$$v = Ri + \frac{d\psi(\theta, i)}{dt} \tag{8.6}$$

Here R is the winding resistance, i is the phase current of the SR motor, and $\psi(\theta, i) = iL(\theta, i)$. The instantaneous electromagnetic torque developed by one phase is given for a linear circuit assumption by

$$T_e = \frac{1}{2} i^2 \frac{dL}{d\theta} \tag{8.7}$$

Here, L is the winding inductance of the SR motor.

As can be seen in equation 8.7, the developed torque is proportional to the square of the current, giving it excellent starting torque, and is independent

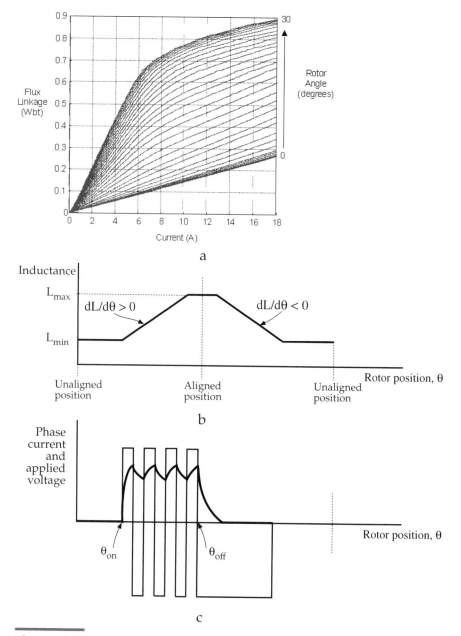

Figure 8-26
(a) A measured flux-linkage versus current characteristic of a four-phase, 8/6 SR motor obtained using the method described in Chapter 5.
(b) Approximated inductance versus rotor position characteristic of an SR motor.
(c) Low-speed current and voltage profiles for one phase of an SR motor.

of the polarity of phase current (hence permitting a single switch per phase in the converter design). Moreover, the torque sign is dependent on the inductance variation with position ($dL/d\theta$).

Since the developed torque in the SR motor depends on the inductance profile and therefore the position of the rotor, it is necessary to obtain rotor position information to find the correct interval for switching. To achieve continuous rotation, the stator phase currents are switched on and off in each phase in a sequence according to the position of the rotor. In other words, a switching sequence that produces a switched magnetic field rotating in one direction will produce a rotation of the rotor in the opposite direction.

From this discussion, one may see that the switching on and off of excitation current to the motor phases is related to the rotor pole positions. This means that some form of position sensor is essential for the effective operation of the SR motor. In addition, it will be seen next that the motor has a wide range of different output characteristics, depending on what strategy is used to switch the stator excitation currents. However, in general, for motoring operation the stator phases are excited in such a way that the stator phase nearest to a pair of rotor poles is magnetized, in order that the rotor poles are attracted into alignment with the magnetized stator poles. Once the aligned position is reached, the stator is deenergized, so that rotation of the motor occurs in one direction only.

However, the simplicity of the SR does not extend to its operation. SR motors cannot be operated directly from a three-phase ac supply or a dc source. They require a complex controller that provides the winding excitation and must be accurately synchronized with the angular position of the rotor for effective operation. Therefore, measurement of the rotor position is necessary to operate the SR motor. Controlling these motors requires complex control algorithms because of their inherent nonlinear properties. Ideally, the control algorithms should take all the nonlinear aspects of the motor and nonideal rotor position sensors into account and smooth the irregularities from the output torque-speed characteristics of the motor.

A diagram of the main components of the drive system is shown in Fig. 8-27. For optimum performance, the motor, power electronic circuits, control circuits, and measuring circuits should be considered together, and not in isolation.

Inverter Topology

As was mentioned, the torque produced in the SR motor is independent of the direction of current flow in each motor phase. This means the inverter is only

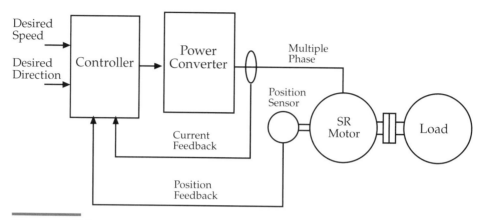

Figure 8-27
Main components of an SR motor drive.

required to supply unidirectional currents into the stator windings. One of the major circuit topologies that have been used for SR motor drives, known as two-switch per-phase inverter type, is shown in Fig. 8-28. This circuit is used successfully in SR motor drives and has the advantage of allowing full reverse voltage across the windings.

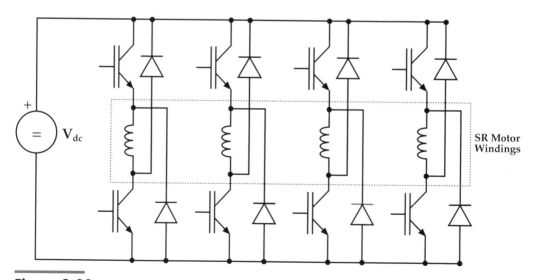

Figure 8-28
Two-switch type inverter topology for four-phase SR motors.

As shown in the figure, this inverter type uses two switching devices and two diodes per phase, hence three modes of operation are possible in this circuit.

- **Mode 1, Positive Phase Voltage**
  A positive phase voltage can be applied by turning both switching devices on. This will cause current to increase in the phase winding.

- **Mode 2, Zero Phase Voltage (soft chopping control)**
  A zero voltage loop can be imposed on the motor phases when either of the switches is turned off while current is flowing through the phase winding. This results in current flow through a freewheeling loop consisting of one switching device and one diode, with no energy being supplied by, or returned to, the dc supply. The current will decay slowly due to the small resistance of the semiconductors and connections, which leads to small conduction losses.

- **Mode 3, Negative Phase Voltage**
  When both switching devices in a motor phase leg are turned off, the third mode of operation occurs. In this mode, the motor phase current will transfer to both of the freewheeling diodes and return energy to the supply. When both of the diodes in the phase circuit are conducting, a negative voltage with amplitude equal to the dc supply voltage level is imposed on the phase windings.

Position Sensors

The need for rotor position information has been traditionally satisfied by the use of electromechanical sensors, which are sometimes called angle transducers or position encoders. These sensors are attached to the motor shaft and measure the mechanical rotation of the rotor. They provide an electrical signal as an output that provides information about the rotor position.

There are various electromechanical position sensors used in SR motor drives. However, we will consider the Gray binary coded disk as a position sensor, which can provide the absolute position of the rotor. The wheel for the Gray coded disk and its associated signals are shown in Fig. 8-29. Note that use of a Gray coded disk removes the potential for error due to multi-bit changes that occur in other types of absolute encoders. In the Gray coded disks, even when a transition is under way, the only possible values that can be read are the before and after positions, either of which is valid.

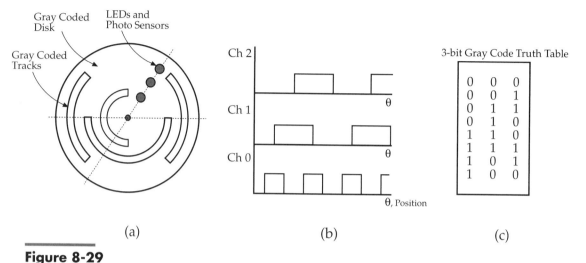

Figure 8-29
(a) Three-bit Gray coded disk, (b) signals for a Gray binary code disk, and (c) its truth table.

A Gray binary coded disk sensor is an optical sensor that consists of a concentrically patterned wheel, a light emitting diode per bit (acts as a light transmitter), and a phototransistor to detect light. The wheel either blocks or passes the emitted light to the phototransistor depending on the shaft position. Hence the phototransistor turns on or off depending on the rotor angle.

The output of the sensor can be directly interfaced to a computer controller that can generate the control signals for the inverter. Furthermore, the current feedback signals can be used to regulate the winding currents for the torque control purpose.

8.5.2 Virtual Instrumentation

The custom-written VI is named SR Motor Control.vi and is located in Section 8.5 of the Chapter 8 folder. The front panel of the VI is shown in Fig. 8-30, which accepts the inputs from a Gray coded position sensor and generates switching signals (without current feedback) for the inverter switches. Fig. 8-31 shows the brief user guides for these front panels.

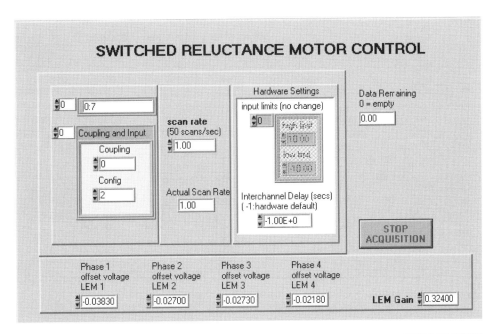
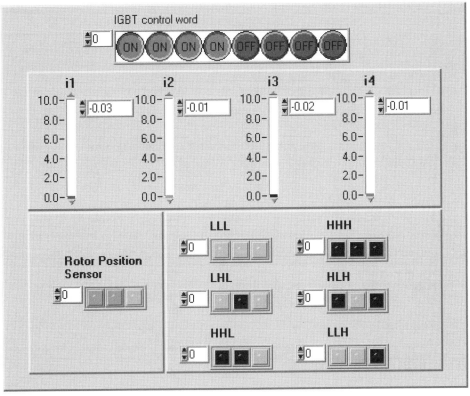

Figure 8-30
The front panels of the main VI and the control sub-VI (LabVIEW 5.1).

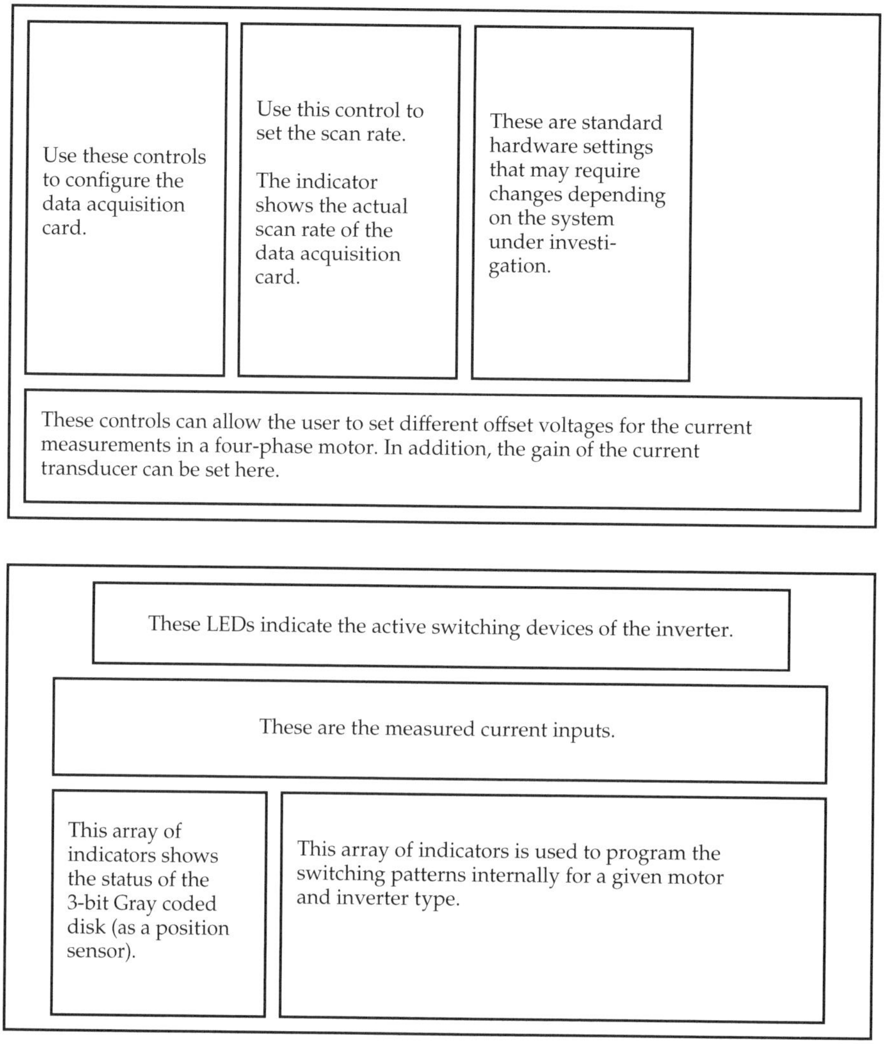

Use these controls to configure the data acquisition card.

Use this control to set the scan rate.

The indicator shows the actual scan rate of the data acquisition card.

These are standard hardware settings that may require changes depending on the system under investigation.

These controls can allow the user to set different offset voltages for the current measurements in a four-phase motor. In addition, the gain of the current transducer can be set here.

These LEDs indicate the active switching devices of the inverter.

These are the measured current inputs.

This array of indicators shows the status of the 3-bit Gray coded disk (as a position sensor).

This array of indicators is used to program the switching patterns internally for a given motor and inverter type.

Figure 8-31
Brief user guides for the front panels shown in Fig. 8-30.

8.6 References

"L297/L297D Stepper Motor Controllers." Data Sheet. SGS Thompson Microelectronics, August 1996.

"L298 Dual Full-Bridge Drive." Data Sheet. SGS Thompson Microelectronics, July 1999.

"LMD18245, 3 A, 55 V DMOS Full-Bridge Motor Drive." Data Sheet. National Semiconductor Corp., 1993.

DC Motors, Speed Controls, Servo Systems. The Electrocraft Engineering Handbook, 5th ed. Rockwell Automation/Electrocraft.

Ertugrul, N. "Position Estimation and Performance Prediction for Permanent-Magnet Motor Drives." Ph.D. thesis, University of Newcastle upon Tyne, 1993.

Ertugrul, N. "Power Electronics Lecture Notes." Department of Electrical and Electronic Engineering. University of Adelaide, 1997.

Ertugrul, N. "The Speed Control of Slip-Ring Induction Motor from Rotor Circuit and the Design of a Static Starting Circuit." M.Sc., Istanbul Technical University, Institute of Science and Technology, 1989.

Handbook of Small Motor, Gearmoter, and Control, 5th ed. Bodine Electric, 1993.

Kenjo, T. *Power Electronics for the Microprocessor Age.* New York: Oxford University Press, 1990.

Sen, P. C. *Principles of Electric Machines and Power Electronics.* New York: Wiley, 1989.

Appendix

The following pages provide a range of additional information that will help potential developers of similar laboratory systems. Note that the circuits provided here are used in various real-time VIs presented in this book. Although these circuits are tested thoroughly and are in use, I strongly advise that for any practical implementation you test the circuits before initiating any high voltage and current measurements, and always follow standard safety precautions.

For Chapter 1

Table A-1 summarizes the ratings of some of the devices available in our laboratory that were utilized in the development stages of the relevant VIs.

Table A-1 *Laboratory devices and their ratings.*

3-phase supply	415 V, 50 Hz, 50 A
DC supply	200 V, 40 A
Single-phase autotransformers (as variable voltage sources)	240 V, 8 A, 50 Hz
3-phase autotransformers	Input: 415 V, 15 A; Output: 0–470 V
Rheostats	50 Ω, 5 A
Capacitors	4 μF, 1000 V
Switched Reluctance motor drive	Qulton, 380/415 V, 50/60 Hz, 4 kW, 9 A

Implementation details of the three-phase current interface circuits are given in Figs. A-1 through A-4 and in Table A-2.

Appendix

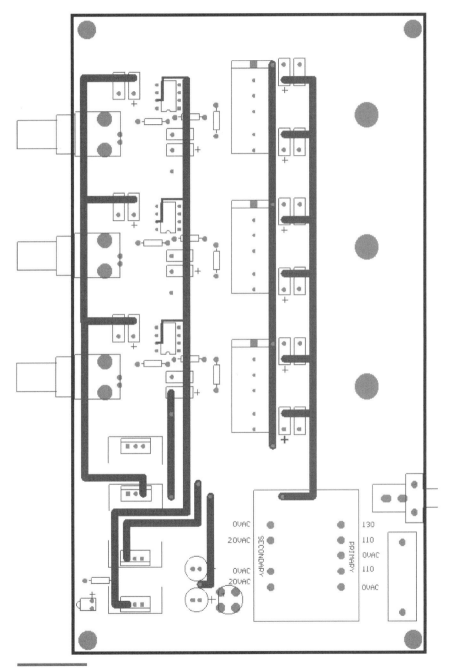

Figure A-1
Top layer of the printed circuit board (PCB) containing three identical Hall-effect transducers.

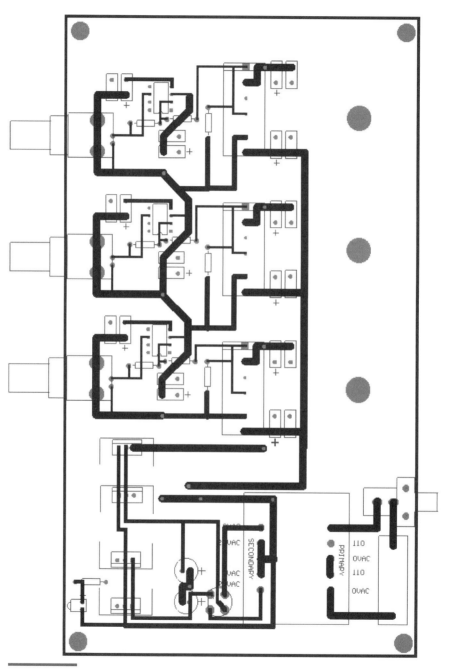

Figure A-2
Bottom layer of the PCB.

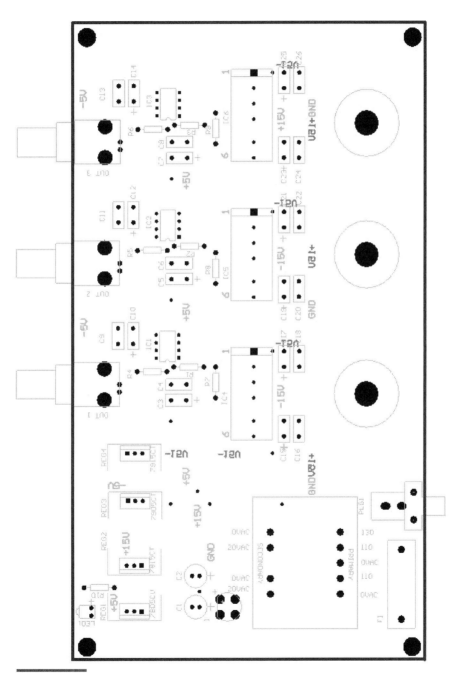

Figure A-3
Overlay of the PCB.

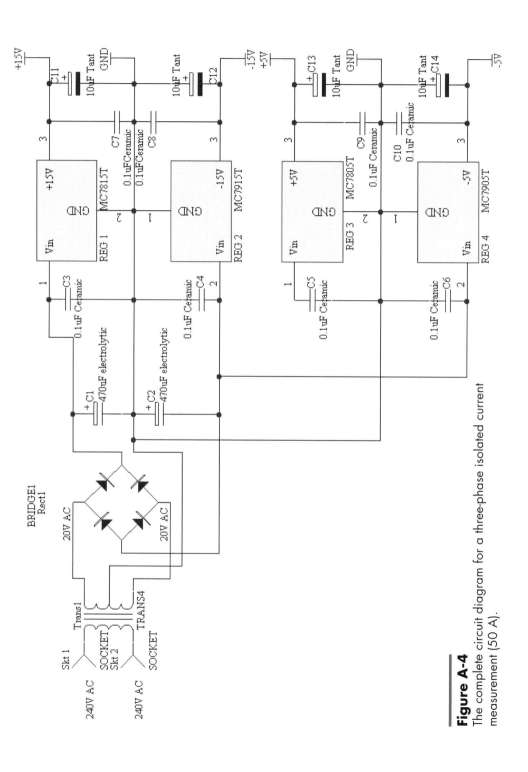

Figure A-4
The complete circuit diagram for a three-phase isolated current measurement (50 A).

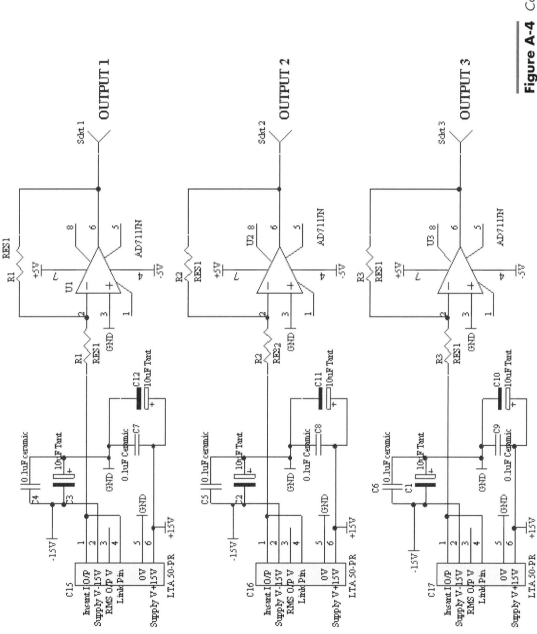

Figure A-4 *Continued*

Table A-2 *List of components used in the three-phase current interface circuit.*

Designators	Comment
C8	0.1 μF 63 V RS126-556
C13	0.1 μF 63 V RS126-556
C26	0.1 μF 63 V RS126-556
C4	0.1 μF 63 V RS126-556
C18	0.1 μF 63 V RS126-556
C24	0.1 μF 63 V RS126-556
C22	0.1 μF 63 V RS126-556
C6	0.1 μF 63 V RS126-556
C11	0.1 μF 63 V RS126-556
C16	0.1 μF 63 V RS126-556
C20	0.1 μF 63 V RS126-556
C9	0.1 μF 63 V RS126-556
C1	100 μF Electrolytic
C2	100 μF Electrolytic
R5	10 K 0.5 W METAL FILM
R6	10 K 0.5 W METAL FILM
R1	10 K 0.5 W METAL FILM
R3	10 K 0.5 W METAL FILM
R2	10 K 0.5 W METAL FILM
R4	10 K 0.5 W METAL FILM
C12	10 μF 25 V RS117-675
C5	10 μF 25 V RS117-675
C25	10 μF 25 V RS117-675
C23	10 μF 25 V RS117-675
C14	10 μF 25 V RS117-675
C7	10 μF 25 V RS117-675
IC6	LEM LTA, 50P/SP1, 1:1000
C21	10 μF 25 V RS117-675
IC5	LEM LTA, 50P/SP1, 1:1000
IC4	LEM LTA, 50P/SP1, 1:1000
C15	10 μF 25 V RS117-675
C3	10 μF 25 V RS117-675
C17	10 μF 25 V RS117-675
C10	10 μF 25 V RS117-675
C19	10 μF 25 V RS117-675
IC2	AD711 FEC 400-970
IC3	AD711 FEC 400-970
1C1	AD711 FEC 400-970
OUT	BNC R/A FEC150-682
OUT	BNC R/A FEC150-682

Table A-2 *Continued*

Designators	Comment
OUT 1	BNC R/A FEC150-682
1	Bridge rectifier, 1.5 A W005fec572743
F1	FUSE FEC319-491 & FEC471-677
LED1	LED Right Angle, FEC264-362
REG1	LM7805 REG +5 V
REG2	LM7815 REG +15 V
REG3	LM7905 REG −5 V
REG4	LM7915 REG −15 V
PLG1	MOLEX 2PIN
R7	R2 (100 r) S.O.T.
R8	R2 (100 r) S.O.T.
R9	R2 (100 r) S.O.T.
R10	RESISTOR 0.5 W 330 R
2	Transformer 20/20VFEC150-075

Implementation details of the three-phase voltage interface circuits are given in Figs. A-5 through A-8 and in Table A-3.

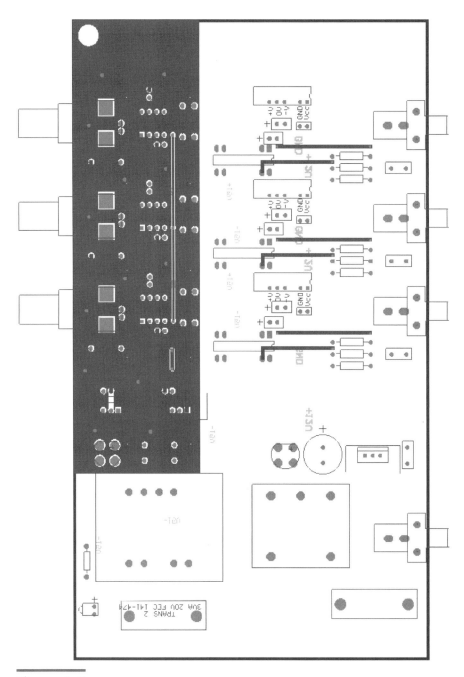

Figure A-5
Top layer of the PCB containing three identical voltage isolation circuits.

Appendix 395

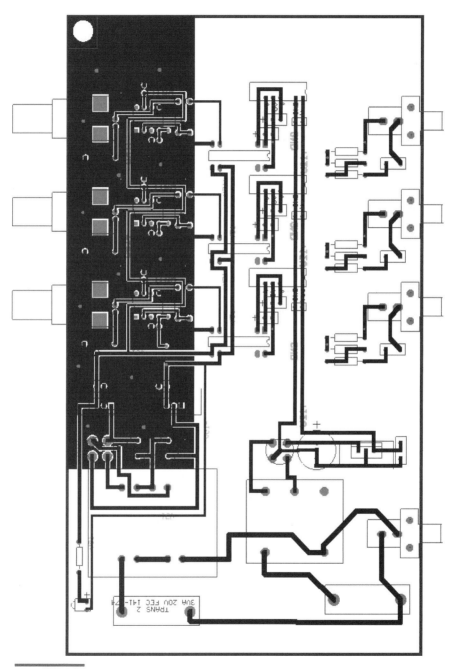

Figure A-6
Bottom layer of the PCB.

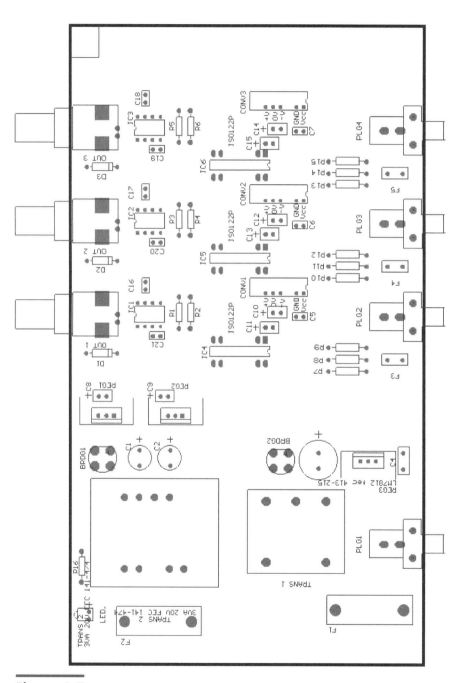

Figure A-7
Overlay of the PCB.

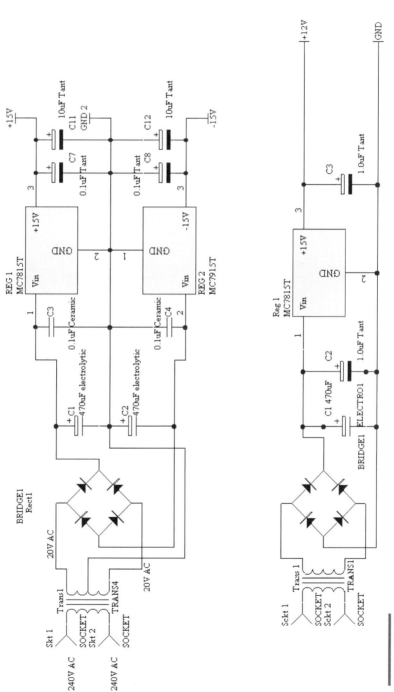

Figure A-8
The complete circuit diagram for isolated high-voltage measurement (500 V).

(continued)

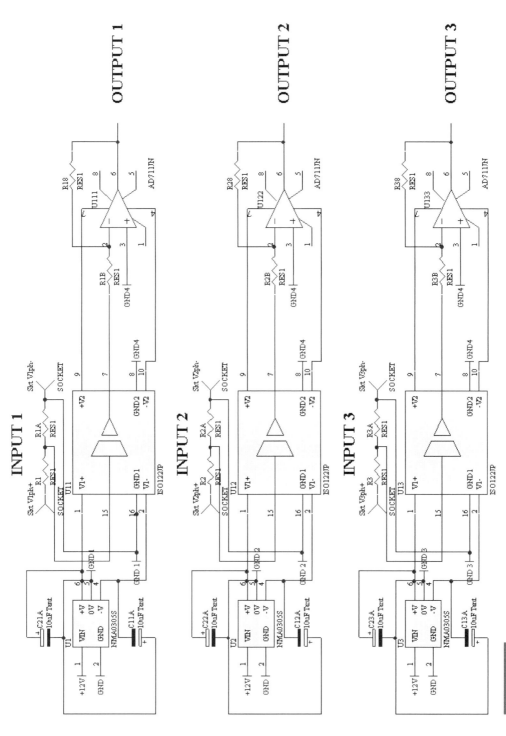

Figure A-8
Continued

Table A-3 *List of components used in the three-phase voltage interface circuit.*

Designators	Comment
BRDG1	BRIDGE RECT FEC706-553
BRDG2	BRIDGE RECT FEC706-553
C1	100 μF ELECTRO 25/50 V
C10	0.1 μF TANT
C11	0.1 μF TANT
C12	0.1 μF TANT
C13	0.1 μF TANT
C14	0.1 μF TANT
C15	0.1 μF TANT
C16	0.1 μF CERAMIC
C17	0.1 μF CERAMIC
C18	0.1 μF CERAMIC
C19	0.1 μF CERAMIC
C2	100 μF ELECTRO 25/50 V
C20	0.1 μF CERAMIC
C21	0.1 μF CERAMIC
C3	100 μF ELECTRO 25/50 V
C4	0.1 μF 2F4 63 V CERAMIC FEC108-993
C5	0.1 μF 2F4 63 V CERAMIC FEC108-925
C6	0.1 μF 2F4 63 V CERAMIC FEC108-925
C7	0.1 μF 2F4 63 V CERAMIC FEC108-925
C8	1 μF TANT
C9	1 μF TANT
CONV1	NMA1215S FEC330-802
CONV2	NMA1215S FEC330-802
CONV3	NMA1215S FEC330-802
F1	FUSE FEC319-491/471-677
F2	FUSE FEC319-491/471-677
F3	FUSE 50 mA FEC151-101/HLDR149259
F4	FUSE 50 mA FEC151-101/HLDR149259
F5	FUSE 50 mA FEC151-101/HLDR149259
IC1	AD711op/ampFEC400-970
IC2	AD711op/ampFEC400-970
IC3	AD711op/ampFEC400-970
IC4	ISO122P, Isolation Amplifier
IC5	ISO122P, Isolation Amplifier
IC6	ISO122P, Isolation Amplifier
LED1	LED R/A FEC264-362
OUT 1	BNC R/A FEC150-682
OUT 2	BNC R/A FEC150-682

(continued)

Table A-3 *Continued*

Designators	Comment
OUT 3	BNC R/A FEC150-682
PLG1	MOLEX 2PIN FEC151-884/151-886/151-890/151-891
PLG2	MLX 2PINFEC151-884/14259/151890
PLG3	MLX 2PINFEC151-884/14259/151890
PLG4	MLX 2PINFEC151-884/14259/151890
R1	RESISTOR 1/4 W 10 K
R10	S.O.T. GAIN TO BE SET
R11	S.O.T. GAIN TO BE SET
R12	S.O.T. GAIN TO BE SET
R13	S.O.T. GAIN TO BE SET
R14	S.O.T. GAIN TO BE SET
R15	S.O.T. GAIN TO BE SET
R16	2 K 1/4 W
R2	RESISTOR 1/4 W 10 K
R3	RESISTOR 1/4 W 10 K
R4	RESISTOR 1/4 W 10 K
R5	RESISTOR 1/4 W 10 K
R6	RESISTOR 1/4 W 10 K
R7	S.O.T. GAIN TO BE SET
R8	S.O.T. GAIN TO BE SET
R9	S.O.T. GAIN TO BE SET
REG1	LM7815fec 413-446
REG2	LM7915fec 413-987
REG3	LM7812 fec 413-215
TRANS 1	1.2 VA 15 V fec432-726
TRANS 2	3 VA 20-0-20 V 0.07 A fec 141-474

For Chapter 5

Implementation details of the interface circuit for the electromechanical device experiment are given in Figs. A-9 through A-13 and in Table A-4.

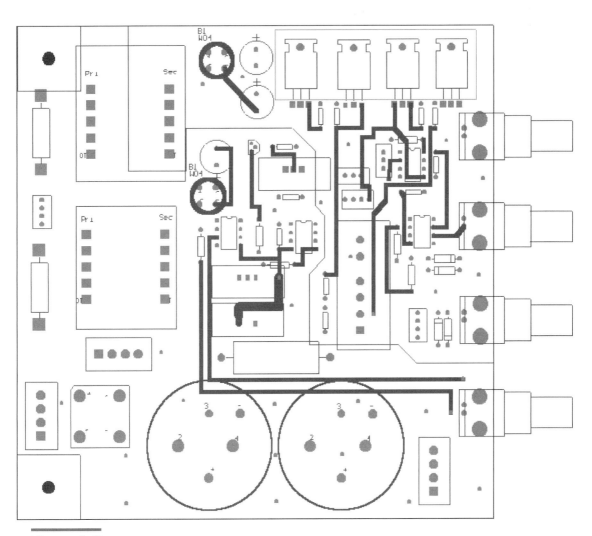

Figure A-9
Top layer of the PCB containing the hardware for the electromechanical device experiment.

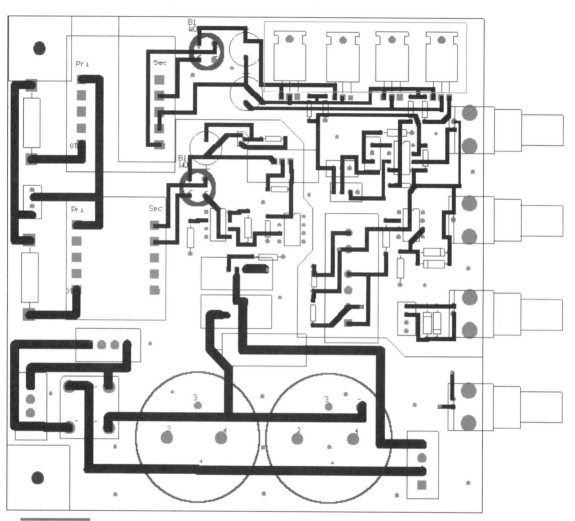

Figure A-10
Bottom layer of the PCB.

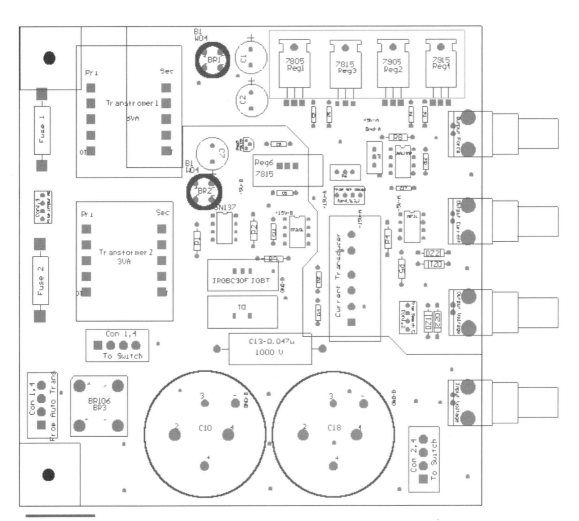

Figure A-11
Overlay of the PCB.

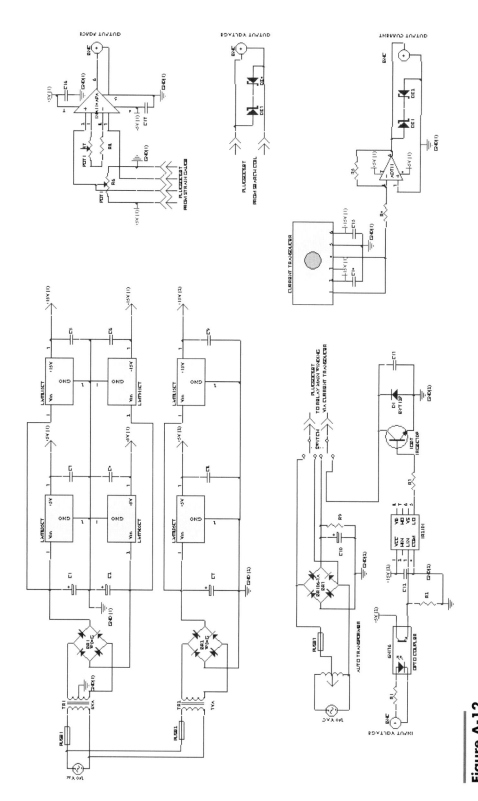

Figure A-12
The complete circuit diagram for the electromechanical device experiment interface.

Table A-4 *List of components used in the interface circuit for the electromechanical device.*

Designators	Comment	Designators	Comment
R1	220 Ω	C16	0.1 μF / 35 V
R2	1.8 KΩ	C17	0.1 μF / 35 V
R3	10 Ω	TR1	6 VA
R4	1 KΩ	TR2	3 VA
R5	1 KΩ	BR1	W04G
R6	10 KΩ	BR2	W04G
R7	100 Ω	BR3	MIC BR106
R8	22 Ω	FUSE 1	250 mA
R9	4.7 KΩ / 4 W	FUSE 2	250 mA
C1	1000 μF / 35 V	FUSE 3	250 mA
C2	1000 μF / 35 V	REG1	78L05
C3	0.1 μF / 35 V	REG2	79L05
C4	0.1 μF / 35 V	REG3	78L15
C5	0.1 μF / 35 V	REG4	79L15
C6	0.1 μF / 35 V	REG5	7805
C7	470 μF / 35 V	REG6	78L15
C8	0.1 μF / 35 V	D1	BYT12P
C9	0.1 μF / 35 V	IGBT	IRGBC30F
C10	470 μF / 400 V	Opto-coupler	6N136
C12	0.1 μF / 35 V	DRIVER	IR2101
C13	0.047 μF / 1000 V	Op-Amp	AD711
C14	0.1 μF / 35 V	Inst-Amp	1NA114P
C15	0.1 μF / 35 V		

Figure A-13
Photo of the custom-built interface circuit for the electromechanical device experiment.

For Chapter 8

Implementation details of the dc motor control are given in Figs. A-14 through A-17 and in Table A-5.

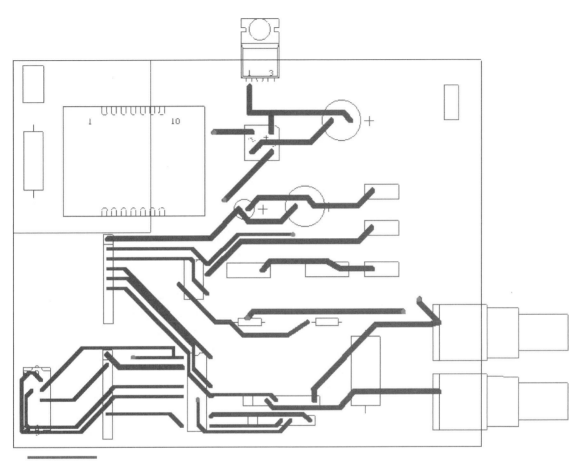

Figure A-14
Top layer of the PCB containing the hardware for the dc motor control VI.

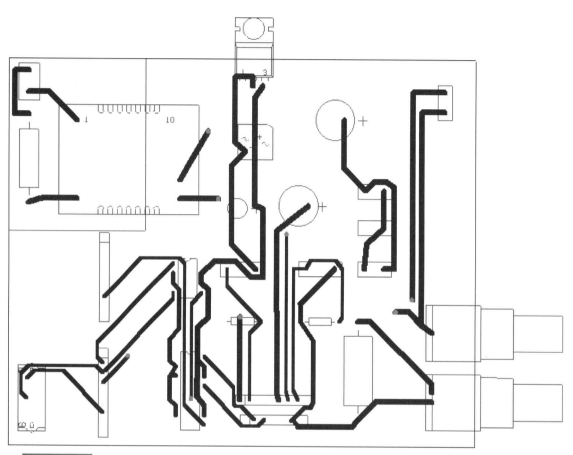

Figure A-15
Bottom layer of the PCB.

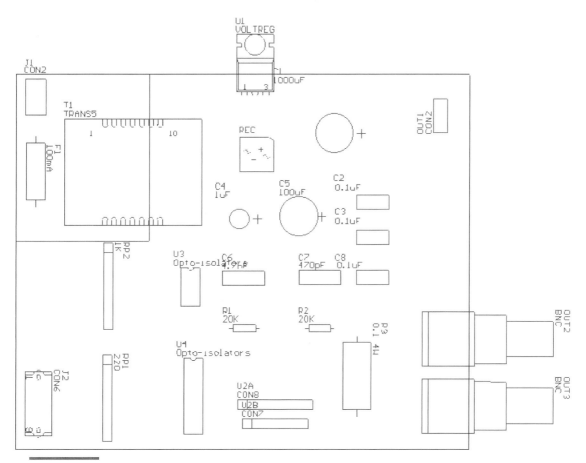

Figure A-16
Overlay of the PCB.

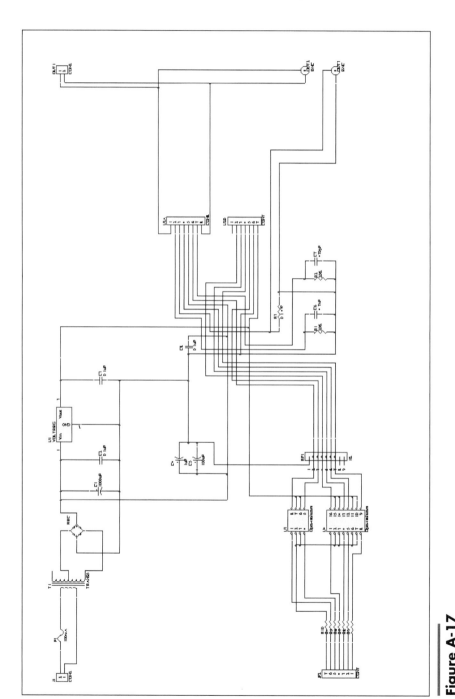

Figure A-17
The complete circuit diagram for the dc motor control.

Table A-5 *List of components used in the dc motor control circuit.*

Designators	Comment
J1	2-pin Molex connector for main input
J2	10-pin connector
OUT1	2-way screwed connector
F1	100 mA Fuse
T1	Transformer, 3 VA, 2×15 V
V1	Voltage regulator, 7805
C1	Electrolytic Capacitor, 1000 μF, 35 V
C5	Electrolytic Capacitor, 100 μF, 25 V
REC	W04 bridge rectifier
R1	20 K, 1/4 W Resistor
R2	20 K, 1/4 W Resistor
U3	Opto-coupler, RS 308-813
U4	Opto-coupler, RS 307-064
R3	0R1, 5 W Power Resistor
C2	0.1 μF
C3	0.1 μF
C4	10 μF
C5	100 μF
C6	4.7 nF
C7	470 pF
C8	0.1 μF
RP1	220 Ω, Resistor pack
RP2	1 K, Resistor pack
OUT2	Right Angle BNC
OUT3	Right Angle BNC
U2A, U2B	Connectors for LMD18245

Implementation details of the stepper motor control are given in Figs. A-18 through A-21 and in Table A-6.

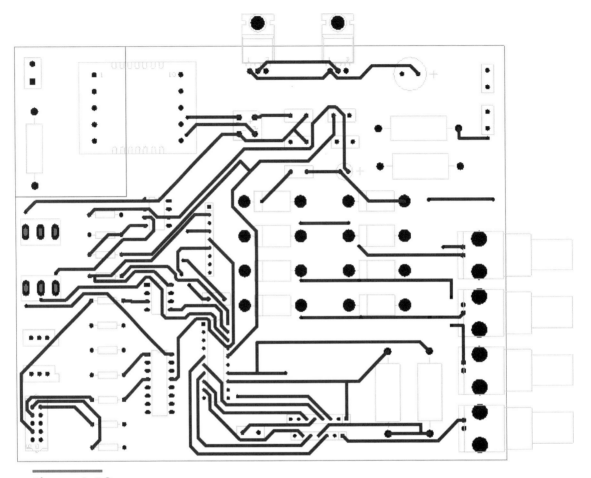

Figure A-18
Top layer of the PCB containing the hardware for the stepper motor control.

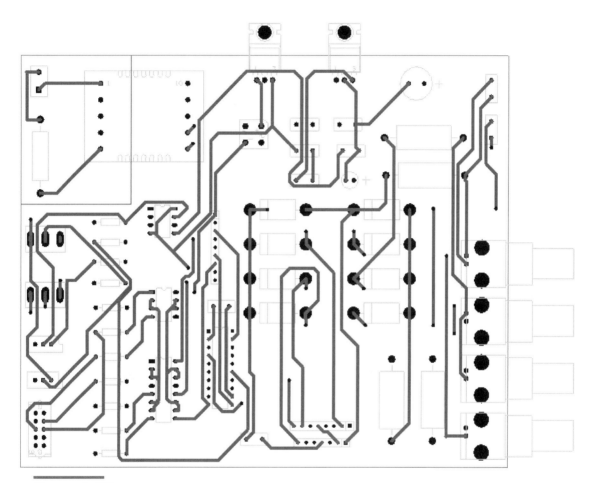

Figure A-19
Bottom layer of the PCB.

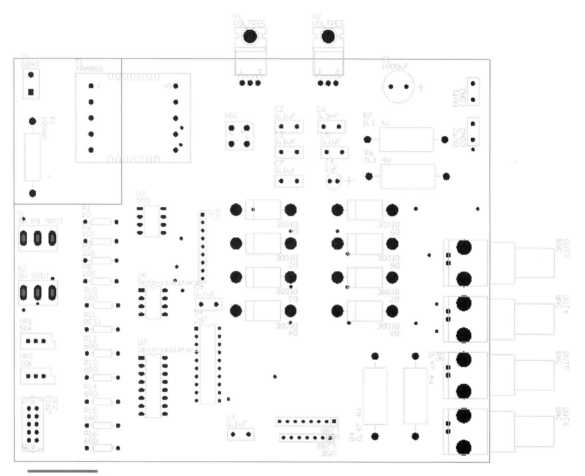

Figure A-20
Overlay of the PCB.

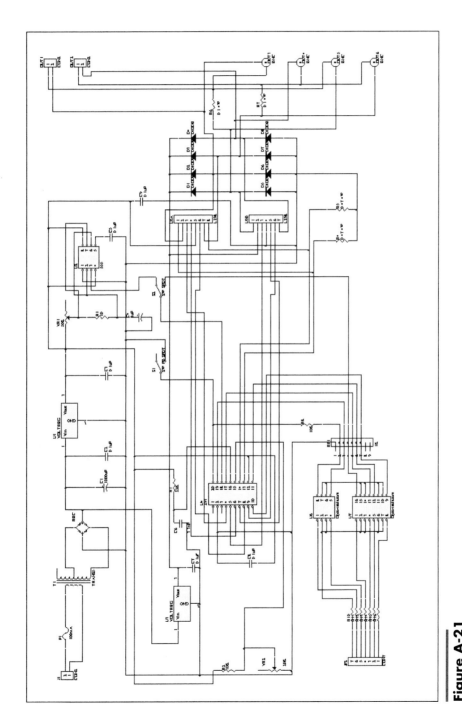

Figure A-21
The complete circuit diagram for the stepper motor control.

Table A-6 *List of components used in the stepper motor control circuit.*

Designators	Comment
J1	2-pin Molex connector for main input
JP2	10-pin connector
OUT1	Connector for stepper motor
OUT2	Connector for stepper motor
F1	100 mA Fuse
T1	Transformer, 3 VA, 2×15 V
V1	Voltage Regulator, 7805
V3	Voltage Regulator, 7805
C1	Electrolytic Capacitor, 1000 μF, 25 V
C4	Electrolytic Capacitor, 1 μF, 63 V
REC	W005G Bridge Rectifier
D1-D6	1N5408 Freewheeling Diode
C2, C3, C5	0.1 μF
C7, C8, C9	0.1 μF
C4	1 μF
C6	3.3 nF
R6, R7	0R1, 4 W Power Resistor
R4, R5	0R47, 4 W Power Resistor
OUT3	Right Angle BNC
OUT4	Right Angle BNC
OUT5	Right Angle BNC
OUT6	Right Angle BNC
U5A, U5B	Connectors for L298N
U4	L297
U6	Opto-coupler, RS 308-813
U7	Opto-coupler, RS 307-064
U2	555 Timer
RP1	1 K, Resistor Pack
R10-R16	680 R, 1/4 W Resistor
R1	10 R, 1/4 W Resistor
R2, R8	10 K, 1/4 W Resistor
R3	22 K, 1/4 W Resistor
S1	Normally closed push button
S2	Single pole switch
VR1	50 K, multi-turn resistor
VR2	20 K, multi-turn resistor

Implementation details of the static-relay interface for the starting circuit of the wound-rotor asynchronous motor experiment are given in Figs. A-22 through A-26 and in Table A-7.

Appendix

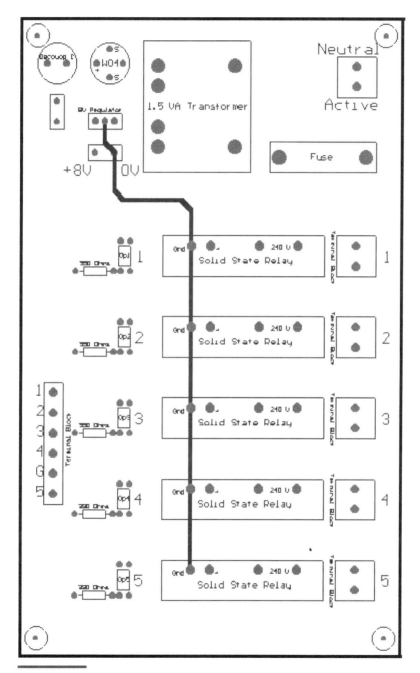

Figure A-22
Top layer of the PCB containing the hardware for the solid-state relay and computer interface.

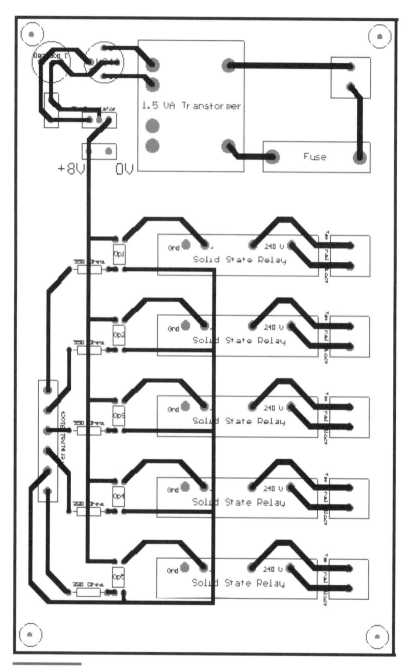

Figure A-23
Bottom layer of the PCB.

Appendix

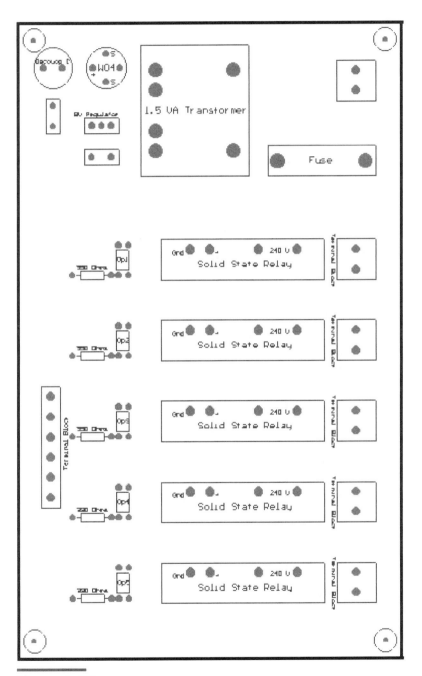

Figure A-24
Overlay of the PCB.

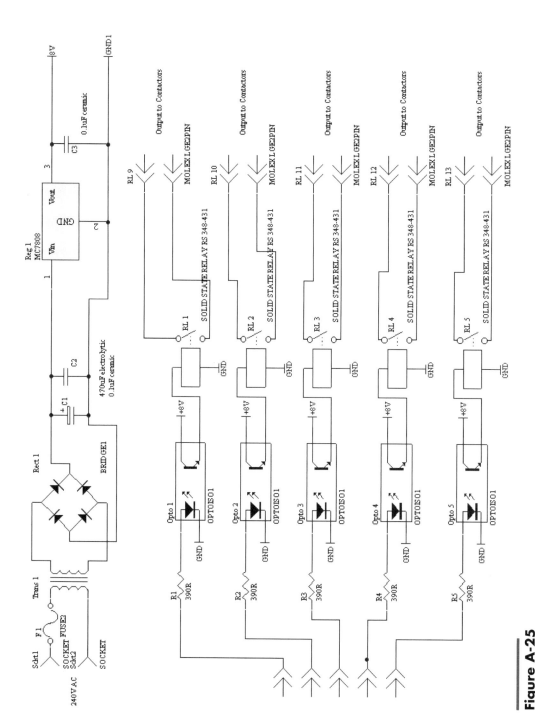

Figure A-25
The complete circuit diagram for the solid-state relay and computer interface via opto-couplers.

Appendix

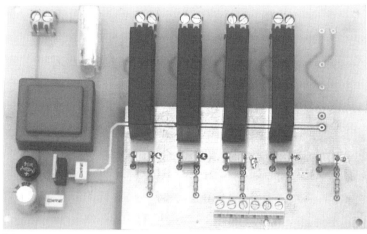

(a)

(b)

Figure A-26
Photos of the custom-built interface circuits for the starting of wound-rotor asynchronous motor: (a) computer interface, and (b) the power circuit.

Table A-7 List of components used in the starting circuit of the wound-rotor asynchronous motor.

Designators	Comment
RS348-431	Solid State Relays, 240 V, 2.5 A, 3-18 V Control
TLP521-1	6 Opto-Couplers
LM7808	Voltage Regulator
W04	Bridge Rectifier
Capacitor 1	1000 µF, 16 V, Electrolytic
Capacitor 2	22 nF, ceramic, 16 V, for the voltage regulator
Capacitor 3	22 nF, ceramic, 16 V, for the voltage regulator
Resistors 1–6	385 R, Input Resistors for the Opto-Couplers
Fuse and holder	240 V, 100 mA
1 Connector	6-way (for digital inputs)
5 Connectors	2-way (240 V, 5 A)

Front Panels of Multiple-Choice Quiz VIs

The front panels of the two VIs presented here (Fig. A-27 and Fig. A-28) are designed to perform preliminary multiple-choice quiz sessions that should run prior to the practical tests in the laboratory.

Sample 1: Synchronization Observer Experiment Multiple-Choice Quiz

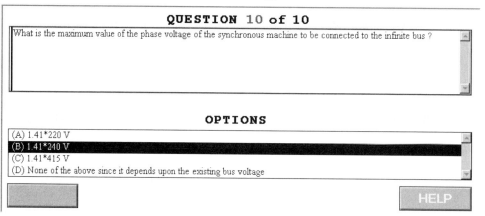

Figure A-27
Front panel of the multiple-choice quiz VI (Sample 1, no-password)

Appendix

Sample 2: Implementation of Password Access

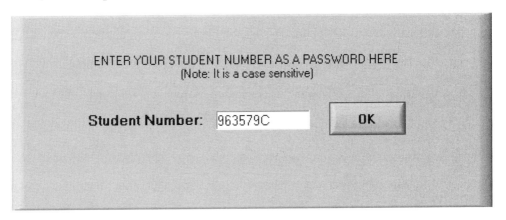

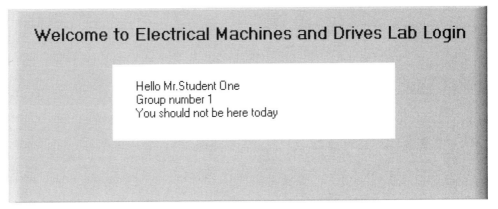

Figure A-28
Front panels of the password VI, which can be integrated into the previous VI.

The complete codes of the VIs and sample input files are provided on the accompanying CD-ROM. To operate these VIs, you will need to prepare custom-written text files (which should be experiment-specific) and define correct path names.

Note that the text files contain questions/multiple choices/answers (Sample 1; Table A-8) and students' list (Sample 2; Table A-9). The student list is designed to integrate logging in with a password. For correct operation, the text files should be written in the format provided.

Table A-8 *A sample tab-delimited text file for the multiple-choice quiz VI.*

What is the synchronous speed of the asynchronous machine? (A) 1400 rpm   (B) 1500 rpm (C) 1410 rpm   (D) 1000 rpm   1   0.102793
What is the maximum speed of the prime drive (asynchronous motor) in this experiment?   (A) 1410 rpm   (B) 1500 rpm   (C) Slightly less than 1500 rpm   (D) Slightly greater than 1500 rpm   2   0.581896
What is the rated phase voltage of the asynchronous machine?   (A) 220 V   (B) 240 V   (C) 415 V   (D) 339 V   1   0.945002
What is the number of poles of the synchronous machine?   (A) 1   (B) 2   (C) 4   (D) 6   2   0.009851
In this test, the slip-ring induction (asynchronous) motor will be started with a resistor bank connected across the rotor windings. What is the starting position of these resistors, and what is done after it is started? (A) At start the rotor resistors are at their maximum values, and they are gradually reduced to the half values.   (B) The motor starts with the minimum values, and they are increased while the speed increases.   (C) No adjustment is done, the resistors are left alone. (D) At start the rotor resistors are at their maximum values, and they are gradually reduced and short-circuited. 3   0.070456
What would happen if synchronization is not performed correctly?   (A) The machine will generate no output voltage.   (B) Dangerously large voltages will be generated.   (C) The machine will draw no current.   (D) Dangerously large currents will flow. 3   0.012279
What is the most important outcome if you were to come in contact with high-voltage in the laboratory?   (A) You would fail the practical and we would need to find a new student.   (B) The lights would dim and the demonstrator would be upset.   (C) This is undesirable since we may be injured.   (D) I cannot answer this question since I did not read the experiment's handout. 2   0.585975
What is the characteristic of the stator in the synchronous machine?   (A) It rotates.   (B) It stays stationary.   (C) It produces a static magnetic field.   (D) It controls the current direction.   1   0.120449
What is the most important concern in running any machine above its maximum speed?   (A) The large voltages on its terminals could cause electric shock. (B) The rotor may disintegrate catastrophically.   (C) It may experience "runaway."   (D) The stator may disintegrate catastrophically.   1   0.076411
What is the maximum value of the phase voltage of the synchronous machine to be connected to the infinite bus?   (A) 1.41*220 V   (B) 1.41*240 V   (C) 1.41*415 V   (D) None of the above since it depends on the existing bus voltage 3   0.644833

Table A-9 *A sample tab-delimited text file for the Login VI.*

963579C	Mr. Student One	1	1/2/2003
952514H	Ms. Student Two	1	1/2/2003
963470I	Mr. Student Three	1	1/2/2003
954475K	Mr. Student Four	1	1/2/2003
968132A	Mr. Student Five	1	1/2/2003
962568N	Ms. Student Six	1	1/2/2003
964195H	Mr. Student Seven	2	1/2/2003
968133S	Mr. Student Eight	2	1/5/2003
968143E	Mrs. Student Nine	2	1/5/2003
944858O	Mr. Student Ten	2	1/2/2003
933777J	Mr. Student Eleven	3	1/2/2003

Index

A

ABC reference frame, 285
Absolute temperature, 225
AC chopper, 245, 249
AC circuits, 37
   analysis, 42
Acceleration, 115
Active circuit, 224
Active diodes, 238
Active (real) power, 50, 76
Amplitude, 14
Analog integrator, 102
Angular frequency, 14
Animation ring, 233
Apparent power, 51, 76
Armature
   copper losses, 128
   current, 123
   resistance, 123
   voltage, 123

Asynchronous motor, 45
   approximate equivalent circuit, 175, 182
   blocked-rotor test, 175, 176, 177, 181
      equivalent impedance, 177
      equivalent leakage reactance, 177
      equivalent resistance, 177
   core loss resistance, 174
   dynamic simulation, 282
   efficiency, 179
   exact equivalent circuit, 174, 175, 179, 182
   full-load, 174
   induced voltage, 282
   input power, 179
   magnetizing branch current, 176
   magnetizing reactance, 174
   no-load test, 175, 176, 177, 181
      equivalent turns ratio, 178
      rotor-induced voltage, 173, 178
   output power, 179

Asynchronous motor (*continued*)
  performance characteristics, 180
  phase current, 174
  phase voltage, 174
  power factor, 179
  rated speed, 174
  rotating field (flux density wave), 171
  rotor, 173
    current, 174
    leakage reactance, 174
    resistance, 174
  shaft torque, 179
  slip, 177
  slip-ring, 173
  squirrel cage, 173
  starting current, 187
  stator, 173
    current, 178
    leakage reactance, 174
    resistance, 174
  synchronous speed, 173, 174
  wound rotor, 173
    acceleration range, 370
    external resistors, 369
    starting, 366
    starting current, 366
    starting torque, 370
  torque-speed, 369
Average (mean or dc) value, 15
Average power, 76

B

Back emf, 288
Battery load, 237
BH curve, 96, 139
BH loop, 142
Bi-directional current, 307
Boltzmann constant, 225
Brushless PM motor, 287
  dc link voltage, 292
  sinusoidal, 288
  star-point voltage, 292
  terminal voltage, 292
  trapezoidal, 288
    control, 358, 365

C

Calibration, 105
Chopping angle, 18
Circular buffer, 350
Clamping device, 105
Clipped angle, 18
Commutation, 237
  angle (overlap angle), 238
  instant, 238
  interval, 238
  period, 238
  time, 243
Complex power, 50, 151
Conductance, 25
Continuous current conduction, 238, 239, 301
Controller characteristics, 340
Copper losses, 121
Core permeability, 155
Current conduction angle, 248
Current phasor, 66
Current sensor, 150
Cycloconverter, 254

D

Data acquisition, 113, 135
Data logging, 75
DC circuit, 23
DC link voltage
DC load, 237
DC motor
  acceleration and deceleration, 340
  armature voltage, 298
  commutators, 338
  control, 336, 337, 348
    closed-loop, 338, 345, 350
    inner loop, 337

outer loop, 337
controller characteristics, 340
demagnetization, 337
dynamic simulation, 297
electromagnetic torque, 298
equivalent circuit, 336
feedback devices, 337
feedback elements, 339
four quadrant operation, 336, 339, 342
inductance of armature, 299
open-loop system, 299
operating quadrants, 340
resistance of armature, 299
speed sensing, 341
speed (velocity) profiles, 340, 341, 342, 347
supply voltage, 299
tacho-generator, 337
winding current, 299
DC offset, 13, 136, 139
DC resistance, 175
Deceleration, 115
period, 116
speed curve, 117
Delta-connected balanced load, 64
Delta-star conversion, 59
Differential amplifier, 75, 136
Diode, 224
characteristic, 225
resistance, 225
voltage drop, 240
Discontinuous current conduction, 238, 240, 301
Distortion, 18
Drop-down menu, 182
Duty cycle, 18, 258, 299

E

Eddy current losses, 143, 155
Efficiency, 156
Electrical isolation, 7
Electrical rotor position, 291

Electric vehicles, 312
accelerator pedal position, 317
aerodynamic drag coefficient, 319
aerodynamic resistance, 318
air density, 319
climbing resistance, 318, 319
drive train options, 314
four-wheel steering, 313, 314, 315
in-hub direct drive, 313, 314, 315
linear and angular velocities, 315, 319
propulsion, 313
resultant acceleration, 320
road wheel angle, 316
rolling resistance, 318
coefficient, 319
skidding, 321
steering, 314
angle, 315, 316, 317
geometry, 316
turning angle, 319
turning point, 316
line, 317
position, 317
turning radius, 315, 316
two-wheel steering, 314
velocity, 315, 317
forward, 318
Electrocution, 115
Electromagnetic torque, 101, 115, 290
Electromechanical device, 100, 134, 135
Electromotive force, 85, 282
Equation of motion, 126
Excitation current, 145, 156, 288
Excited winding, 101

F

Faraday's Law, 139
Fault current, 327
Fault scenarios, 327
Fault-tolerant factor, 326
Fault-tolerant motor drive, 324, 325

Ferromagnetic
  core, 155
  material, 88
Field intensity, 140
Finite Element Analysis, 86
Flux
  density, 139
  linkage, 97, 102, 107, 139, 173
    current characteristics, 143
Flux density
  maximum, 142
  minimum, 14
  vectors, 281
Forward-bias region, 224
Fourier series, 21
Freewheeling diode, 230, 233, 307, 379
Frequency, 14
  bandwidth, 113
Frequency modulation (FM), 258
Friction and windage loss, 117, 121
Fringing, 87,
Front panel, 2

G

Gray code truth table, 380

H

Harmonic, 18, 97
  distortion, 75
  negative sequence, 20
  positive sequence, 20
  triplens, 20
  zero sequence, 20
H-bridge, 260, 307, 338, 354, 355
  bipolar driving method, 338
  control methods, 258
  power amplifier, 343
    threshold voltage, 345
  unipolar driving method, 339
Hysteresis, 98
  bandwidth, 330
  current control, 296, 330
  effect, 139
  loop, 96, 100, 139
    critical points, 141
  losses, 96, 98, 143, 155

I

Ideal transistor, 259
Impedance, 41, 148, 156
  angle, 52, 248
Induced voltage, 98
Inductance-current characteristics, 143
Inductive load, 149, 150
Infinite bus, 189
Instantaneous
  angular velocity, 115
  current measurement, 101
  power, 76
  reactive power, 51
  voltage measurement, 101
Integration, 102
Inverter, 360
Iron losses, 121
Isolation amplifier, 9
Isolation circuit, 371

K

Kinetic energy, 115
Kirchhoff's Current Law, 28
Kirchhoff's Voltage Law, 28

L

Leakage flux, 86, 155
Line
  current, 64
  voltage, 64
Load line, 225

M

Magnetic
  core loss, 155
  device, 146

field, 282, 305
field intensity, 86, 90, 97
flux, 86
flux density, 86
saturation, 96, 97
Magnetic circuits, 85, 89, 91
dc characteristics, 143
force, 144
Magnetization
characteristics (flux-current curves), 100, 101
cycle, 140
Magnetomotive force, 88
Mechanical rotation, 278
Mesh analysis, 26
Moment of inertia, 115
Motor torque, 116
Mutual
flux magnetic circuit, 156
inductance, 290
magnetic field, 155

N

Negative coercive force, 142
Nodal analysis, 28
Noise, 20
immunity, 9
No-load iron losses, 128, 131
Norton's equivalent, 30, 31, 32

O

Offset error, 105
One-wattmeter method, 72
Open-circuit, 26
Open-loop system, 299
Operating point, 225
Optical isolators, 371, 372

P

Passive impedance, 200
Period, 14

Permeability, 88
Per-unit value, 42, 243
Phase
angle, 76
hypothetical, 162
current, 64
difference, 162
impedance, 66
shift, 18
voltage, 64
Phasor, 40
diagram, 151
Picture ring, 5, 182
Picture/text ring, 234
Piecewise rectangular waveforms, 296
Polar moment of inertia, 129
Position sensor, 377
encoders, 380
Hall-effect, 360, 364
logic circuit, 361
resolver, 358
Positive coercive force, 142
Power, 50
angle, 52
equation, 51
measurement, 70, 75
triangle, 50, 150
loss, 121
total, 143
loss curves, 132
Power electronics, 223
Power factor, 39, 162
correction, 55, 153
nonsinusoidal waveform, 162
Practical capacitor, 150
Practical transformer, 156
Proportional integral (PI), 332
Pseudo-code, 234, 264
PWM, 258, 260, 262
current control, 296, 330
duty cycle, 299

Q

Quiz, 112

R

Reactance, 41, 149
Reactive power, 50, 76
Real-time control, 335
Real-time experimental system, 7
Rectification, 20, 236
Rectifiers, 236
   efficiency, 239
   ripple factor, 239
   single-phase, 237
   three-phase, 237
Redundancy, 324, 328
Relative permeability, 88
Relative velocity, 201
Reluctance, 87, 197
Residual flux density (remnant magnetization), 96
   negative, 142
   positive, 142
Resistance, 25, 41
Resistance circuits, 23
Retardation test, 116, 120, 129
   method, 117
Retarding torque, 173
Reverse-bias region, 224
Reverse saturation current, 224
RMS power, 76
Root-mean-square (rms) value, 15
Rotating field, 278, 358
Rotational inertia, 115
Rotor field, 278, 280
Runge-Kutta, 245, 275, 284

S

Saliency, 101
Sampling time interval, 102
Saturation, 96, 107, 162
   time, 137
Schockley diode equation, 224
Search coil, 97, 145
Secondary effects, 101
Short-circuit, 26, 156
Signal conditioning devices, 152
Single-phase circuit, 150
Single-phase inverter, 325
Single-quadrant controller, 299
Sinusoidal voltage, 14
Skin effect, 175
Soft starting, 246
Solid-state relay, 372
Speed profile, 115
Square wave generator, 261
Star-connected
   balanced load, 64
   resistance, 133
Star-delta conversion, 59
Step-down converter, 299
Stepper motors, 301
   bipolar, 305
   control, 306, 352
      design criteria, 353
   current conduction path, 311
   CW/CCW rotation, 310
   full-step mode, 307
   half-stepping, 304, 307
   hybrid, 303
   initial state, 311
   linear, 303
   micro-stepping, 304, 307
   number of phases, 304
   number of rotor teeth, 304
   number of steps, 304
   permanent magnet, 302
   rotor position, 311
   rotor state diagrams, 304
   step angle, 304
   step size, 304
   timing diagram, 306
   variable reluctance, 302
   wave-drive mode, 307

Stray load losses, 121
Supply inductance, 239
Switched Reluctance motor, 374
   aligned/unaligned position, 374, 375
   concentrated windings, 374
   control, 375
      soft chopping, 379
   electromagnetic torque, 375
   flux-current characteristics, 376
   inductance profile, 376
   inverter type, 378
   saliency, 375
   voltage equation, 375
Switching
   device, 103
   pattern, 360
   signal, 361
   soft/hard, 273
Synchronization, 188, 189
   conditions, 189, 190
   phase angle, 193
   phase difference, 189
   phasor graph, 193
   phasor rotation, 193
   synchronoscope, 193
   test, 191
Synchronous generator, 189
   armature
      current, 195, 197, 202
         leakage reactance, 197
         reaction reactance, 195, 199
         resistance, 195, 197
   equivalent circuit, 195, 196
   field current, 189, 196
   frequency, 189
   induced voltage, 195
   leakage reactance, 195, 198, 199
   open-circuit characteristics, 196, 205
   phase sequence, 189
   phasors, 195, 199, 201
   Potier triangle, 198, 199, 211
   power angle, 202
      characteristic, 203
   power factor, 202
   reactive power, 203
   real power, 203
   round rotor, 195
   round rotor phasor diagrams, 196
   saturation, 195
   short-circuit characteristics, 197, 207
   slip test, 200, 212
   stator resistance, 205
   synchronizing lamps, 200
   synchronous impedance, 195, 197, 198, 207
   synchronous reactance, 195, 198
      direct-axis, 201
      quadrature-axis, 201
   synchronous speed, 189, 197
   terminal voltage, 195
   unsaturated operation, 201
   zero power factor, 198
      test, 209
Synchronous machine, 194
   damping, 194
   field excitation
      normal, 194
      overexcitation, 195
      underexcitation, 195
   magnetic field, 194
   rotating field, 194
   synchronous speed, 282
   test, 194

T

Tacho-generator, 116, 121, 345
Thermal voltage, 224
Thevenin resistance, 31
Thevenin's equivalent, 30, 31, 32
Three-wattmeter method, 71
Three-phase ac, 20
   circuit test, 151
Thyristor (SCR), 228
   control (gate) current, 229
   di/dt, 230

Thyristor (SCR) (*continued*)
  dv/dt, 230
  forward blocking state, 229
  gate pulse, 233
  gate signal, 233
  holding current, 230, 232, 233, 236
  latching current, 229, 233, 236
  on-state, 229
  regenerative feedback, 228
  reverse blocking state, 229
  threshold voltage, 229
  triggering, 229
    pulse duration, 229
  two-transistor model, 228
Time-varying ideal signals, 15
Torque
  ripple, 20, 330, 358
  speed characteristic, 132, 133
  transducer, 132
Torque imbalance, 115
Torque pulsation, 288
Total harmonic distortion, 40
Transducer, 113
Transformation
  inverse, 285
  rotor, 285
  stator, 285
Transformer
  efficiency, 160, 165
  equivalent circuits, 156, 159, 164
  equivalent impedance, 159
  equivalent leakage reactance, 159, 160
  equivalent resistance, 159
  leakage reactance, 156
  load impedance, 165
  magnetizing branch, 156
  magnetizing current, 159, 162
  magnetizing current rush, 156
  nameplate, 169
  phasors, 165
  primary, 155
  rated primary current, 159, 160
  secondary, 155
  test, 153, 161
    full-load, 160, 165, 169
    open-circuit, 157, 162, 169
    short-circuit, 157, 159, 162, 169
  voltage regulation, 160, 165
  winding resistance, 156
Trapezoidal back emf, 288
Trapezoidal method (integration), 140, 142, 231, 312
Triggering angle, 247, 248
Turn-off angle, 247, 253
Two-segment motor, 329
Two-wattmeter method, 70
Two-way conduction, 246

V

Vibration, 20
Voltage
  drop, 102
  phasor, 66, 149
  regulation, 156
  source, 24
Voltage divider, 9, 25

W

Warm-up time, 123
Waveform
  chart, 240
  chopped sine, 15
  clipped sine, 15
  distorted, 97
  exponential, 15
  logarithmic, 15
  periodic, 15
  programmed harmonics, 15
  ramp, 15
  rectangular, 15
  sine, 15
  square, 15
  trapezoidal, 15
  triangular, 15
Winding resistance, 155

About the Author

Nesimi Ertugrul received his B.S. and M.S. degrees in electrical engineering and electronic and communication engineering from Istanbul Technical University in 1985 and 1989, respectively. In 1993, he received his Ph.D. from the University of Newcastle upon Tyne and in 1994 he joined Adelaide University, where he is now a senior lecturer.

His research topics include rotor position sensorless operation of brushless permanent magnet and switched reluctance motors, real-time control of electrical machines, fault tolerant motor drives, predictive maintenance, power electronics systems, and electric vehicles. He is currently researching the development of interactive computer-based teaching and learning systems that involve object-oriented programming and data acquisition.

Dr. Ertugrul is a member of IEEE and serves on the editorial advisory board for the International Journal of Engineering Education (IJEE). He coordinated and edited two special issues of IJEE, one on LabVIEW Applications in Engineering Education and one on Remote Access and Distance Learning Laboratories.

He also prepared the courseware "LabVIEW-based Course on Electrical Machines and Circuits" for National Instruments (Part No.: 322765A-01), which is available at http://www.ni.com/academic/experiments.htm. In addition, he contributed "LabVIEW-Based Interactive Teaching Laboratory" to *LabVIEW for Automotive, Telecommunications, Semiconductor, Biomedical, and Other Applications*, edited by H. T. Martin and M. L. Martin, Prentice Hall, 2000.

He has won the following awards for his work with LabVIEW:

1. Outstanding Application in Academic Category, National Instruments, NI Week, August 1999.
2. Winning Application Paper at VIP Days 2000 paper contest.

Get the Most Out of LabVIEW...

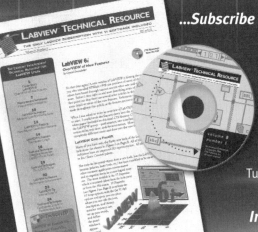

...Subscribe to LabVIEW™ Technical Resource!

Receive a new issue each quarter full of technical solutions to help you get the most out of LabVIEW. Each LTR issue includes a Resource CD packed with LabVIEW VIs, Utilities and Source Code. Learn LabVIEW Tips, Techniques, Tutorials and more.

Increase your LabVIEW programming knowledge!

To Order your Subscription with Resource CD or other LTR products, visit the **LTR Online Store** or fax the attached order form.

WEB: ltrpub.com • **FAX:** 214 706 0506 • **EMAIL:** ltr@ltrpub.com

---

"LabVIEW Technical Resource... a must-have training tool for LabVIEW programmers"
Jeff Kodosky – NI Fellow and LabVIEW Inventor, National Instruments

Free LabVIEW Tips

LabVIEW Style Guides

"Saved me hours if not days of development time!"
Brad Hedstrom – Senior Systems Engineer, Applied Biosystems

CD Library of Back Issues

"LTR is a valuable resource for the LabVIEW community, providing highly technical information and useful tools to improve LabVIEW users' productivity."
Tamra Kerns – Director of Software Marketing, National Instruments

LTR VI Code Paks

Online & On Demand   **ltrpub.com**

PH0202

LABVIEW™ TECHNICAL RESOURCE
THE ONLY LABVIEW SUBSCRIPTION WITH VI SOFTWARE INCLUDED

ORDER FORM

TEL: 214-706-0587   FAX: 214-706-0506

WHAT IS LTR?

LabVIEW Technical Resource (LTR) is a quarterly journal for LabVIEW users and developers available by subscription from LTR Publishing, Inc.

Each LTR issue presents powerful LabVIEW tips and techniques and includes a Resource CD packed with LabVIEW VIs, Source Code, Utilities, and Documentation. Technical articles on LabVIEW programming methodology, in-depth tutorials, and time-saving tips and techniques address everyday programming issues in LabVIEW.

In its ninth year of publication, LTR has subscribers in over 50 countries and is well-known as a leading independent source of LabVIEW-specific information.

Purchase the LabVIEW Technical Resource CD-ROM Library of Back Issues Version 3.0 and browse this searchable CD-ROM for easy access to over 250 VIs from LTR Volumes 1-8.

To subscribe to the LabVIEW Technical Resource or to order the CD-ROM Library of Back Issues, fax this form to LTR Publishing at **(214) 706-0506** or visit the LTR web page at **ltrpub.com** to download a free sample issue.

CONTACT INFORMATION

Name _____ Company _____
Address _____
City _____ State _____ Zip/Post Code _____
Country _____ E-mail (required) _____
Tel _____ FAX _____

LTR PRODUCT LISTING

QTY	LTR Subscriptions and Products • Single-User License	U.S.	INTL.	EXT. PRICE
	1 year subscription (4 Issues / 4 Resource CDs)	$95	$120	
	2 year subscription (8 Issues / 8 Resource CDs)	$175	$215	
	Individual Back Issues with Resource CDs (Article Index at ltrpub.com)	$25	$30	
✗	VI Code Paks – **Available ONLY at ltrpub.com** (priced per pak) Download bundled paks of LTR VIs & accompanying articles by topic.	$50	$50	Order Online

QTY	LTR Library of Back Issues on CD-ROM • Single-User License	U.S.	INTL.	EXT. PRICE
	CD-ROM Library of Back Issues Ver. 3.0 (28 issues / over 250 VIs)	$350	$375	
	Upgrade CD-ROM to Ver. 3.0 (requires Version 2.0)	$89	$99	

QTY	LTR Library of Back Issues on CD-ROM • Multi-User License	U.S.	INTL.	EXT. PRICE
	Server Version CD-ROM Library of Back Issues Ver. 3.0 (5 user license)	$495	$530	

◯ Please send me information on Multi-User Add-On Pak Discounts for Servers with > 6+ users.

◯ Please send me information on Group Subscription discounts.

TOTAL _____

PAYMENT INFORMATION

✔	PAYMENT METHOD		
	VISA / MC / AMEX Card Number		Exp.
	Signature		Date
	Bill company (**U.S. ONLY**) / Fax Purchase Orders to (214) 706-0506 / P.O. Number		
	Check enclosed (**U.S. BANK ONLY** – Make check payable to LTR Publishing) (Texas residents please add 8.25% sales tax)		
	Wire / TT (**INTERNATIONAL ORDERS**) – Contact LTR for Banking Information		

ORDER ONLINE or Fax this order form to 214-706-0506. Please include a signature on all credit card orders. Mail all check orders (US bank only) to:

LTR Publishing, Inc., 860 Avenue F, Suite 100 Plano, Texas 75074
Tel: 214-706-0587 • Fax: 214-706-0506 • email: ltr@ltrpub.com

Standard shipping is included on all orders. If you prefer, you may include your own Airborne, Federal Express, or UPS Air Collect #. Texas residents please include Texas Sales Tax at 8.25%.

© Copyright 2002 LTR Publishing, Inc. All rights reserved. LabVIEW Technical Resource is an independently produced publication of LTR Publishing, Inc. LabVIEW is a registered trademark of National Instruments Corporation.

LTRPUB.COM

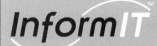

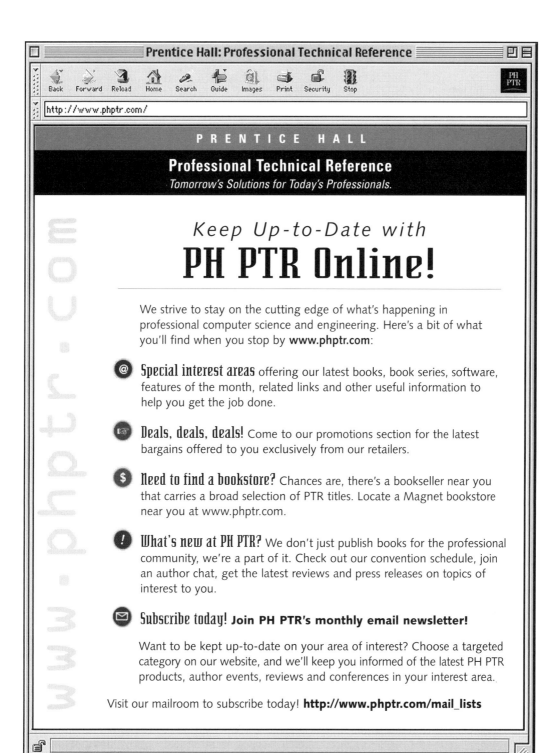

LICENSE AGREEMENT AND LIMITED WARRANTY

READ THE FOLLOWING TERMS AND CONDITIONS CAREFULLY BEFORE OPENING THIS SOFTWARE PACKAGE. THIS LEGAL DOCUMENT IS AN AGREEMENT BETWEEN YOU AND PRENTICE-HALL, INC. (THE "COMPANY"). BY OPENING THIS SEALED SOFTWARE PACKAGE, YOU ARE AGREEING TO BE BOUND BY THESE TERMS AND CONDITIONS. IF YOU DO NOT AGREE WITH THESE TERMS AND CONDITIONS, DO NOT OPEN THE SOFTWARE PACKAGE. PROMPTLY RETURN THE UNOPENED SOFTWARE PACKAGE AND ALL ACCOMPANYING ITEMS TO THE PLACE YOU OBTAINED THEM FOR A FULL REFUND OF ANY SUMS YOU HAVE PAID.

1. **GRANT OF LICENSE:** In consideration of your payment of the license fee, which is part of the price you paid for this product, and your agreement to abide by the terms and conditions of this Agreement, the Company grants to you a nonexclusive right to use and display the copy of the enclosed software program (hereinafter the "software") on a single computer (i.e., with a single CPU) at a single location so long as you comply with the terms of this Agreement. The Company reserves all rights not expressly granted to you under this Agreement.

2. **OWNERSHIP OF SOFTWARE:** You own only the magnetic or physical media (the enclosed software) on which the software is recorded or fixed, but the Company retains all the rights, title, and ownership to the software recorded on the original software copy(ies) and all subsequent copies of the software, regardless of the form or media on which the original or other copies may exist. This license is not a sale of the original software or any copy to you.

3. **COPY RESTRICTIONS:** This software and the accompanying printed materials and user manual (the "Documentation") are the subject of copyright. You may not copy the Documentation or the software, except that you may make a single copy of the software for backup or archival purposes only. You may be held legally responsible for any copying or copyright infringement which is caused or encouraged by your failure to abide by the terms of this restriction.

4. **USE RESTRICTIONS:** You may not network the software or otherwise use it on more than one computer or computer terminal at the same time. You may physically transfer the software from one computer to another provided that the software is used on only one computer at a time. You may not distribute copies of the software or Documentation to others. You may not reverse engineer, disassemble, decompile, modify, adapt, translate, or create derivative works based on the software or the Documentation without the prior written consent of the Company.

5. **TRANSFER RESTRICTIONS:** The enclosed software is licensed only to you and may not be transferred to any one else without the prior written consent of the Company. Any unauthorized transfer of the software shall result in the immediate termination of this Agreement.

6. **TERMINATION:** This license is effective until terminated. This license will terminate automatically without notice from the Company and become null and void if you fail to comply with any provisions or limitations of this license. Upon termination, you shall destroy the Documentation and all copies of the software. All provisions of this Agreement as to warranties, limitation of liability, remedies or damages, and our ownership rights shall survive termination.

7. **MISCELLANEOUS:** This Agreement shall be construed in accordance with the laws of the United States of America and the State of New York and shall benefit the Company, its affiliates, and assignees.

8. **LIMITED WARRANTY AND DISCLAIMER OF WARRANTY:** The Company warrants that the software, when properly used in accordance with the Documentation, will operate in substantial conformity with the description of the software set forth in the Documentation. The Company does not warrant that the software will meet your requirements or that the operation of the software will be uninterrupted or error-free. The Company warrants that the media on which the software is delivered shall be free from defects in materials and workmanship under normal use

for a period of thirty (30) days from the date of your purchase. Your only remedy and the Company's only obligation under these limited warranties is, at the Company's option, return of the warranted item for a refund of any amounts paid by you or replacement of the item. Any replacement of software or media under the warranties shall not extend the original warranty period. The limited warranty set forth above shall not apply to any software which the Company determines in good faith has been subject to misuse, neglect, improper installation, repair, alteration, or damage by you. EXCEPT FOR THE EXPRESSED WARRANTIES SET FORTH ABOVE, THE COMPANY DISCLAIMS ALL WARRANTIES, EXPRESS OR IMPLIED, INCLUDING WITHOUT LIMITATION, THE IMPLIED WARRANTIES OF MERCHANTABILITY AND FITNESS FOR A PARTICULAR PURPOSE. EXCEPT FOR THE EXPRESS WARRANTY SET FORTH ABOVE, THE COMPANY DOES NOT WARRANT, GUARANTEE, OR MAKE ANY REPRESENTATION REGARDING THE USE OR THE RESULTS OF THE USE OF THE SOFTWARE IN TERMS OF ITS CORRECTNESS, ACCURACY, RELIABILITY, CURRENTNESS, OR OTHERWISE.

IN NO EVENT, SHALL THE COMPANY OR ITS EMPLOYEES, AGENTS, SUPPLIERS, OR CONTRACTORS BE LIABLE FOR ANY INCIDENTAL, INDIRECT, SPECIAL, OR CONSEQUENTIAL DAMAGES ARISING OUT OF OR IN CONNECTION WITH THE LICENSE GRANTED UNDER THIS AGREEMENT, OR FOR LOSS OF USE, LOSS OF DATA, LOSS OF INCOME OR PROFIT, OR OTHER LOSSES, SUSTAINED AS A RESULT OF INJURY TO ANY PERSON, OR LOSS OF OR DAMAGE TO PROPERTY, OR CLAIMS OF THIRD PARTIES, EVEN IF THE COMPANY OR AN AUTHORIZED REPRESENTATIVE OF THE COMPANY HAS BEEN ADVISED OF THE POSSIBILITY OF SUCH DAMAGES. IN NO EVENT SHALL LIABILITY OF THE COMPANY FOR DAMAGES WITH RESPECT TO THE SOFTWARE EXCEED THE AMOUNTS ACTUALLY PAID BY YOU, IF ANY, FOR THE SOFTWARE.

SOME JURISDICTIONS DO NOT ALLOW THE LIMITATION OF IMPLIED WARRANTIES OR LIABILITY FOR INCIDENTAL, INDIRECT, SPECIAL, OR CONSEQUENTIAL DAMAGES, SO THE ABOVE LIMITATIONS MAY NOT ALWAYS APPLY. THE WARRANTIES IN THIS AGREEMENT GIVE YOU SPECIFIC LEGAL RIGHTS AND YOU MAY ALSO HAVE OTHER RIGHTS WHICH VARY IN ACCORDANCE WITH LOCAL LAW.

ACKNOWLEDGMENT

YOU ACKNOWLEDGE THAT YOU HAVE READ THIS AGREEMENT, UNDERSTAND IT, AND AGREE TO BE BOUND BY ITS TERMS AND CONDITIONS. YOU ALSO AGREE THAT THIS AGREEMENT IS THE COMPLETE AND EXCLUSIVE STATEMENT OF THE AGREEMENT BETWEEN YOU AND THE COMPANY AND SUPERSEDES ALL PROPOSALS OR PRIOR AGREEMENTS, ORAL, OR WRITTEN, AND ANY OTHER COMMUNICATIONS BETWEEN YOU AND THE COMPANY OR ANY REPRESENTATIVE OF THE COMPANY RELATING TO THE SUBJECT MATTER OF THIS AGREEMENT.

Should you have any questions concerning this Agreement or if you wish to contact the Company for any reason, please contact in writing at the address below.

Robin Short
Prentice Hall PTR
One Lake Street
Upper Saddle River, New Jersey 07458

About the CD-ROM

The CD-ROM included with *LabVIEW® for Electric Circuits, Machines, Drives, and Laboratories* contains the following files that can be run using LabVIEW®:

CHAPTER 2: Waveforms Generator.llb, Waveforms and Harmonic Analysis.llb, Current Divider Circuit.vi, Equivalent Resistances.vi, Mesh Analysis Circuit.vi, Series and Parallel Circuit .vi, Voltage Divider Circuit.vi, Norton_1.vi, Norton_2.vi, Norton_3.vi, Thevenin Norton Circuits.vi, Thevenin_1.vi, Thevenin_2.vi, Thevenin_3.vi

CHAPTER 3: AC Current Divider Circuit.llb, AC Series and Parallel Circuit.llb, Dual Supply Mesh Analysis.llb, Impedance Voltage Divider.llb, Parallel Impedances.llb, Series Impedances.llb, Single Phase AC Definitions.llb, Power Triangles.llb, Single Phase Power and Power Factor Correction.llb, Star Delta Transformations.llb, Voltage and Currents in Delta Star loads.llb, Three Phase Phasors.llb, 3 phase power measurements.llb, Data Logging.llb

CHAPTER 4: Magnetics 1.llb, Magnetics 2.llb, BH Hysteresis.llb, RealTimeVoltageCurrent.txt, Flux Linkage Test.llb

CHAPTER 5: Retardation Test.llb, Dcloss, Dcloss2, Dcloss3, Dcloss4, D.C. Losses Experiment.llb, Dcloss, Dcloss1, Dcloss2, Dcloss3, Dcloss4, Dcloss5, Dcloss6, Dcloss7, Dcloss8, Dcloss9, Dcloss10, Dcloss11, Dcloss12, Dcloss13, Dcloss14, Electromechanical Device Exp.llb, SinglePhase.llb, ThreePhase.llb, Power Factor Correction.llb, Transformer Experiment.llb, Induction Motor Experiment.llb, Synchronisation Observer.llb, Synchronous Machines Experiment.llb

CHAPTER 6: Diode Simulation.llb, diodevals.txt, lookupf.txt, lookupr.txt, Thyristor.llb, Three phase halfway uncontrolled rectifier.llb, Single phase AC Chopper.llb, Cycloconverter.llb, Part 1.llb, Part 2.llb

CHAPTER 7: Rotating Field Demonstration.llb, Rotating Field Induction.llb, Induction Motor.llb, Brushless PM Motor Simulation.llb, Brush DC Motor Simulation.llb, Stepper Motor Simulation.llb, Four Wheel Steer and Drive.llb, Fault Tolerant Drive.llb

CHAPTER 8: DC Motor Control.llb, SpeedProfilingPanel.llb, Stepper Motor Control.llb, Brushless PM Simulation.llb, Brushless real-time.llb, Wound Rotor Induction Starting.llb, SR Motor Control.llb

APPENDIX: Login.llb, Question Time.llb, student list.txt, Synchronisation.txt

The CD-ROM can be used on Microsoft Windows® 95/98/NT®

Limitations

National Instruments LabVIEW 6i Evaluation Version is intended to expose full functionality of LabVIEW and evaluate VIs written by the author. Evaluation version has the following limitations to provide adequate opportunity for a customer to evaluate the software while protecting against misuse.

1) 5-minute continuous execution VI timeout
2) 60-minute cumulative execution LabVIEW timeout
3) 30-day usability timeout
4) Hard coded date for expiration
5) 15 days of "friendly reminders"
6) Unable to integrate external code

License Agreement

Use of the software accompanying *LabVIEW® for Electric Circuits, Machines, Drives, and Laboratories* is subject to the terms of the License Agreement and Limited Warranty, found on the previous two pages.

Technical Support

Prentice Hall does not offer technical support for any of the programs on the CD-ROM. However, if the CD-ROM is damaged, you may obtain a replacement copy by sending an email that describes the problem to: disc_exchange@prenhall.com.